Inverse Canonical Energy Flow in Laser Beams

Victor Kotlyar • Alexey Kovalev • Sergey Stafeev

Inverse Canonical Energy Flow in Laser Beams

Springer

Victor Kotlyar
Image Processing Systems Institute
NRC Kurchatov Institute
Samara, Russia

Alexey Kovalev
Image Processing Systems Institute
NRC Kurchatov Institute
Samara, Russia

Sergey Stafeev
Samara National Research University
Samara, Russia

ISBN 978-3-032-27288-1 ISBN 978-3-032-27289-8 (eBook)
https://doi.org/10.1007/978-3-032-27289-8

This Springer imprint is published by the registered company Springer Nature Switzerland AG
The registered company address is: Gewerbestrasse 11, 6330 Cham, Switzerland

Preface

This book examines two-dimensional light fields (sinc beams, Airy beams, Hankel beams, and Bessel beams of fractional order), in which reverse energy flow is formed near isolated intensity nulls, that is, near those points where the energy (power) density is zero. A reverse canonical energy flow has also been found in three-dimensional nonparaxial vector light fields with linear, azimuthal, and hybrid polarizations. The reverse energy flow is formed in the near field, in the presence of evanescent waves, as well as in the region of sharp focus, where there are no evanescent waves. Interestingly, reverse energy flow takes place in the near field of a Hertzian linear point dipole.

This book is devoted to an interesting and largely unexplored effect in optics: the reverse energy flow of light. It examines two-dimensional and three-dimensional nonparaxial vector light fields, in which reverse canonical energy flow is formed in certain spatial regions. The propagation of coherent monochromatic light is described in several ways: by decomposition into a plane-wave spectrum and by Rayleigh-Sommerfeld vector integrals. Projections of the electric vector in sharp focus are described using Richards-Wolf theory and Debye integrals. The reverse energy flow effect is closely related to the well-known superoscillation effect and the formation of giant local wave vectors in nonparaxial light fields.

This book presents precise analytical expressions for the longitudinal projections of the canonical energy flow vector, which strictly prove the presence of reverse flow in the light fields under consideration.

This book includes the results that the authors obtained in 2024–2025.

This book will be of interest to a wide range of scientists and engineers working in the fields of optics, photonics, laser physics, opto-information technologies, and optical instrumentation. It can also be useful for bachelor's and master's students in the specialties of applied mathematics and physics, applied mathematics and informatics, and optics, as well as for graduate students specializing in these areas.

Samara, Russia — Victor Kotlyar
Samara, Russia — Alexey Kovalev
Samara, Russia — Sergey Stafeev

Introduction

In modern optics and photonics, despite their rapid recent development, some theoretical problems remain unresolved. One such problem is the occurrence of negative values for the longitudinal projection of the Poynting vector. As early as 1919, V.S. Ignatovsky, and later in 1959, B. Richards and E. Wolf discovered that in the sharp focus of coherent light, the Poynting vector takes on negative values. They called this effect reverse energy flow. However, this effect has not been studied in detail. The Poynting vector is included in the differential law of conservation of energy as its divergence. Therefore, the Poynting vector is defined only up to the curl of any continuous function, since the divergence of the curl is zero. Therefore, the question arises: Does the Poynting vector always determine the magnitude and direction of the energy flow at each point in space? In 2007, A. Ya. Bekshaev and M.S. Soskin showed that the Poynting vector is the sum of two vectors: the orbital (canonical) energy flux vector and the spin flux vector. The spin flux cannot influence the energy flux. Therefore, a true reverse energy flow occurs when the axial projection of the canonical energy flux vector takes on negative values. In other words, the canonical energy flux is proportional to the local wave vector, which is equal to the phase gradient at that point. If the component of the wave vector parallel to the optical axis is negative, then the energy flux has a reverse direction.

Reverse canonical energy flow occurs in nonparaxial light fields that diverge strongly or are focused into a sharp focus. Typically, reverse canonical energy flow occurs near intensity zeros. In two-dimensional light fields, a longitudinal optical vortex (a 2D vortex) forms near these intensity zeros, and the light energy circulates around the intensity zeros in the longitudinal plane. In some parts of this circulating energy flow, the direction is reversed (toward the light source). The local wave vector near the intensity zeros exceeds the wavenumber of the light. Therefore, the reverse energy flow effect is closely related to the superoscillation effect discussed by M. Berry in 2009 (optical currents). The region where the energy circulates in the longitudinal plane is only a tenth of a wavelength in size, that is, the region with the reverse flow is essentially subwavelength.

In three-dimensional nonparaxial light fields, a reverse flow forms near lines of zero intensity. For example, if a 3D vector field has a curve of zero intensity in the transverse plane that forms a ring centered on the optical axis, then two isolated zero intensity points will appear in any longitudinal plane section of that ring. Around these two zero intensity points, two energy flows will form, circulating in opposite directions: around one zero intensity point, the energy flow will rotate clockwise, while around the other zero intensity point, the energy flow will rotate counterclockwise. In some parts of these trajectories along which light energy circulates, the direction of the energy flow will be reversed.

Although the fraction of light energy directed in the opposite direction is only 1 or 2% of the total beam energy, studying the backward canonical flux allows us to better understand the complex nature of light.

At the end of the Introduction, we would like to mention several works that are predecessors to this work, in which reverse energy flow is tightly related to the phase singularities of light fields. For instance, the work by J.F. Nye et al. [1] demonstrated for the first time that near an optical vortex (near a screw phase dislocation) in 2D waves, a phase saddle point should exist. In another work [2], M.V. Berry showed that in the waist of a 2D nonparaxial Gaussian beam without evanescent waves, two points of phase singularity (screw dislocations) exist with topological charges of $+1$ and -1, symmetrical with respect to the optical axis, and two phase saddle points exist near these singularities. It was also shown that around the phase singularity points (intensity zeros), energy flow circulates, which is proportional to the wave vector. This follows directly from the definition of the wave vector as the gradient of the complex amplitude describing the 2D nonparaxial light field. However, M.V. Berry's work [2] was inspired by the works of G.P. Karman et al. [3, 4], where the authors illustrated that near the paraxial focus of 3D rotationally symmetric beams, bounded by a circular aperture or having spherical aberration, a zero-intensity ring in the focus generates in the longitudinal section an isolated intensity zero and thus generates a phase singularity point. However, since the beam is paraxial (numerical aperture is 0.1), there is no phase saddle point near the phase singularity point, whereas the reverse flow can be generated in the interval between the phase singularity point and the phase saddle point in 2D beam [5]. Paraxial light fields cannot have a reverse energy flow.

References

1. J.F. Nye, J.V. Hajnal, J.H. Hannay, Proc. R. Soc. Lond. Ser. A Math. Phys. Eng. Sci. **417**, 7 (1988).
2. M.V Berry, J. Mod. Opt. **45**, 1845 (1998).
3. G.P. Karman, M.W. Beijersbergen, A. van Duijl, J.P. Woerdman, Opt. Lett. **22**, 1503 (1997).

4. G.P. Karman, A. Van Duijl, J.P. Woerdman, Opt. Lett. **23**, 403 (1998).
5. M.V. Berry, J. Phys. A Math. Theor. **43**, 415302 (2010).

The authors would like to thank Dr. Anton Nalimov for assistance with the computer modeling.

The results presented in this monograph were obtained during research supported by Russian Science Foundation Grant No. 26-12-00117.

Competing Interests The authors have no competing interests to declare that are relevant to the content of this manuscript.

Contents

About the Authors

Victor Kotlyar is the Head of the Laboratory at the Image Processing Systems Institute of the National Research Center "Kurchatov Institute" and a Professor of Computer Science at Samara National Research University. He received his MS, PhD, and DrSc degrees in physics and mathematics from Samara State University (1979), Saratov State University (1988), and the Moscow Central Design Institute of Unique Instrumentation, Russian Academy of Sciences (1992). He is the co-author of 400 scientific papers, 10 books, and 7 inventions. His current interests are diffractive optics and singular optics. E-mail: Kotlyar@ipsiras.ru

Alexey Kovalev graduated in 2002 from Samara National Research University, majoring in applied mathematics. He obtained his DrSc in Physics and Maths degree in 2012. He is a senior researcher at the Laser Measurements Laboratory at the Image Processing Systems Institute of the National Research Center "Kurchatov Institute" and an Associate Professor of Computer Science at Samara National Research University. He is the co-author of more than 270 scientific papers and 4 books. His research interests are mathematical diffraction theory and optical vortices. E-mail: Alanko@ipsiras.ru

Sergey Stafeev received his master's degree in applied mathematics and physics from Samara National Research University in 2009. He received his PhD in 2012. He is a researcher at the Laser Measurements Laboratory at the Image Processing Systems Institute, NRC "Kurchatov Institute." His scientific interests include diffractive optics, nonuniform polarization, and tight focusing. E-mail: stafeev@ipsiras.ru. ORCID: 0000-0002-7008-8007.

Chapter 1
Canonical Energy Flow in 2D Laser Beams

1.1 Reverse Flow During Propagation of Half a Plane Wave

Ever since the seminal works by Ignatovsky [1] and Richards and Wolf [2], optical researchers have been pondering over why at certain conditions, the Poynting vector [3] takes negative values. Being a constituent part of integral and differential equations describing the energy conservation law [4], the Poynting vector is associated with the energy flow. On the other hand, the Poynting vector is defined up to the curl of any analytic function, because taking divergence of curl always yields zero [5]. The energy backflow (negative Poynting vector value) has been discovered to occur for Bessel vector beams [6] and other vector beams [7–9] and on the optical axis at the sharp focus [10, 11]. However, rigorous simulations have shown the energy backflow to be incapable of displacing a microparticle in the negative direction (an optical 'tractor') [12, 13]. In the work by Bekshaev [14], the Poynting vector was shown to be composed of the sum of two vectors: a (canonical) orbital energy flow and a spin flow. The canonical energy flow is proportional to the intensity multiplied by the phase gradient, thus being co-directed with the wave-vector of a light field [15]. Meanwhile, the spin flow is proportional to the curl of the spin angular momentum and, so, is incapable of taking part in the energy transfer. Negative values of the energy flow have been shown to occur [16] when the spin flow is counter-directed to the optical axis and larger in magnitude than the direct canonical energy flow.

In this section, we analyze 2D diffraction of a TE-polarized plane wave by an opaque rectangular screen [4]. This problem can be analytically solved using Fresnel integrals, meaning that an analytic expression for a canonical energy flow can be derived and parameters leading to the formation of the canonical energy backflow identified. We reveal that the canonical energy backflow is generated in the diffraction pattern areas where values of diffraction-enabled intensity and phase modulation on the screen edge are smaller than the incident wavelength. In those areas the local wave-vector is larger than the incident wave vector. There is nothing unusual

V. Kotlyar et al., *Inverse Canonical Energy Flow in Laser Beams*,
https://doi.org/10.1007/978-3-032-27289-8_1

about this, for it is known that in the vicinity of a vortex center (intensity null) of a light field, the azimuthal component of the wave vector can take huge values [17–22].

1.1.1 Fresnel Diffraction by a Rectangular Edge

Scalar Fresnel diffraction of a plane wave by a rectangular edge is described by a complex amplitude [4]:

$$\begin{aligned} E_y(x,z) &= \frac{-ie^{ikz}}{2z}[(C(w)-S(w))+i(1+C(w)+S(w))] \\ &= \frac{e^{ikz}}{2z}\left(1+\sqrt{2}e^{-i\frac{\pi}{4}}C(w)+\sqrt{2}e^{i\frac{\pi}{4}}S(w)\right), \end{aligned} \tag{1.1}$$

where k is the wave-number, the x-coordinate is in the plane of an opaque screen, the z-coordinate is perpendicular to the screen, with magnitudes remaining unchanged along the y-coordinate, E_y defines linear (TE)-polarization along the y-coordinate, and $w = x\sqrt{\frac{2}{\lambda z}}$, λ being the wavelength. Fresnel integral in (1.1) takes the form:

$$C(w) = \int_0^w \cos\left(\frac{\pi t^2}{2}\right)dt, \quad S(w) = \int_0^w \sin\left(\frac{\pi t^2}{2}\right)dt. \tag{1.2}$$

While describing the Fresnel diffraction, light field amplitude (1.1) remains valid at any small z, as with z tending to zero, field amplitude (1.1) tends to a constant value, meaning that half the constant-amplitude plane wave gets obstructed by the screen. For light field (1.1), the intensity distribution is given by

$$\begin{aligned} I(x,z) = \left|E_y(x,z)\right|^2 = \frac{1}{4z^2} \times \\ \left[(C(w)-S(w))^2+(1+C(w)+S(w))^2\right]. \end{aligned} \tag{1.3}$$

From (1.3), the intensity pattern is seen to remain unchanged at any z, changing only in scale and decreasing in magnitude with increasing distance z. In view of $C(\infty) = S(\infty) = 1/2$ and Eq. (1.3), we infer that at $x \to \infty$ the intensity tends to a constant value of $I(x \to \infty, z) = z^{-2}$ at a given z; while being always four-times lower than that at infinity at $x = 0$. The intensity extremums are found by equating to zero the derivative of intensity (1.3):

$$\begin{gathered}\frac{\partial}{\partial w}I(w)=\frac{1}{2z^2}\frac{\partial}{\partial w}\left[\left(\frac{1}{2}+C(w)\right)^2+\left(\frac{1}{2}+S(w)\right)^2\right]=\\ \frac{1}{z^2}\left[\left(\frac{1}{2}+C(w)\right)\cos\frac{\pi w^2}{2}+\left(\frac{1}{2}+S(w)\right)\sin\frac{\pi w^2}{2}\right]=0.\end{gathered} \tag{1.4}$$

Equation (1.4) leads to a transcendental equation for deriving the extremes:

$$\frac{2C(w)+1}{2S(w)+1}=-\mathrm{tg}\frac{\pi w^2}{2}. \tag{1.5}$$

The first root of (1.5) can be approximately assessed from the following considerations. The left-hand side of (1.5) equals 1 at $C(w) = S(w)$, which occurs approximately at $w = 1.26$ (see Tables of Integrals in [23]). And on the right-hand side of (5), the tan function takes a value of -1 at $w = 1.22$. Hence, the coordinate x_m of the first (larger) local intensity maximum is approximately defined by

$$x_m \approx 1.22\sqrt{\frac{\lambda z}{2}}. \tag{1.6}$$

1.1.2 Canonical Energy Flow Generated by the Diffraction of Light by a Rectangular Edge

Disregarding inessential constants, the Poynting vector is composed of the sum of two vectors [15]: the (canonical) orbital energy flow, P_{or} and the spin flow, P_{sp}:

$$\mathbf{P}=\mathbf{P}_{or}+\mathbf{P}_{sp}=\mathrm{Im}(\mathbf{E}^*\cdot(\nabla)\mathbf{E})+\frac{1}{2}(\nabla\times\mathbf{S}),\quad \mathbf{S}=\mathrm{Im}(\mathbf{E}^*\times\mathbf{E}). \tag{1.7}$$

Based on (1.7), the longitudinal component of the orbital energy flow can be expressed as

$$P_{or,z}=\mathrm{Im}\left(E_x^*\frac{\partial}{\partial z}E_x+E_y^*\frac{\partial}{\partial z}E_y+E_z^*\frac{\partial}{\partial z}E_z\right). \tag{1.8}$$

Let us remind that under the 2D diffraction of a TE-polarized wave there is a single non-zero E-vector projection of Eq. (1.1), which produces, after being substituted into (1.8):

$$P_{or,z} = \mathrm{Im}\left(E_x^* \frac{\partial}{\partial z} E_x\right) = \frac{1}{2z^3} \times \left\{kz\left[\left(\frac{1}{2} + C(w)\right)^2 + \left(\frac{1}{2} + S(w)\right)^2\right] - \frac{w}{2}\left[\sin\frac{\pi w^2}{2}\left(\frac{1}{2} + C(w)\right) - \cos\frac{\pi w^2}{2}\left(\frac{1}{2} + S(w)\right)\right]\right\}. \tag{1.9}$$

An analysis of (1.9) shows that with the first term in the square brackets being always positive, for the canonical energy flow to take negative values the second term in the square brackets should also be positive and larger in magnitude than the first one. Let us find at least a single value of the parameter w in (1.9) which makes the entire expression in (1.9) negative. For instance, assuming that the sine function equals 1 and the cosine—zero, we find that $\pi w^2/2 = \pi/2$, i.e. $w = 1$. Hence, the value of the x-coordinate at which the energy flow in (1.9) can be negative is

$$x = \sqrt{\frac{\lambda z}{2}}. \tag{1.10}$$

Since the diffraction pattern of a plane wave by a rectangular edge is a result of interference between a wave propagating past the screen and secondary diverging waves generated by the edge diffraction, the wave-vector projection on the z-axis away from the screen can take negative values only at negative z, with the screen located at $z = 0$. In this case, the light diffracted from the screen edge is propagating in the negative z-axis direction. Rather than being reverse, this is a direct energy flow propagating backwards. Behind the screen (at $z > 0$), both the plane wave and the diffracted secondary waves are propagating in the positive z-axis direction. So, it is only thanks to surface waves near the $z = 0$ plane that a canonical energy backflow can occur at $z > 0$, leading us to choose $z = \lambda/30$. In view of Eq. (1.10), the transverse coordinate is $x = 0.13\lambda$. At $w = 1$, table values of Fresnel integrals can be employed [23]: $C(w) = 0.78$, $S(w) = 0.44$. Substituting them into (1.6) yields:

$$P_{or,z}(x=0.13\lambda, z=\lambda/30) = \frac{1}{2z^3} \times \left\{0.21\left[\left(\frac{1}{2}+0.78\right)^2 + \left(\frac{1}{2}+0.44\right)^2\right] - \frac{1}{2}\left(\frac{1}{2}+0.78\right)\right\} \approx -\frac{0.11}{2z^3} \approx -\frac{1500}{\lambda^3} \approx 10000nm^{-3} \tag{1.11}$$

There are a countable number of x-coordinate values where the canonical energy flow in (1.9) has a local negative extremum. An increase in x leads to increasing w, and the second term is becoming ever larger in magnitude than the first term in the square brackets in (1.9). Actually, at large $w>>1$, Fresnel integrals take values C $(\infty) = S(\infty) = 1/2$ and Eq. (1.9) is rearranged to

$$P_{or,z}(x \to \infty, z) = \frac{1}{2z^3}\left[2kz - \frac{w}{2}\left(\sin\left(\frac{\pi w^2}{2}\right) - \cos\left(\frac{\pi w^2}{2}\right)\right)\right]. \tag{1.12}$$

Equation (1.12) suggests that with increasing distance from the edge, the negative energy flow also increases in magnitude, for $w > 0$ is growing. From (1.12) it also follows that given a unit-value sine function and zero cosine, i.e. at points x_m, localized minima of energy flow (1.9) are observed at

$$x_m = \sqrt{\lambda z(2m+1)}, \quad m = 0, 1, 2, \ldots, \tag{1.13}$$

if the inequality holds:

$$x_m > 4\pi z^{3/2}\sqrt{\frac{2}{\lambda}}. \tag{1.14}$$

From (1.14) it is seen that at small z ($kz << 1$), the first localized negative minimum of energy flow is near the origin, $x_m < \lambda$, whereas with increasing distance z, the transverse coordinate in (1.14) at which the first local negative localized maximum is moving away from the screen edge.

1.1.3 Profile of the Longitudinal Wave-Vector Projection

It can be shown that the longitudinal projection of the wave-vector of canonical energy flow (1.9) is defined by the product of intensity by the phase derivative of the light field complex amplitude (1.1), thus

$$P_{or,z} = \left|E_y\right|^2 \frac{\partial}{\partial z}\arg\left(E_y\right) =$$
$$\left|E_y\right|^2\left(k + \frac{w}{2z}\frac{\cos\frac{\pi w^2}{2}\left(\frac{1}{2}+S(w)\right) - \sin\frac{\pi w^2}{2}\left(\frac{1}{2}+C(w)\right)}{\left(\frac{1}{2}+C(w)\right)^2 + \left(\frac{1}{2}+S(w)\right)^2}\right). \tag{1.15}$$

The derivative of the argument of function (1.1) relative to the longitudinal z-coordinate in (1.15) equals the longitudinal wave-vector projection. Thus, given a positive-valued second term in the round brackets in (1.15), the longitudinal wave-vector projection will be larger than the wave-vector magnitude, which is possible if evanescent waves are present in the field. If the second term in the round brackets in

(1.15) is negative and larger in magnitude than the first term k, a canonical energy backflow occurs at those points. Thus, an energy backflow is generated in those intensity pattern areas where the local intensity and phase modulation period is smaller than the incident wavelength. We note that the smaller the local period of (subwavelength) phase and amplitude modulation, the larger (in magnitude) is the canonical energy backflow. From (1.15), it is seen that with growing w or the transverse x-coordinate, the contribution of evanescent waves with a larger-than-k local wave-vector will grow and so will the magnitude of the canonical energy backflow.

1.1.4 Canonical Energy Backflow for a Gaussian Beam

By way of demonstrating that the canonical energy backflow is a universal phenomenon, below we conduct a comparative analysis of two seemingly very different problems: diffraction of light by an edge and propagation of a paraxial Gaussian beam. It appears that a canonical energy backflow can also be found on the periphery of a conventional linearly polarized Gaussian beam, being generated far from the optical axis in areas with near-zero intensity. The complex amplitude of a Gaussian beam has been well known [24]:

$$E_g(r,z) = \frac{1}{q(z)} \exp\left(ikz - \frac{r^2}{w^2 q(z)}\right), \quad q(z) = 1 + i\frac{z}{z_0}. \tag{1.16}$$

The longitudinal component of the canonical energy flow in (1.8) and (1.15) is defined as a product of beam intensity (1.16) by the phase derivative with respect to the longitudinal coordinate:

$$\begin{gathered} P_{g,z} = \left|E_g\right|^2 \frac{\partial}{\partial z} \arg\left(E_g\right) = \frac{z_0^2}{z^2 + z_0^2} \exp\left(-\frac{2r^2}{w^2(z)}\right) \times \\ \left(k - \frac{z_0}{z^2 + z_0^2} - \frac{2r^2 z_0^2 z}{w^2\left(z^2 + z_0^2\right)^2}\right), \quad w^2(z) = w^2\left(1 + \frac{z^2}{z_0^2}\right), \end{gathered} \tag{1.17}$$

where $z_0 = kw^2/2$ is the Rayleigh range. From (1.17), it follows that the projection of the phase derivative of Gaussian beam (1.16) with respect to the longitudinal coordinate will be negative if the expression in the round brackets is negative:

$$k - \frac{z_0}{z^2 + z_0^2} - \frac{2r^2 z_0^2 z}{w^2\left(z^2 + z_0^2\right)^2} < 0. \tag{1.18}$$

Condition (1.18) is valid at some distance z from the waist if the circle radius r is larger than a certain value:

$$r > \frac{w}{z_0}\left[\frac{\left(z^2+z_0^2\right)\left(k\left(z^2+z_0^2\right)-z_0\right)}{2z}\right]^{1/2}. \tag{1.19}$$

Simplifying (1.19) and putting, for clarity, $z = z_0$ yields:

$$r > kw^2. \tag{1.20}$$

May the radius of the Gaussian beam waist equal the least value at which the paraxial approximation still holds [25]: $w = 2\lambda$. Then, the threshold value of the transverse coordinate r marking the birth of a canonical energy backflow is $r > 8\pi\lambda = 4.4w(z_0)$. This relation suggests that the radius behind which a canonical energy backflow is starting to occur is about four times larger than the radius of the Gaussian beam waist at the Rayleigh range.

1.1.5 Edge Diffraction: A Non-Paraxial Consideration

Let us consider for comparison the non-paraxial propagation of light after an opaque screen with a rectilinear edge. The complex amplitude of light in the 2D case is described by means of expansion over the two-dimensional plane waves spectrum

$$E_{np}(x,z) = \int_{-\infty}^{\infty} A(\xi)\exp\left(ikz\sqrt{1-\xi^2}+ikx\xi\right)d\xi, \tag{1.21}$$

To find the amplitude $A(\xi)$, it is necessary to perform the inverse Fourier transform from the initial complex amplitude of the light field:

$$A(\xi) = \frac{1}{\lambda}\int_{-\infty}^{\infty} E(x,z=0)\exp(-ikx\xi)d\xi, \tag{1.22}$$

where the amplitude of the initial field (1.22) for the case of an opaque screen is equal to the Heaviside function:

$$E(x,z=0) = \begin{cases} 1, x>0, \\ 0, x<0. \end{cases} \tag{1.23}$$

Substituting (1.23) in (1.22) we derive the amplitude of the plane wave spectrum for the 2D opaque screen:

$$A(\xi) = \frac{1}{\lambda}\int\limits_0^{\infty} \exp(-ikx\xi)d\xi = \frac{\delta(x)}{2} - \frac{i}{2\pi\xi}, \tag{1.24}$$

where $\delta(x)$ is a Dirac delta function. Substituting (1.24) in (1.21) we finally obtain a non-paraxial description of 2D diffraction at a rectilinear edge:

$$E_{np}(x, z) = \frac{1}{2}e^{ikz} - \frac{i}{2\pi}\int\limits_{-\infty}^{\infty} \frac{\exp\left(ikz\sqrt{1-\xi^2} + ikx\xi\right)}{\xi}d\xi. \tag{1.25}$$

Integral in (1.25) in limits $-\infty < \xi < -1$ и $1 < \xi < \infty$ describes contribution to the light field of evanescent waves, which are noticeable at $z < \lambda$. In this case, the longitudinal projection of the wave vector of the evanescent wave is greater than the wave vector of the incident wave:

$$k_z = k\sqrt{\xi^2 - 1} > k, \quad |\xi| > 1. \tag{1.26}$$

Therefore, near the initial plane ($z = 0$) negative values of the longitudinal projection of the canonical energy flux have to be expected. Substituting (1.25) into (1.8) and considering that the amplitude (1.25) is a linearly polarized TE-wave with the polarization vector along the y coordinate, only one term remains in (1.8). Let us find an expression for the longitudinal projection of the canonical flux of non-paraxial diffraction at the edge of a plane TE-wave in the initial plane ($z = 0$):

$$\begin{aligned}
P_{or,z} &= \mathrm{Im}\left(\frac{1}{2} + \frac{i}{2\pi}\int\limits_{-\infty}^{\infty}\frac{e^{-ikx\xi}}{\xi}d\xi\right)\left(\frac{ik}{2} + \frac{k}{2\pi}\int\limits_{-\infty}^{\infty}\sqrt{1-\xi^2}\frac{e^{ikx\xi}}{\xi}d\xi\right) = \\
&\frac{k}{4}\left(1 + \frac{2}{\pi}\int\limits_0^1\sqrt{1-\xi^2}\frac{\sin(kx\xi)}{\xi}d\xi + \frac{2}{\pi}\int\limits_0^{\infty}\frac{\sin(kx\xi)}{\xi}d\xi + \right. \\
&\left.\frac{4}{\pi^2}\int\limits_0^1\sqrt{1-\xi^2}\frac{\sin(kx\xi)}{\xi}d\xi\int\limits_0^{\infty}\frac{\sin(kx\xi)}{\xi}d\xi\right) = \\
&= \frac{k}{4}(1 + kx)\left(1 + \frac{2}{\pi}\int\limits_0^1\sqrt{1-\xi^2}\frac{\sin(kx\xi)}{\xi}d\xi\right).
\end{aligned} \tag{1.27}$$

The last integral in (1.27) is less than $\pi kx/2$, because the integrand $\sin(kx\xi)/\xi$ is multiplied by a function $\sqrt{1-\xi^2}$ less than 1, and since the integral sine without this function is equal to:

$$\int_0^\infty \frac{\sin(kx\xi)}{\xi} d\xi = \frac{\pi kx}{2}. \tag{1.28}$$

Therefore, for negative transverse coordinates x satisfying the condition $x < -1/k$, the longitudinal canonical flow (1.27) is negative. In this case, the edge of the truncated plane wave becomes the source of a cylindrical wave, which, propagating from the edge of the plane wave along the initial plane, forms a light field with a wave vector directed in the opposite direction to the optical axis.

1.1.6 Numerical Modeling Results

Figure 1.1 depicts 2D color patterns of intensity (1.3) (a, c) and the longitudinal component of canonical energy flow (6) (b, d) in the (x, y)-plane and their respective profiles at distances $z = \lambda/30$ (a, b) and $z = 2\lambda$ (c, d) for wavelength $\lambda = 532$ nm. The intensity was calculated using (1.3) and the energy flow — using (1.8).

From Fig. 1.1a, c the intensity pattern is seen to comply with Eq. (1.3). With decreasing distance z, the intensity modulation frequency in the brighter region is seen to be growing. We note that the distance z is much smaller than the wavelength, as is the intensity oscillation period. For instance, in Fig. 1.1a about 13 intensity modulation half-periods fit in a wavelength. From Fig. 1.1b, it can be seen that at small z, negative values of the canonical energy flow start to be observed almost

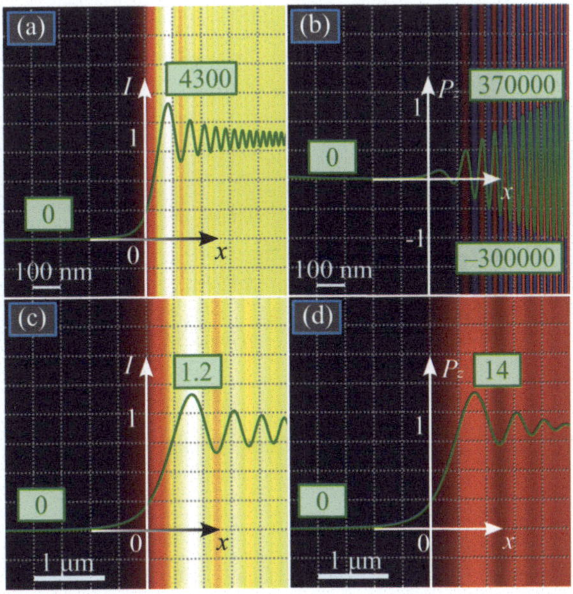

Fig. 1.1 2D color patterns of intensity (1.3) (**a**, **c**) and the longitudinal component of canonical energy flow (1.8) (**b**, **d**) in the (x, y)-plane and their respective profiles at distances $z = \lambda/30$ (**a**, **b**) and $z = 2\lambda$ (**c**, **d**)

immediately (100 nm from the edge), with the localized negative minima increasing in magnitude with increasing transverse x-coordinate, whereas the local recurrence interval is decreasing. At a larger value of $z = 2\lambda$, the intensity in Fig. 1.1c decreases in magnitude thanks to a diverging diffraction pattern, and the localized intensity modulation period also increases. Although in Fig. 1.1d an energy backflow is nowhere to be seen but it does occur, appearing at large values of w or x, as suggested by (1.9). An energy backflow can be expected to take place at any z, though appearing increasingly farther from the rectangular screen edge at distance x proportional to $\sqrt{\lambda z}$.

Figures 1.2 and 1.3 depict numerical simulation results of the diffraction of a 3D plane half-wave using the FDTD method. Simulation parameters are as follows: λ is 532 nm, simulation grid is λ/30 in all three coordinates, computational domain is 10 × 10 μm in transverse plane, incident wave is a plane wave E_y. The observation plane (first monitor) is located within the coordinates $-2 < x < 0$ μm, $-2.5 < y < 2.5$ μm, x, y are transverse coordinates, $z = 0$ is the initial plane. The plane wave source is located within the coordinates $0 < x < 8$ μm, $5.5 < y < 5.5$ μm, $z = 0$. The second monitor for taking the derivative with respect to z is located within the same boundaries along the x, y axes and at a grid step distance of λ/30 along the z axis from the first monitor. There are PML regions with ideal absorption around the all calculation area, the PML width is 0.5 μm. Figure 1.2 depicts the intensity $|\mathbf{E}|^2$,

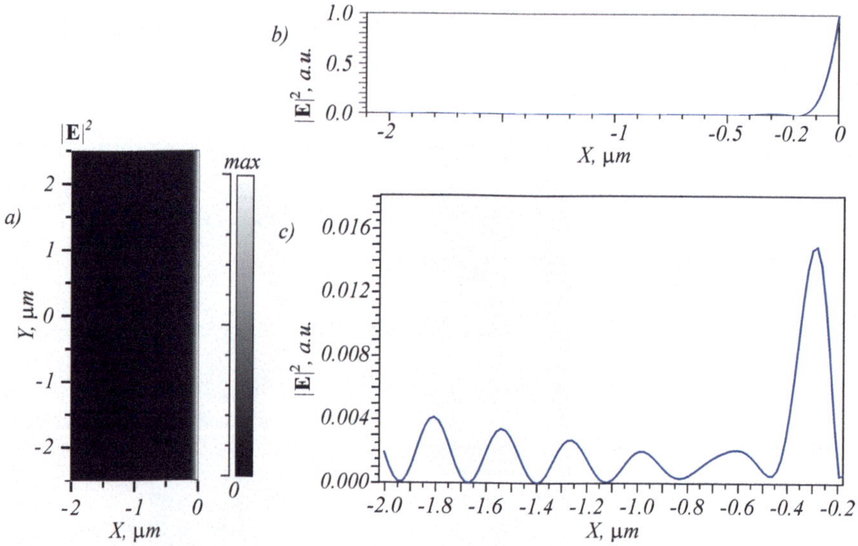

Fig. 1.2 2D grayscale intensity distribution in relative units in the shadow region (**a**) and its profile along the x-axis at $y = 0$ (**b**) and its enlarged view (**c**)

and Fig. 1.3 shows the longitudinal projection of the canonical energy flux, which is calculated using the expression:

$$\mathbf{P}_{or} = \sum_{p=x,y,z} \operatorname{Im}\left(E_p^* \nabla E_p\right). \tag{1.29}$$

It is seen from Fig. 1.2 that there are intensity lobes with a period of approximately 300 nm (less than the wavelength) in the shadow region and the intensity value in local maxima is less than 1% of the incident wave intensity (from unity). Figure 1.3 depicts the 2D distribution of the longitudinal coordinate of the canonical energy flux in the shadow region in the same dimensions as the intensity in Fig. 1.2 and its profiles along the x axis. It is seen from Fig. 1.3 that there is a reverse canonical energy flux in the shadow region (i.e. at negative values of the transverse coordinate x). Moreover, the approximate period of its modulation is the same as the period of intensity modulation in Fig. 1.3c. The magnitude of the reverse flux, i.e. the tilt of the wave vector in the negative direction of the z axis, is small. The results in Fig. 1.3 confirm the conclusions of the theory that at negative values of the transverse coordinate (in the shadow region) there is a reverse canonical energy flux (1.27).

We have deduced an analytic relationship, Eq. (1.9), to describe the longitudinal component of a canonical TE-polarized light energy flow that occurs after the light beam is diffracted by a rectangular screen blocking half the space. The relationship derived suggests that at any distance from the screen along the optical axis, a screen-parallel diffraction pattern is formed with a multitude of fringes parallel to the screen edge on which the canonical energy flow is negative. Thus, we have concluded that a localized canonical energy backflow occurs there. An analysis of Eq. (1.9) has shown that owing to the field phase derivative relative to the longitudinal coordinate entering Eq. (1.15), the wave-vector is counter-directed to the optical axis. Also,

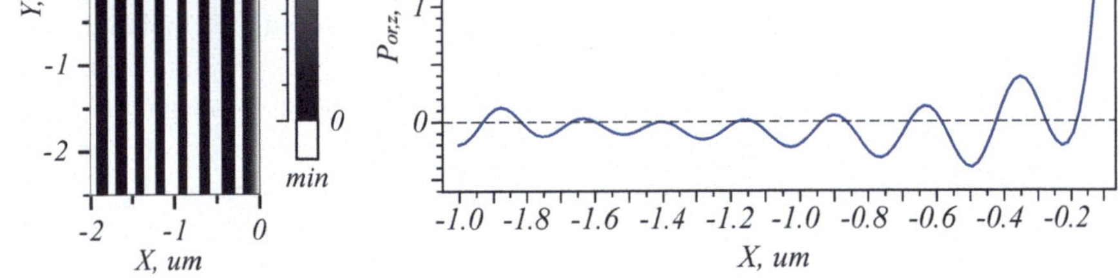

Fig. 1.3 2D distribution of the canonical energy flux in the shadow region (**a**) and its profile along the x-axis at $y = 0$ (**b**)

Eq. (1.15) suggests that the canonical energy backflow occurs in those areas where localized intensity and phase modulation periods are smaller than the wavelength. Similar to evanescent waves, in the afore-mentioned areas, the local wave-number is larger than the incident field wave-number. The propagation of light through such areas can be interpreted as travelling through an aperiodic subwavelength grating, leading to the generation of evanescent waves and reverse waves with a backward directed wave-vector. In the case of non-paraxial propagation of a TE-polarized half-plane wave in the shadow region, where the intensity is almost zero, the reverse canonical energy flux also takes place (1.27).

1.2 Canonical Energy Backflow in Airy Beams

Accelerating Airy beams are described by a complex amplitude, which is an exact solution of the 2D paraxial propagation equation [25–27]. They propagate on parabolic trajectories, although there are also accelerating beams propagating along hyperbolic paths [28]. The Airy beams and their modifications [29–34] are used for filamentation [35], micromanipulation [36, 37], in micromachines [38] and plasmonics [39–41]. It has been demonstrated [9] that in the cross-section of a 2D nonparaxial Airy beam, a backward canonical energy flow arises. However, the nonparaxial Airy beam in [9] was only studied numerically, with no analytical estimations given. An energy backflow was also found to occur in other types of the optical beams. For instance, the work [42] analyzed superposition of differently polarized vector vortex Bessel beams, having numerically shown that the longitudinal Poynting vector component can take negative values. In [43], presence of negative values of the Poynting vector was numerically shown for 2D Bessel beams, whose amplitude contains a product of two Bessel functions of the order −1/4 with different arguments. A three-dimensional analogue of this beam has previously been considered [44]. It has been numerically shown [45] that inside a dielectric microsphere with a nanohole, a backward canonical energy flow takes place. It has been shown [10] that on the optical axis of a tightly focused circularly polarized optical vortex, the longitudinal Poynting vector component can be negative. In the sense of the negative values of longitudinal Poynting vector component, the backward energy flow was also found in other beam types [6, 7].

In this paper, we demonstrate both theoretically and numerically that in the far sidelobes of the intensity pattern of paraxial 2D Airy beams with TE-polarization, the longitudinal component of the canonical energy flow [15] has negative values. In this area, the wavevector is directed oppositely to the beam propagation axis. Besides, in these areas of the backward energy flow, the local period of intensity and phase modulation is less than the wavelength. A 2D TE-polarized light field does not have the spin and the canonical energy flow, thus coinciding with the Poynting vector. In this work, we show that the energy backflow is proportional to the negative longitudinal component of the wavevector, suggesting that the wavevector is directed oppositely to the propagation direction of the light field.

We demonstrate for the first time that this phenomenon arises not only in nonparaxial Airy beams, but also in paraxial beams, which do not have evanescent waves. The phenomenon of energy backflow in a nonparaxial Airy beam can be explained by the presence of local superoscillations regions. The superoscillation effect arises when a paraxial light field with a limited spatial spectrum can have local amplitude modulations with a period smaller than the boundary values of the Fourier spectrum [46]. It is also known [17] that near the center of the optical vortex $(x + iy)^n$, the modulus of the wave vector K, equal to the phase gradient $K = |\nabla(n\arctan(y/x))| = n/\sqrt{x^2+y^2}$, can be greater than the wave number of the light wave $k = 2\pi/\lambda$.

1.2.1 Two-Dimensional Accelerating Airy Beam

At an arbitrary distance z from the initial plane, similar to a Bessel beam or a plane wave, the complex amplitude of a 2D Airy beam with infinite energy is given by [25]

$$E(x,z) = \mathrm{Ai}\left(s - \frac{\xi^2}{4}\right)\exp\left(ikz - \frac{i\xi^3}{12} + \frac{is\xi}{2}\right), \quad (1.30)$$

where s and ξ are the dimensionless coordinates, $s = x/x_0$, $\xi = z/(kx_0^2)$, x and z are the transverse and longitudinal coordinates, k is the wavenumber of monochromatic light, x_0 is an arbitrary scaling parameter (further, we suppose that $x_0 = \lambda$ with λ being the wavelength), $Ai(x)$ is the Airy function [23]:

$$\begin{aligned} \mathrm{Ai}(x) &= \frac{1}{\pi}\left(\frac{x}{3}\right)^{1/2} K_{1/3}\left(\frac{2x^{3/2}}{3}\right), \\ \mathrm{Ai}(x) &= \frac{1}{2\pi}\int\limits_{-\infty}^{\infty}\exp\left(\frac{it^3}{3} + itx\right)dt. \end{aligned} \quad (1.31)$$

The Airy function in Eq. (1.31) is expressed in the integral form through a modified Bessel function of the second kind $K_{1/3}(x)$. The complex amplitude of a finite-energy Airy beam is also known and given by [27]

$$\begin{aligned} E_a(x,y) = \mathrm{Ai}\left(s - \frac{\xi^2}{4} + ia\xi\right) \times \\ \exp\left(ikz - \frac{i\xi^3}{12} + \frac{is\xi}{2} + \frac{ia^2\xi}{2} - \frac{a\xi^2}{2} + as\right), \end{aligned} \quad (1.32)$$

with $a > 0$ being a dimensionless parameter that ensures the exponential decay of amplitude (1.32) at negative values x. Below, we use an asymptotic representation of the Airy function at large argument values and therefore, we can write down [23] ($x \to \infty$):

$$\mathrm{Ai}(-x) \approx \frac{1}{\sqrt{\pi}x^{1/4}} \sin\left(\frac{2}{3}x^{3/2} + \frac{\pi}{4}\right). \tag{1.33}$$

The intensity of the Airy beam (1.30) is determined only by the Airy function:

$$I(x,z) = \left(\mathrm{Ai}\left(s - \frac{\xi^2}{4}\right)\right)^2. \tag{1.34}$$

The diffraction pattern of a 2D Airy beam consists of a set of parabolas. The main parabola, where the greatest of the local maxima resides, can be easily obtained by equating the argument of the intensity (1.34) to zero:

$$x = \frac{z^2}{4k^2x_0}. \tag{1.35}$$

1.2.2 Canonical Energy Flow of a 2D Airy Beam

In work [15], it was shown that the Poynting vector includes, up to insignificant constants, two vectors that describe the orbital (canonical) energy flow P_{or} and the spin flow P_{sp}:

$$\mathbf{P} = \mathbf{P}_{or} + \mathbf{P}_{sp} = \mathrm{Im}(\mathbf{E}^* \cdot (\nabla)\mathbf{E}) + \frac{1}{2}(\nabla \times \mathbf{S}), \quad \mathbf{S} = \mathrm{Im}(\mathbf{E}^* \times \mathbf{E}). \tag{1.36}$$

Based on Eq. (1.36), the formula for deriving the longitudinal component of the orbital energy flow can be written as:

$$P_{or,z} = \mathrm{Im}\left(E_x^* \frac{\partial}{\partial z}E_x + E_y^* \frac{\partial}{\partial z}E_y + E_z^* \frac{\partial}{\partial z}E_z\right). \tag{1.37}$$

We will consider here a 2D Airy beam with linear TE-polarization, i.e. the beam (1.30) has only one component of the electric vector, directed along the coordinate y, $E_y(x, z)$. Along this coordinate, nothing changes. Thus, in our case, only the second term is nonzero in the expression for the energy flow (1.37). From Eq. (1.37), it can be shown that the canonical energy flow equals the intensity multiplied by the derivative of the phase of field (1.30) with respect to the longitudinal coordinate. Thus, an expression for the longitudinal component of the canonical energy flow of the Airy beam (1.30) is given by

$$P_{or,z} = \mathrm{Im}\left(E_y^* \frac{\partial}{\partial z} E_y\right) = |E(x,z)|^2 \frac{\partial}{\partial z} \arg(E(x,z)) = \\ \left(\mathrm{Ai}\left(s - \frac{\xi^2}{4}\right)\right)^2 \frac{\partial}{\partial z}\left(kz + \frac{s\xi}{2} - \frac{i\xi^3}{12}\right). \tag{1.38}$$

At a given distance z, a canonical energy backflow arises at such values x of the transverse coordinate where the derivative of phase in Eq. (1.38) is negative:

$$k + \frac{x}{2kx_0^3} - \frac{z^2}{4k^3x_0^6} < 0. \tag{1.39}$$

Condition (1.39) is fulfilled for the transverse coordinates satisfying the following inequality:

$$x < -2k^2x_0^3 + \frac{z^2}{2k^2x_0^3}. \tag{1.40}$$

As seen from Eq. (1.40), the energy backflow occurs at the coordinates (x, z) on parabolas similar to those formed by the intensity maxima and minima of the Airy beam. Thus, with increasing distance from the initial plane, the transverse coordinates at which the energy backflow occurs are shifted in the positive x-axis direction. We can obtain transverse coordinates x at which the energy backflow occurs in the initial plane ($z = 0$). At $z = 0$, the energy backflow occurs at

$$x < -2k^2x_0^3. \tag{1.41}$$

We note that beams (1.30) and (1.32) are assumed to be paraxial. Thus, such beams do not produce subwavelength intensity patterns in the transverse planes, i.e. the scaling parameter x_0 should exceed the wavelength λ. As shown in [9], the paraxial regime of an Airy beam propagation at small value $a = 0.01$ starts from the value of $x_0 > 2\lambda$ [9]. For instance, at $x_0 = 2\lambda$, the transverse coordinate starting from which an energy backflow occurs is defined by the inequality $x < -64\pi^2\lambda \approx -631\lambda$. Based on Eq. (1.33), we find that at this distance from the first, and largest, local intensity maximum of 0.3—at which the Airy function equals $\mathrm{Ai}(-1) = 0.54$—the intensity is about 0.017. We note that for the negative coordinates x satisfying inequality (1.41), the canonical energy flow in (1.38) is everywhere negative, as far as minus infinity. Equation (1.40) suggests that with increasing distance z from the initial plane, the boundary of the energy backflow region is being shifted towards increasing negative transverse coordinates. It is interesting that the energy backflow persists at any distance from the initial plane z, suggesting that it is generated not due to evanescent waves, which are present in the nonparaxial case near the initial plane. Our guess is that in a paraxial Airy beam, the energy backflow is generated due to a specific term in the beam phase, proportional to xz. This term causes an unlimited growth of the longitudinal derivative of phase as the negative transverse coordinate x is decreasing indefinitely. We can estimate a period of the Airy beam intensity modulation in the domain of negative coordinates x where the energy backflow occurs. Thus, we will find how many periods of intensity modulation are contained in 631λ? To do so, we consider the asymptotic relationship in (1.33). Positive maxima of amplitude (1.33) are at points where the sine equals 1:

$$x_p = x_0\left(\frac{3}{2}\pi\left(p+\frac{1}{4}\right)\right)^{2/3}, \quad p = 0, 1, 2, \ldots \tag{1.42}$$

Then, the local distance between two neighboring amplitude maxima is approximately equal to:

$$\Delta x = x_{p+1} - x_p \approx \frac{2}{3}x_0\left(\frac{3}{2}\pi\right)^{2/3} p^{-1/3} \approx \frac{3.78\lambda}{p^{1/3}}, \quad p = 1, 2, 3, \ldots \tag{1.43}$$

Equation (1.43) suggests that with increasing negative values of the transverse coordinate, the magnitude of the local period of intensity and phase modulation is decreasing. Using Eq. (1.43), we obtain an approximate number of periods of intensity modulation p, after which the energy backflow is generated:

$$\begin{aligned} p\Delta x &= \frac{3.78\lambda}{p^{1/3}} p = 631\lambda, \\ p &= \frac{(631)^{3/2}}{3.78} \approx 4193. \end{aligned} \tag{1.44}$$

Notwithstanding a significant number of intensity modulation periods in (1.44) along the transverse coordinate at $x < 0$, prior to the energy backflow occurrence, the intensity of the energy backflow amounts nearly to 6% of the maximum value. Thus, the energy backflow can be measured experimentally. It is worth noting that at the threshold coordinate x in (1.41) of the energy backflow occurrence, the intensity modulation period is about one fourth of the wavelength. We also note that starting approximately from $p > (3.78)^3 \approx 54$, the local intensity modulation period in (1.43) is smaller than the wavelength. Therefore, the energy backflow occurs at negative transverse coordinates $x < 0$ where the period of intensity and phase modulation is subwavelength. The magnitude of the longitudinal component of the wavevector of the light field (1.30) is equal to the z-derivative of phase of field (1.30). Hence, from Eqs. (1.38) and (1.39) it follows that at $z = 0$, the wave-vector of the Airy beam is larger in magnitude than the wave-number k:

$$k - \frac{|x|}{2kx_0^3} < -k, \quad |x| > 4k^2x_0^3 \approx 1263\lambda. \tag{1.45}$$

From inequalities (1.45), it follows that the threshold coordinate x at which the wavevector of light field (1.30) is larger than k is twice as large as threshold coordinate (1.41) of the energy backflow occurrence. Interestingly, amplitudes and phases of subwavelength period, as well as large wavenumbers, are uncharacteristic of paraxial light fields. Subwavelength areas in the light field (1.30) can be explained by the fact that the absolute value of the amplitude of Fourier-spectrum (1.31) is the same for all spatial frequencies. Therefore, as seen from asymptotic relationship (1.33), at large absolute values of the negative transverse coordinate x, the local amplitude modulation period of the Airy beam tends to zero. This can account for the energy backflow of a paraxial Airy beam.

1.2.3 Canonical Energy Flow for a 2D Nonparaxial Airy Beam

It has been shown numerically [9] that in the neighborhood of the initial plane, a nonparaxial 2D Airy beam has a negative canonical energy flow. Here, we derive analytical expressions for such beams, making up for their absence in [9]. With the Airy function being a Fourier transform of the cubic phase function in (1.31), the complex amplitude of a nonparaxial Airy beam in the initial plane can be written as

$$E(x, z=0) = 2\pi \mathrm{Ai}(kx) = \int_{-\infty}^{\infty} A(\xi)\exp(ikx\xi)d\xi = \int_{-\infty}^{\infty} \exp\left(i\frac{\xi^3}{3} + ikx\xi\right)d\xi. \tag{1.46}$$

The function $A(\xi)$ in Eq. (1.46) describes the amplitude of a 2D plane wave spectrum of the Airy beam. Then, the complex amplitude of a nonparaxial Airy beam at an arbitrary propagation distance z from the initial plane reads as:

$$E(x, z) = = \int_{-\infty}^{\infty} \exp\left(i\frac{\xi^3}{3} + ikx\xi + ikz\sqrt{1-\xi^2}\right)d\xi. \tag{1.47}$$

Differentiation of Eq. (1.47) with respect to the longitudinal coordinate yields:

$$\frac{\partial}{\partial z}E(x,z)= ik\int_{-\infty}^{\infty}\sqrt{1-\xi^2}\exp\left(i\frac{\xi^3}{3}+ikx\xi+ikz\sqrt{1-\xi^2}\right)d\xi. \tag{1.48}$$

Supposing that, like in a TE-wave, field (1.47) is the y-component of the electric field strength vector, the longitudinal component of the canonical energy flow in the initial plane ($z=0$) can be written as:

$$\begin{aligned}P_{or,z}&=\operatorname{Im}\left(E^*\frac{\partial}{\partial z}E\right)\Bigg|_{z=0}=\\&=2\pi k\mathrm{Ai}(kx)\left(\int_{-1}^{1}\sqrt{1-\xi^2}\cos\Phi(\xi)d\xi-\right.\\&\left.\int_{-\infty}^{-1}\sqrt{\xi^2-1}\sin\Phi(\xi)d\xi-\int_{1}^{\infty}\sqrt{\xi^2-1}\sin\Phi(\xi)d\xi\right),\end{aligned} \tag{1.49}$$

with

$$\Phi(\xi)=\frac{\xi^3}{3}+kx\xi.$$

Since the function $\Phi(\xi)$ in Eq. (1.49) is odd, two last integrals cancel each other and Eq.(1.49) is reduced to

$$P_{or,z}=4\pi k\mathrm{Ai}(kx)\left(\int_{0}^{1}\sqrt{1-\xi^2}\cos\Phi(\xi)d\xi\right). \tag{1.50}$$

The Airy function in (1.50) is alternating and the integral can have negative values. Therefore, near the initial plane of the nonparaxial Airy beam (1.46) or (1.47), an energy backflow occurs in the regions where the Airy function and integral (1.50) are of different sign. An energy backflow in nonparaxial Airy beams was also discovered in [9], but the expression (1.50) for the canonical flow in the initial plane was not obtained. Besides, the main contribution into the backflow in the initial plane was supposed [9] to be given by evanescent waves. Expression (1.50) confirms the said supposition, as from Eq. (1.50), the complex amplitude of the Airy beam is seen to also consist of evanescent waves, because the plane waves spectrum in (1.50) is integrated from negative to positive infinity. Separating in (1.50) the intensity and the z-derivative of phase yields:

$$P_{or,z} = |E(x,z)|^2 \frac{\partial}{\partial z} \arg E(x,z) =$$
$$4\pi k |\mathrm{Ai}(kx)|^2 \frac{\int\limits_0^1 \sqrt{1-\xi^2} \cos \Phi(\xi) d\xi}{\mathrm{Ai}(kx)}, \tag{1.51}$$
$$k_z = \frac{\partial}{\partial z} \arg E(x,z) = \frac{\int\limits_0^1 \sqrt{1-\xi^2} \cos \Phi(\xi) d\xi}{\mathrm{Ai}(kx)}.$$

From the second equation in (1.51), it is seen that near the zero of the Airy function in the denominator, the longitudinal projection of the wave-vector k_z will tend to infinity when x tends to x_0, $\mathrm{Ai}(kx_0) = 0$. On passing through the zero intensity point, the wave-vector projection changes sign, which means that an energy backflow occurs near the intensity null.

In work [9], a finite-energy Airy beam was discussed. In this case, instead of Eq. (1.47), the complex amplitude reads as:

$$E_a(x,z) = \int\limits_{-\infty}^{\infty} \exp\left(a^3 - 3a\xi^2 - i3a^2\xi + \right.$$
$$\left. i\frac{\xi^3}{3} + ikx\xi + ikz\sqrt{1-\xi^2}\right) d\xi, \quad a > 0. \tag{1.52}$$

In Eq. (1.51), a is a dimensionless constant that makes the energy of the Airy beam finite. Then, instead of Eq. (1.50), we obtain the following longitudinal component of the canonical energy flow:

$$P_{or,z}^{(a)} = 4\pi k \mathrm{Ai}(kx) \exp\left(kxa + \frac{a^3}{3}\right) \times$$
$$\left(\int\limits_0^1 \exp(-a\xi^2) \sqrt{1-\xi^2} \cos \Phi_a(\xi) d\xi\right), \tag{1.53}$$
$$\Phi_a(\xi) = \frac{\xi^3}{3} + kx\xi - a^2\xi.$$

Similarly to Eq. (1.50), Eq. (1.52) also indicates that near the origin, the canonical energy flow can have negative values when the Airy function and the integral in (1.52) are of different sign.

1.2.4 Numerical Simulation Results

Using Eq. (1.30) for a paraxial 2D Airy beam, we computed its intensity distribution (Fig. 1.4a) along the transverse coordinate x and the distribution of the longitudinal component of the canonical energy flow (Fig. 1.4b) in the initial plane for the wavelength $\lambda = 532$ nm and scaling parameter $x_0 = 2\lambda$. Figure 1.1a shows that at $x = -\lambda$, the maximum intensity is close to 0.3 in relative units. The canonical flow (Fig. 1.4b) is seen to be positive and very close in shape to the shape of the intensity. The field amplitude and intensity at different distances from the initial plane were computed by formula (1.30), whereas the energy backflow was computed by using the general expression in (1.37), with the phase being differentiated numerically using two complex amplitudes in two nearby planes.

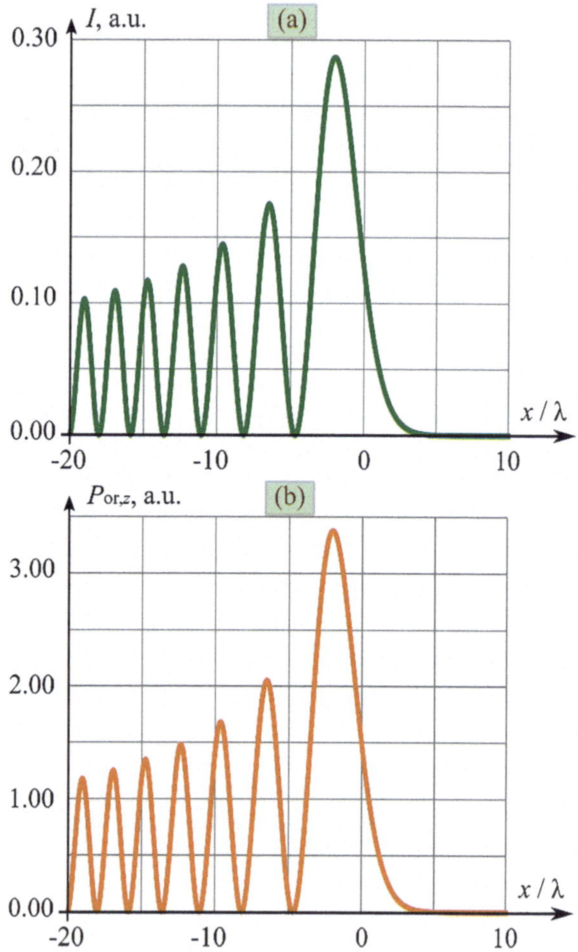

Fig. 1.4 Profiles of (**a**) the intensity (1.34) and (**b**) the longitudinal canonical energy flow component in (1.37) for an unbounded ($a = 0$) 2D Airy beam in (1) in the initial plane ($z = 0$) at the wavelength $\lambda = 532$ nm and $x_0 = 2\lambda$

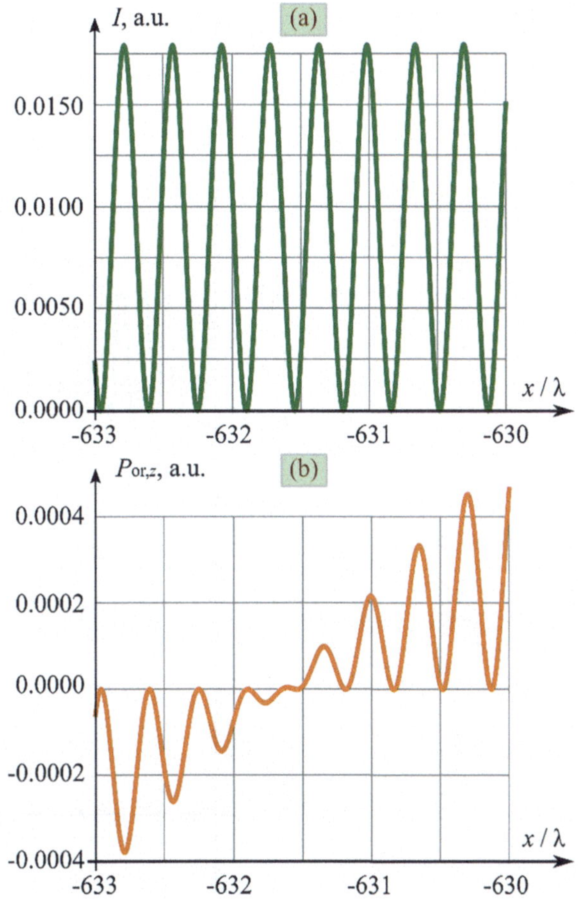

Fig. 1.5 Profiles of (**a**) intensity (1.34) and (**b**) the longitudinal component of canonical energy flow (1.37) for 2D Airy beam (1.30) with unlimited energy ($a = 0$) at $z = 0$, simulated at the wavelength $\lambda = 532$ nm and $x_0 = 2\lambda$. The cross-sections sections of the intensity and the energy flow are illustrated in the interval from -633λ to -630λ

Figure 1.5 also illustrates profiles of the intensity (Fig. 1.5a) and the canonical energy flow (Fig. 1.5b) in the initial plane of the same beam and at the same parameters as in Fig. 1.4, but at a greater transverse distance from the origin for the negative transverse coordinates. In Fig. 1.5, transverse sections of the intensity and the energy flow are shown in the interval from -633λ to -630λ. As seen in Fig. 1.5, the intensity at such a large distance from the origin decreases and amounts to nearly 5% of the maximal intensity. The transverse coordinate $x = -631.5\lambda$ is a threshold after which a backward canonical energy flow occurs. The energy flow remains negative at any negative coordinate x smaller than the threshold value. The numerically calculated threshold coordinate is almost the same as the threshold coordinate predicted theoretically in Eq. (1.41): $x < -64\pi^2\lambda \approx -631\lambda$. The numerical simulation has shown that the energy backflow occurs at distances larger than several wavelengths from the initial plane. Thus, we can infer that the energy backflow in a paraxial Airy beam occurs not due to evanescent waves, whose magnitude is significant at a distance of about one wavelength, but due to a

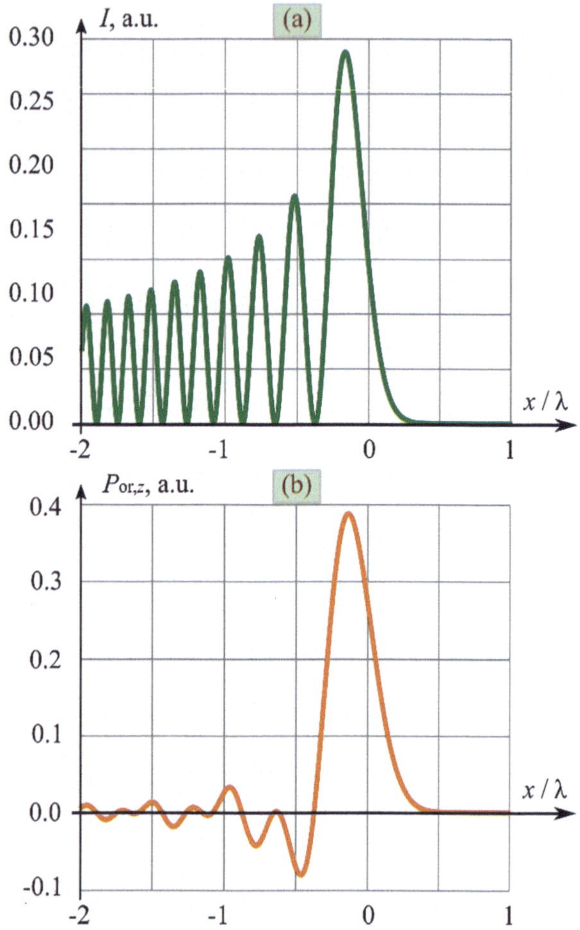

Fig. 1.6 Profiles of (**a**) intensity (1.47) at $x_0 = 0.2\lambda$, $z = 0.02\lambda$, $\lambda = 532$ nm and (**b**) the longitudinal canonical energy flow component in (1.37)

peculiarity of the phase function of Airy beam (1.30). It can be said that the Airy beam with unlimited energy is actually 'limited' as only its part (up to a threshold coordinate) propagates in the positive direction of the optical axis.

Shown in Fig. 1.6 are simulation results of a nonparaxial Airy beam. The scaling parameter was chosen to be less than the wavelength, $x_0 = 0.2\lambda$. The beam has a finite energy, $a = 0.1$. The amplitude and intensity distribution of the nonparaxial Airy beam at distance $z = 0.02\lambda$ from the initial plane was computed by Eq. (1.47). Figure 1.6 illustrates distributions of the intensity (Fig. 1.6a) and the longitudinal energy flow (Fig. 1.6b) near the initial plane of the Airy beam, computed using expressions (1.47) and (1.37). From Fig. 1.6, the period of intensity modulation is seen to be about one tenth of that in Fig. 1.4, as is the parameter x_0. It is also seen that the energy flow pattern is near the same as that in Fig. 1.4. As seen in Fig. 1.6b, the energy backflow at the chosen parameters already occurs in the first and the second sidelobes, which is consistent with the results of work [9]. Besides, the energy backflow is four times smaller than the forward energy flow. From Fig. 1.6, the energy backflow is seen to first occur exactly at the zero-intensity point, with its coordinates derivable from (1.50). With the first zero of the Airy function being known to be approximately defined by $kx = -2.34$, the intensity null is at $x = -0.37\lambda$, which is seen to comply with Fig. 1.6. The presence of the intensity null allows us to devise a mechanism of generating the reverse energy flow, which we see as follows. On passing through the intensity null point, the phase experiences a phase jump by π. On further propagation along the z-axis, an intensity minimum will be found at the same point x, instead of the intensity null. On passing through this minimum, the phase will change smoothly by a value less than π, suggesting that with increasing z, the phase is decreasing rather than increasing. This scenario is only possible if the light wave propagates backwards. From Fig. 1.6, the maximum backward energy flow is also seen to comprise one fifth of the maximum direct energy flow.

To demonstrate a more realistic situation, we numerically simulated the free-space propagation of a 2D Airy beam with linear polarization along the y-axis. Along this axis, the Airy beam remained unchanged, $\mathrm{Ai}(x, z = 0)$. The numerical simulation was conducted by using vector Rayleigh-Sommerfeld integrals. For the field component E_y, the Rayleigh-Sommerfeld integral is given by [47]:

$$E_y(z>0) = -\frac{z}{2\pi} \iint\limits_{u,v} E_y(u, v, z=0) \frac{\exp(ikL)}{L^2} \left(ik - \frac{1}{L}\right) dudv, \tag{1.54}$$

where $L = [(u - x)^2 + (v - y)^2 + z^2]^{1/2}$, z is the longitudinal coordinate, k is the wavenumber of light of wavelength λ. The field component E_z can be derived from the initial transverse field component E_y:

$$E_z(x, y, z>0) = \frac{1}{2\pi} \iint\limits_{u,v} E_y(u, v, 0)(u - y) \frac{\exp(ikL)}{L^2} \left(ik - \frac{1}{L}\right) dudv. \tag{1.55}$$

The numerical simulation was performed with the following parameters: a 1200×1200-pixel initial field of 10.6×10.6 μm (~ $\lambda/60$ per pixel), the wavelength $\lambda = 532$ nm, $E_y = \mathrm{Ai}(x/x_0)$, and $x_0 = 0.2\lambda$. A 600×50-pixel output field with the computing domain of -1 μm $< X < 0.5$ μm, -0.024 μm $< Y < 0.024$ μm. The field was computed twice, at a 10-nm and 20-nm distance from the initial plane, in order to obtain the z-derivative. Figure 1.7 shows cross-sections of the resulting field: intensity $|\mathbf{E}|^2$ (Fig. 1.7a) and the longitudinal canonical energy flow component, $P_{or,z}$ (Fig. 1.7b). The energy flow was computed using Eq. (1.37). Accounting for the longitudinal component of field (1.54) left the intensity pattern (Fig. 1.7a) almost unchanged, when compared with the pattern in the initial field (Fig. 1.6a). Figure 1.4b shows that a canonical backflow occurs (in the first and third sidelobes). In addition, the energy backflow amplitude is seen to be nearly 6 times lower than that of the direct energy flow. Considering that the intensity modulation period is

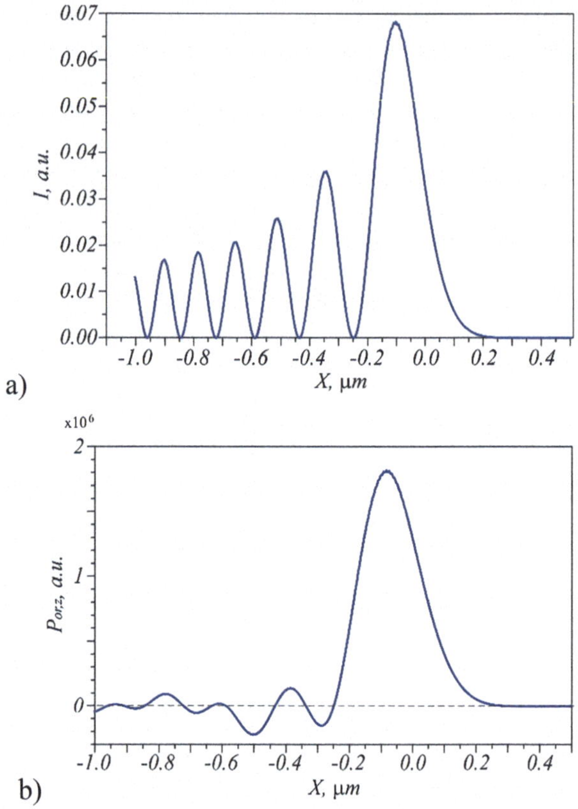

Fig. 1.7 Profiles along the axis x of (**a**) the intensity distribution $|\mathbf{E}|^2$ of the Airy beam $E_y = \mathrm{Ai}(x/x_0)$ at $x_0 = 0.2\lambda$ at a 10-nm distance from the initial plane and (**b**) the cross-section of the longitudinal canonical energy flow component, $P_{\mathrm{or},z}$ in (1.37) at the same distance from the initial plane

about half the wavelength, the energy backflow is generated due to evanescent waves near the initial plane. A comparison of energy backflows in the nonparaxial 2D Airy beam (Fig. 1.6b) and the vector 3D Airy beam (Fig. 1.7b) shows them to be similar.

Summing up, we have analyzed a well-known two-dimensional paraxial Airy beam with unlimited-energy, Eq. (1.30), having derived a simple analytical expression for the longitudinal component of the canonical energy flow, Eq. (1.38). The Airy beam under analysis was assumed to be TE-polarized. We have derived expression (1.40) that defines a threshold transverse coordinate such that for all smaller negative transverse coordinates, a negative canonical energy flow starts to occur near the initial plane. The intensity sidelobes of the Airy beam with unlimited-energy have been shown to decay slowly. We have shown that in the local region where the energy backflow begins, the intensity is about 6% of the maximal intensity of the first lobe, but the distance to this region from the origin is nearly 630 wavelengths, Eq. (1.44), and the number of sidelobes exceeds 4000. The scaling factor ($x_0 = 2\lambda$) of the Airy beam was chosen such that the beam would remain paraxial. The energy backflow has been revealed to occur at an arbitrary distance from the initial plane, since the energy backflow in a paraxial Airy beam is generated not due to evanescent waves, but due to a specific term in the beam phase that is proportional to the product xz. This term leads to the unlimited increase of the longitudinal phase derivative at the unlimited decrease of the negative transverse coordinate x. In a nonparaxial finite-energy beam with a scaling factor $x_0 = 0.2\lambda$, the energy backflow occurs only near the initial plane, starting as early as the second sidelobe [9]. We have reproduced the numerical simulation reported in [9], having validated the results (Fig. 1.6). In addition, we have theoretically predicted, Eq. (1.52), that near the initial plane of a nonparaxial 2D Airy beam, a backward canonical energy flow occurs near the central lobes of the Airy function. To numerically simulate a more realistic situation, we employed the Rayleigh-Sommerfeld integrals and computed the free-space propagation of a 2D linearly polarized Airy beam, Eq. (1.53), while accounting for the longitudinal field component in (1.54). In this case, a backward canonical energy flow (Fig. 1.7b) was also observed near the initial plane. Although the evanescent waves are present in the near-field of the Airy beams in Fig. 1.6, these are not the reason why a backward energy flow occurs. Instead, the reverse energy flow occurs due to isolated intensity nulls, as in a local subwavelength region in their vicinity, the beam propagates in the negative direction to compensate for a phase jump by π.

For the reverse energy flow in the near field to be measured experimentally, a non-paraxial Airy beam needs to be generated. The amplitude (1.46) of a non-paraxial Airy beam is a complex function. To implement the Airy beam phase, phase jumps by π need to be realized. This can be done through the use of a glass plate with a step-wise relief, with the step depth providing the designed π-phase delay. For the glass plate, the required step depth would be about a wavelength of incident light. After travelling over the required near-wavelength distance providing the phase delay of π, the near-field condition is fulfilled no more. Besides, for the intensity to be measured with a subwavelength accuracy, a near-field scanning optical microscope needs to be utilized. The aforesaid factors make experimental measurements of the reverse canonical energy backflow in the near field of the Airy beam challenging.

It is known that other curvilinear beams exist with sharp bending paths, such as half-Bessel, Weber and Mathieu beams. It can be supposed and investigated in future that the canonical energy backflow also exists in such beams and conditions of this backflow are also to be determined both in paraxial and nonparaxial limits.

1.3 Backward Canonical Energy Flow at the Near Field of a Non-Paraxial 2D Sinc-Beam

The canonical energy flux vector of light introduced into optics in [12, 48, 49] is only a part of the Poynting vector associated with the orbital energy flux without taking into account the spin flux. The reverse canonical energy flux found, for example, in a 2D non-paraxial Airy beam in [9] differs from the reverse energy flux found using the Poynting vector in a sharp focus [1, 2, 10] and in diffraction-free beams [6, 7]. For vector fields, negative values of the Poynting vector longitudinal component occur when the longitudinal projection of the orbital canonical flux vector, which is directed in the forward direction, is smaller in modulus than the longitudinal component of the spin flux vector, that is directed in the opposite direction [16]. The inverse canonical energy flux in a non-paraxial Airy beam [9] occurs near the initial plane, where inhomogeneous evanescent waves are present. Theoretical works [50, 51] are also related to the canonical reverse flow, in which the probabilities of the reverse flow of a superposition of several plane waves are considered. In [15] it is shown that the canonical energy flow in propagating paraxial vortex beams is positive and the azimuthal part of the canonical flow is directed in one direction, determined by the sign of the topological charge and does not depend on the polarization state of the beam. In [43] the reverse flow was found for 2D Bessel beams, the amplitude of which consists of the product of two Bessel functions of the order of $-1/4$ with different arguments.

In this section, several non-paraxial 2D TE-polarized beams are considered. Using plane wave expansion, explicit expressions are obtained for the longitudinal projection of the canonical energy flux vector in the initial plane. These expressions are the product of the light field amplitude in the initial plane and the longitudinal derivative of this amplitude. For a sinc beam, this derivative can be obtained as a cylindrical function. For all beams, a reverse canonical flux is found and shown to arise in the presence of evanescent waves. Moreover, the reverse flux appears in the odd side lobes (antinodes) of the standing evanescent wave. If the beam parameters are chosen such that they do not have evanescent waves, then the reverse energy flux does not appear near the initial plane.

1.3.1 Canonical Energy Flow of a 2D Sinc Beam

Let us consider a 2D TE-polarized non-paraxial sinc beam whose plane wave spectrum amplitude is described by the rect function (by a step). Then the complex amplitude of such a beam in the initial plane ($z = 0$) is equal to:

$$E(x, z=0) = \int_{-\infty}^{\infty} A(\xi) \exp(ikx\xi) d\xi = \int_{-\infty}^{\infty} \mathrm{rect}\left(\frac{\xi}{a}\right) \exp(ikx\xi) d\xi = 2a \frac{\sin(kxa)}{kxa}. \tag{1.56}$$

In (1.56) x is the transverse coordinate, z is the longitudinal coordinate, k is the wave number of light, a is a dimensionless parameter that determines the plane waves spectrum half-width (if $a > 1$, then the spectrum also contains evanescent waves).

$A(\xi) = rect(2\xi/a)$ is the amplitude of the plane waves spectrum. We say "in the initial plane", although there is only one transverse coordinate x in (1.56). The light field (1.56) also describes a 3D problem, which has two Cartesian coordinates in the transverse plane (x, y), but nothing changes along the y coordinate, only the linear polarization vector is directed along this y coordinate. The complex amplitude of the sinc beam at any distance z can be expressed as:

$$E(x,z) = \int_{-\infty}^{\infty} A(\xi)\exp\left(ikx\xi + ikz\sqrt{1-\xi^2}\right)d\xi = \int_{-a}^{a} \exp\left(ikx\xi + ikz\sqrt{1-\xi^2}\right)d\xi. \tag{1.57}$$

The derivative of field (1.57) with respect to the longitudinal coordinate z in the initial plane can be written as:

$$\frac{\partial}{\partial z}E(x,z)\bigg|_{z=0} = ik\int_{-a}^{a}\sqrt{1-\xi^2}\exp(ikx\xi)d\xi. \tag{1.58}$$

Next, we find an expression for the longitudinal coordinate of the canonical energy flux in the initial plane for the sinc beam (1.57). The Poynting vector $\mathbf{P}$, up to insignificant constants, consists of the sum of two vectors [12, 48, 49], which describe the orbital energy flux (canonical energy flux) $\mathbf{P}_{or}$ and the spin flux $\mathbf{P}_{sp}$:

$$\begin{aligned}\mathbf{P} &= \mathbf{P}_{or} + \mathbf{P}_{sp} = \mathrm{Im}(\mathbf{E}^*\cdot(\nabla)\mathbf{E}) + \frac{1}{2}(\nabla\times\mathbf{S}),\\ \mathbf{S} &= \mathrm{Im}(\mathbf{E}^*\times\mathbf{E}).\end{aligned} \tag{1.59}$$

In (1.59) $\mathbf{S}$ is the spin angular momentum. The longitudinal projection of the orbital energy flow, according to (1.59), is described as follows:

$$P_{or,z} = \mathrm{Im}\left(E_x^*\frac{\partial}{\partial z}E_x + E_y^*\frac{\partial}{\partial z}E_y + E_z^*\frac{\partial}{\partial z}E_z\right). \tag{1.60}$$

Since in our 2D case of diffraction with TE polarization there remains only one projection of the electric field strength vector (1.57), we substitute it into (1.60), and obtain ($a > 1$):

$$
\begin{gathered}
P_{or,z} = \mathrm{Im}\left(E_y^* \frac{\partial}{\partial z} E_y\right)\Bigg|_{z=0} = \mathrm{Im}\left[\frac{2ak\sin(kxa)}{kxa} \times \right.\\
\left.\left(i\int_{-1}^{1}\sqrt{1-\xi^2}e^{ikx\xi}d\xi - \int_{-a}^{-1}\sqrt{\xi^2-1}e^{ikx\xi}d\xi - \int_{1}^{a}\sqrt{\xi^2-1}e^{ikx\xi}d\xi\right)\right] = \\
2ak\frac{\sin(kxa)}{kxa} \times \\
\left(\int_{-1}^{1}\sqrt{1-\xi^2}\cos(kx\xi)d\xi - \int_{-a}^{-1}\sqrt{\xi^2-1}\sin(kx\xi)d\xi - \right.\\
\left.\int_{1}^{a}\sqrt{\xi^2-1}\sin(kx\xi)d\xi\right) = 4ak\frac{\sin(kxa)}{kxa}\int_{0}^{1}\sqrt{1-\xi^2}\cos(kx\xi)d\xi.
\end{gathered}
\tag{1.61}
$$

In (1.61) two integrals in symmetric limits on $-a$ to -1 and from 1 to a of the odd function $\sin(kx\xi)$ compensate each other. The last integral in (1.61) is reference [51], so instead of (1.61) we obtain:

$$
\begin{gathered}
P_{or,z} = 4ak\frac{\sin(kxa)}{kxa}\int_{0}^{1}\sqrt{1-\xi^2}\cos(kx\xi)d\xi = \\
4ak\frac{\sin(kxa)}{kxa}\left(\frac{2\pi}{kx}\right)J_1(kx).
\end{gathered}
\tag{1.62}
$$

The longitudinal projection of the canonical energy flux in the initial plane is seen from (1.62) to be negative when the $\sin(kxa)$ function is negative and the $J_1(kx)/(kx)$ function is positive. Conversely, there is a reverse flux where the $\sin(kxa)$ function is positive and the $J_1(kx)/(kx)$ function is negative. The period of the sinc function in the initial plane is less than the wavelength, $x = \lambda/a$, if $a > 1$. The reason for the formation of the reverse flux (1.62) is that evanescent waves contribute to (7) at $a > 1$. Let us consider a simplified plane waves spectrum $A(\xi)$ to get rid of the integrals in (1.57). Let the symmetric (even) plane waves spectrum consist of three delta functions ($a > 1$):

$$
A(\xi) = A\delta(\xi) + B\delta(\xi - a) + B\delta(\xi - a). \tag{1.63}
$$

A and B in (1.63) are real constants. Then the complex amplitude of the non-paraxial 2D light field (1.57) takes the form:

$$E(x,z)=A\exp(ikz)+2B\exp\left(-kz\sqrt{a^2-1}\right)\cos(kxa). \quad (1.64)$$

The light field (1.64) is a plane wave propagating along the z axis and a standing wave decaying along the z axis, formed by two evanescent waves propagating along the x axis in both directions. The period of the standing wave along the x axis is a times smaller than the wavelength, λ/a. It is easy to find the longitudinal canonical flux (1.60) of the field (1.64) in the initial plane ($z = 0$):

$$P_{or,z}=k(A+2B\cos(kxa)). \quad (1.65)$$

The reverse flow is seen from (1.65) to occur if $A < 2B$. The nature of the canonical reverse flow near the initial plane is seen from (1.65). The reverse flow is formed in the presence of evanescent waves in those places where the amplitude of these waves is negative (in the odd antinodes of the standing wave). The evanescent waves themselves do not transfer energy along the z-axis, but, strangely enough, they participate in the formation of the canonical flow along the z-axis. This happens because the complex amplitude (1.64) is a linear combination of propagating and evanescent waves. Therefore, the evanescent waves participate in formation of the phase function of the entire field, and affect the sign of the longitudinal derivative of the phase function. The conclusion that the evanescent waves are responsible for the reverse flow formation is also confirmed by the fact that there is no reverse flow, when $a < 1$ and no evanescent waves present. Therefore, we can agree with the explanation of the reason for the reverse flow formation given in work [9], that the reverse flow is formed due to the interaction of evanescent and propagating waves.

Note that since for a monochromatic 2D light field with TE polarization there is only one projection of the magnetic field:

$$\begin{aligned} H_x(x,z) &= \frac{i}{k}\frac{\partial}{\partial z}E_y(x,z) = \\ &- \int_{-\infty}^{\infty}\sqrt{1-\xi^2}A(\xi)\exp\left(ikx\xi+ikz\sqrt{1-\xi^2}\right)d\xi, \end{aligned} \quad (1.66)$$

then the longitudinal component of the canonical energy flux (1.60) for 2D fields with TE polarization coincides with the longitudinal projection of the Poynting vector (up to a constant):

$$\begin{aligned} P_z &= -\mathrm{Re}\left(E_y^* H_x\right) = \mathrm{Re}\left(E_y^* \frac{-i}{k}\frac{\partial}{\partial z}E_y\right) = \\ &\frac{1}{k}\mathrm{Im}\left(E_y^*\frac{\partial}{\partial z}E_y\right) = \frac{1}{k}P_{or,z}. \end{aligned} \quad (1.67)$$

1.3.2 Canonical Energy Flux of a 2D Gaussian Beam in the Initial Plane

For a 2D Gaussian beam, we can obtain an expression for the canonical energy flux in the initial plane, similar to expression (1.62) for a sinc beam. In this case, the amplitude of the Gaussian beam in the initial plane do not have negative values, but the reverse flux is present, since the Gaussian amplitude is multiplied by the alternating function $J_1(kx)/(kx)$. Indeed, it is well known that the Fourier transform of a Gaussian function is a Gaussian function. Therefore, instead of expression (1.56) for the Gaussian beam, we obtain:

$$\begin{gathered} E(x, z=0) = \int_{-\infty}^{\infty} A(\xi) \exp(ikx\xi) d\xi = \\ \int_{-\infty}^{\infty} \exp\left(-\frac{\xi^2}{a^2}\right) \exp(ikx\xi) d\xi = a\sqrt{\pi} \exp\left(-\frac{k^2x^2a^2}{4}\right). \end{gathered} \tag{1.68}$$

In (1.68) a is a dimensionless parameter that determines the width of the plane wave spectrum. Evanescent waves are always present in the Gaussian beam spectrum, and the larger a, the greater the contribution to the beam amplitude is made by the evanescent waves, and the narrower the Gaussian beam is in the initial plane. The radius of the Gaussian beam (1.68) in the initial plane is equal to:

$$w = \frac{2}{ka} = \frac{\lambda}{\pi a}. \tag{1.69}$$

If $\pi a > 1$, then the radius of the Gaussian beam is clear from (1.69) to be less than the wavelength, i.e. the Gaussian beam (1.68) is non-paraxial. The derivative with respect to z in the initial plane can be written similar to (1.58):

$$\left.\frac{\partial}{\partial z} E(x, z)\right|_{z=0} = ik \int_{-\infty}^{\infty} \sqrt{1-\xi^2} \exp\left(-\frac{\xi^2}{a^2} + ikx\xi\right) d\xi. \tag{1.70}$$

Next, similarly to (1.61), it can be shown that the evanescent waves are not contribute to the canonical flux in the initial plane, and the final expression for the canonical flux for a non-paraxial 2D Gaussian beam is:

$$P_{or,z} = 2ak\sqrt{\pi} \exp\left(-\frac{k^2a^2x^2}{4}\right) \int_0^1 \sqrt{1-\xi^2} \exp\left(-\frac{\xi^2}{a^2}\right) \cos(kx\xi) d\xi. \tag{1.71}$$

For large values of the parameter $a > > 1$, the integral in (1.71) is similar to the integral in (1.62) and have both positive and negative values. That is, the Gaussian beam in the initial plane also have a reverse canonical flux, like the sinc beam (1.62). This reverse flux is formed due to the fact that the complex amplitude of the Gaussian beam is contributed by evanescent waves, since in the integral (1.68) the integration occurs from minus infinity to plus infinity, although the evanescent waves are not included in the integral in (1.71). For large $a >> 1$, the radius of the non-paraxial Gaussian beam is much smaller than the wavelength, and the Gaussian exponent under the integral in (1.71) can be replaced by 1 in the integration interval. Then the integral in (1.71) is equal to the integral in (1.62) and is equal to $2\pi J_1(kx)/(kx)$. Therefore, the negative values of the canonical energy flux in (1.71) is at a distance greater than half the wavelength from the center of the initial plane, although the radius of the Gaussian beam in the initial plane is much less than half the wavelength and the intensity value is almost zero where there is a reverse flux. Numerical simulation has revealed a very weak reverse flux in the non-paraxial Gaussian beam (1.70).

1.3.3 *Canonical Energy Flux of a 2D Hankel Beam in the Initial Plane*

A 2D non-paraxial Hankel beam is also known, similar to the sinc beam. The difference is that the Hankel beam has a singularity in the initial plane ($z = 0$) at the center ($x = 0$). This is a point source of a cylindrical wave that propagates in the positive direction of the z-axis. For such a beam, it is also possible to obtain an expression for the canonical energy flux in the initial plane, close in structure to expression (1.62), but with a singularity. Here we use reference integrals [52]:

$$\begin{aligned} &\int_1^\infty \frac{\sin(kx\xi)}{\sqrt{\xi^2-1}}d\xi = \frac{\pi}{2}J_0(kx), \\ &\int_1^\infty \frac{\cos(k|x|\xi)}{\sqrt{\xi^2-1}}d\xi = \frac{\pi}{2}Y_0(k|x|), \\ &\int_0^1 \frac{\cos(kx\xi)}{\sqrt{1-\xi^2}}d\xi = \frac{\pi}{2}J_0(kx). \end{aligned} \tag{1.72}$$

In (1.72) $Y_0(x)$ is the Neumann function. Then the complex amplitude in the initial plane for a beam with an angular spectrum equal to

$$A(\xi)=\begin{cases}\dfrac{1}{\sqrt{1-\xi^2}}, & |\xi|<1,\\[2mm] \dfrac{1}{\sqrt{\xi^2-1}}, & |\xi|>1,\end{cases} \tag{1.73}$$

can be expressed as:

$$\begin{gathered}E(x,z=0)=\int_{-\infty}^{\infty}A(\xi)\exp(ikx\xi)d\xi=\int_{-1}^{1}\frac{\exp(ikx\xi)}{\sqrt{1-\xi^2}}d\xi-\\ i\int_{-\infty}^{-1}\frac{\exp(ikx\xi)}{\sqrt{\xi^2-1}}d\xi-i\int_{1}^{\infty}\frac{\exp(ikx\xi)}{\sqrt{\xi^2-1}}d\xi=2\int_{0}^{1}\frac{\cos(kx\xi)}{\sqrt{1-\xi^2}}d\xi-\\ 2i\int_{1}^{\infty}\frac{\cos(kx\xi)}{\sqrt{\xi^2-1}}d\xi=\pi(J_0(kx)-iY_0(k|x|))=\pi H_0^{(2)}(k|x|).\end{gathered} \tag{1.74}$$

In (1.74) $H_0^{(2)}(x)$ is the Hankel function of the second kind. The derivative with respect to z in the initial plane have the form (1.58). Then for the longitudinal projection of the canonical energy flux vector for field (1.68) in the initial plane we obtain an expression similar to (1.61):

$$\begin{gathered}P_{or,z}=\mathrm{Im}\left(E_y^*\frac{\partial}{\partial z}E_y\right)\Bigg|_{z=0}=\\ =\mathrm{Im}\left[\pi kH_0^{(1)}(k|x|)\left(i\int_{-1}^{1}\sqrt{1-\xi^2}A(\xi)e^{ikx\xi}d\xi-\right.\right.\\ \left.-\int_{-\infty}^{-1}\sqrt{\xi^2-1}A(\xi)e^{ikx\xi}d\xi-\int_{1}^{\infty}\sqrt{\xi^2-1}A(\xi)e^{ikx\xi}d\xi\right)\Bigg]=\\ \mathrm{Im}[\pi k(J_0(kx)+iY_0(k|x|))\times\\ \left(i\int_{-1}^{1}\exp(ikx\xi)d\xi-\int_{-\infty}^{-1}\exp(ikx\xi)d\xi-\int_{1}^{\infty}\exp(ikx\xi)d\xi\right)=\\ \pi k\frac{\sin(kx)}{kx}(J_0(kx)+Y_0(k|x|))-2\pi^2kY_0(k|x|)\delta(kx).\end{gathered} \tag{1.75}$$

In (1.75) $\delta(x)$ is a delta function. The Hankel beam (1.74) is clear from (1.75) to have a singular point source in the center of the initial plane. But the second term near the initial plane is finite and the reverse flow arises in the center of the beam, since the

Neumann function $Y_0(kx)$ in (1.75) has a negative value near zero ($kx << 1$) and is larger in absolute value. Modeling confirms this conclusion.

1.3.4 Numerical Simulation

The simulation of the sinc beam propagation in the near zone is performed using the Rayleigh-Sommerfeld integral. The field (1.57) is considered TE-polarized, i.e. the initial field has a linear polarization along the y-axis: $E_y = 2a \ \ \mathrm{sinc}\,(kxa)$, $a = 3$. The initial field is 1000 × 1000 samples (sample size is approximately $\lambda/53$), the wavelength is $\lambda = 532$ nm. Figure 1.8 depicts the initial field amplitude (1.56). Shown in Fig. 1.9 are the intensity $|\mathbf{E}|^2$ at a distance of 10 nm from the initial plane (a) and the longitudinal projection of the canonical energy flux $P_{\mathrm{or},z}$ (1.62) also at a distance of 10 nm from the initial plane. The domain size is 1000 points within the boundaries (−0.5 μm < x < 0.5 μm). The field is calculated twice, at a distance of 10 nm from the initial plane and at a distance of 20 nm from it for the difference taking of the derivative with respect to z.

At $a = 3$ the period of the side lobes is seen from Fig. 1.8 to be approximately 3 times shorter than the wavelength. And it follows from Fig. 1.9b that the reverse flow arises already in the first side lobe, and its magnitude is approximately 20% of the maximum forward energy flow. It is clear from a comparison of Figs. 1.8 and 1.9b that the negative flow almost coincides with the first negative side lobe of the sinc field amplitude. This modeling confirms the conclusions made on the basis of Eqs. (1.64) and (1.65) that the reverse flow near the initial plane is formed where the amplitude of the evanescent waves has negative values. Note that the nature of the reverse flow formation is similar to the formation of the reverse flow on the output surface of a gradient microlens, which was discovered in [53]. Interestingly, measuring the energy flow instead of measuring the intensity (energy density) makes it

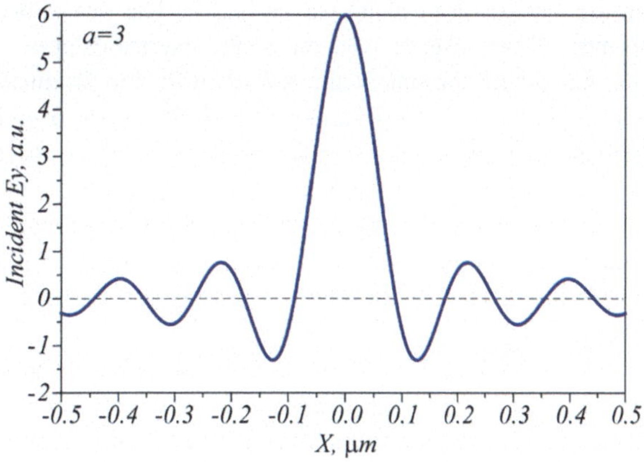

Fig. 1.8 The value of the E_y component of the initial field along the x axis

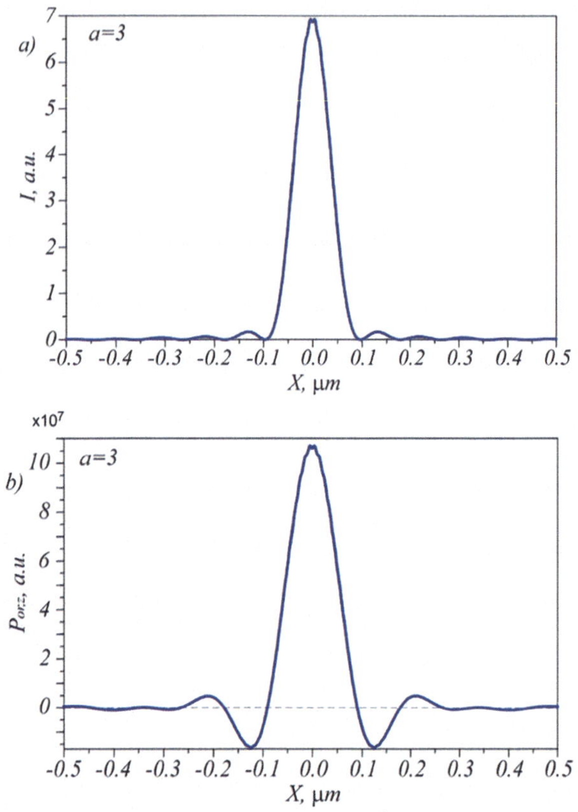

Fig. 1.9 Profiles of intensity $|\mathbf{E}|^2$ along the x-axis at a distance of 10 nm from the initial plane (**a**) and a canonical flux projection onto the z-axis $P_{or,z}$, $a = 3$ (**b**)

possible to increase the resolution of gradient optics, which constructs an image near its output surface.

Figure 1.10 depicts the numerical simulation results for the sinc beam (1.57), obtained in a different way, not using the Rayleigh-Sommerfeld integrals, but based on the expansion in plane waves (1.57). The flux was calculated using the general formula (1.60). Figure 1.10 shows the intensities at $a = 0.8$ (a) and $a = 3$ (c) in the initial plane for the sinc beam, as well as the longitudinal projections of the canonical energy flux at $a = 0.8$ (b) and $a = 3$ (d). It is clear from Fig. 1.10 that in the absence of evanescent waves at $a = 0.8$ there is no reverse flux (b). And if evanescent waves are present near the initial plane at $a = 3$ (d), then there is a reverse flux. Moreover, the reverse flux arises in the first side lobes of the amplitude and intensity, where the evanescent standing waves have odd antinodes (compare Figs. 1.9b and 1.10d), or where the electric field vector y component of the evanescent standing waves is negative.

Shown in Fig. 1.11 is the simulation result of a non-paraxial Gaussian beam propagation (1.68) using the plane wave expansion (1.57). Figure 1.11a shows the distribution of the longitudinal projection of the canonical energy flux in the initial plane of the Gaussian beam (1.68) with a waist radius of $\lambda/4$. Note that the intensity

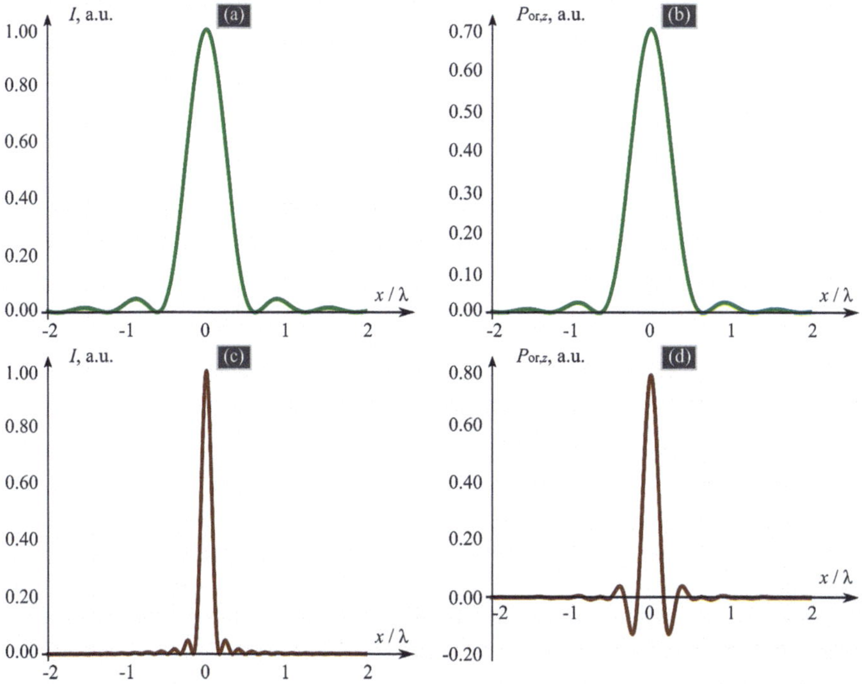

Fig. 1.10 Intensity profiles along the x-axis for $a = 0.8$ (**a**) and $a = 3$ (**c**) in the initial plane for the sinc beam (1.57), as well as longitudinal projections of the canonical energy flux for $a = 0.8$ (**b**) and $a = 3$ (**d**). The wavelengths are plotted along horizontal axis, and relative units along vertical axis

distribution of the Gaussian beam is almost the same as the distribution in Fig. 1.11a. Such a beam has a reverse flux, which coincides in place with the odd side lobes of the standing evanescent wave near the initial plane. The places where there is a reverse flux are highlighted by vertical lines in Fig. 1.11b. Where the line coincides with 0, the energy flux is positive, and where the line coincides with 1, the flux is reverse. The magnitude of this reverse flux is extremely small and is equal to 10^{-6} of the direct flux. This occurs because the contribution to the total field of the evanescent waves is different. For example, for a sinc beam, the amplitude modulus of the plane wave spectrum is 1 for both propagating and evanescent waves. In this case, the contribution of evanescent waves to the total field is significant. But for a Gaussian beam, the amplitude of the plane wave spectrum decreases exponentially for evanescent waves. Therefore, the contribution of such waves is weak, and the reverse flow is also weak.

Below are the results of numerical modeling the Hankel beam propagation (1.74) using the Rayleigh-Sommerfeld integral [47]. The initial field is taken to be 1000×1000 counts in size, 10×10 μm (the count size is approximately $\lambda/53$), the wavelength is $\lambda = 532$ nm. The output field has a size of 1000×10 counts within

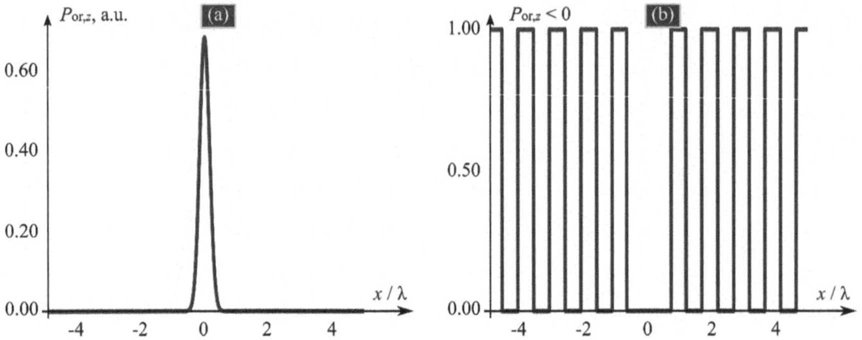

Fig. 1.11 Poynting vector longitudinal projection distribution for a Gaussian beam (1.68) with a waist radius of $\lambda/4$ (where $a = 4/\pi > 1$) in the initial plane (**a**) and the distribution of areas in which a reverse energy flow occurs (where the black line is equal to 1) (**b**). The wavelengths are plotted along the horizontal axis

the boundaries of (-1 μm $< x <$ 1 μm), (-10 nm $< y <$ 10 nm). The field is calculated at a distance of 10 nm from the initial plane. The intensity of the Hankel beam near the initial plane is clear from Fig. 1.12 to rapidly decrease and the reverse flow occurs in the center (at the point of maximum intensity). The magnitude of the reverse flow is 2 times greater than the maximum value of the forward flow.

The following results are obtained in this section. For a 2D non-paraxial TE-polarized sinc beam (1.56), an explicit expression for the longitudinal component of the Poynting vector is obtained (in this case, the canonical energy flux) (1.62). It follows that near the initial plane, such a beam has a reverse canonical energy flux, which is about 20% of the maximum value of the forward flux. It is shown that the reverse flux arises already in the first side lobes of the sinc beam complex amplitude, where the amplitude of the evanescent standing waves is negative (in odd antinodes) (1.65). For a Gaussian beam (1.68), the canonical flux cannot be reduced to explicit functions, and the expression for the flux includes an integral describing only the propagating waves (1.71). The evanescent waves are included in the amplitude of the Gaussian beam in the initial plane (1.68). The contribution of the evanescent waves to the total field is much weaker than for a sinc beam. Therefore the reverse canonical energy flow in the Gaussian beam is very weak and is a millionth of the forward flow. The reverse flow still occurs in those places where the z-projection of the evanescent waves flow is negative (Fig. 1.11b). For the Hankel beam (1.74), an explicit expression for the energy flow near the initial plane (1.75) is also obtained, and the reverse flow is clear from it to occur in the

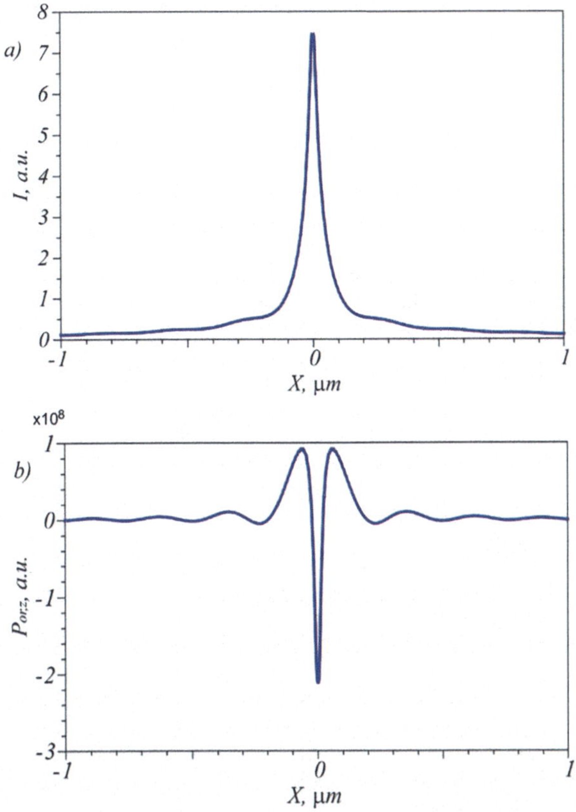

Fig. 1.12 Intensity profile of the Hankel beam (1.74) at a distance of 10 nm from the initial plane (**a**) and the longitudinal component profile of the canonical energy flux (1.60) for this beam at the same distance from the initial plane (**b**)

center of the beam. It is larger in magnitude than the forward flow (Fig. 1.12b). The reverse flow in nanoscale regions of two-dimensional non-paraxial light fields was also considered in [54–56].

References

1. V.S. Ignatowsky, Trans. Opt. Inst. Petrogr. **1**, 4 (1920)
2. B. Richards, E. Wolf, Proc. R. Soc. A Math. Phys. Eng. Sci. **253**, 358 (1959)
3. J.H. Poynting, Philos. Trans. R. Soc. Lond. A **175**, 343 (1884)
4. M. Born, E. Wolf, *Principles of Optics: Electromagnetic Theory of Propagation, Interference and Diffraction of Light* (Elsevier, 2013)
5. K.Y. Bliokh, E.A. Ostrovskaya, M.A. Alonso, O.G. Rodríguez-Herrera, D. Lara, C. Dainty, Opt. Express **19**, 26132 (2011)
6. A.V. Novitsky, D.V. Novitsky, J. Opt. Soc. Am. A **24**, 2844 (2007)
7. M.A. Salem, H. Bağcı, Opt. Express **19**, 8526 (2011)
8. I. Rondón-Ojeda, F. Soto-Eguibar, Wave Motion **78**, 176 (2018)
9. P. Vaveliuk, O. Martinez-Matos, Opt. Express **20**, 26913 (2012)

10. V.V. Kotlyar, S.S. Stafeev, A.G. Nalimov, Phys. Rev. A **99**, 033840 (2019)
11. V.V. Kotlyar, A.A. Kovalev, A.G. Nalimov, Opt. Lett. **43**, 2921 (2018)
12. M.V. Berry, J. Opt. A Pure Appl. Opt. **11**, 094001 (2009)
13. A.Y. Bekshaev, J. Opt. **15**, 044004 (2013)
14. A.Y. Bekshaev, M.S. Soskin, Opt. Commun. **271**, 332 (2007)
15. B. Ghosh, A. Daniel, B. Gorzkowski, A.Y. Bekshaev, R. Lapkiewicz, K.Y. Bliokh, J. Opt. Soc. Am. B **41**, 1276 (2024)
16. V.V. Kotlyar, S.S. Stafeev, Optik (Stuttg) **245**, 167703 (2021)
17. N.I. Zheludev, G. Yuan, Nat. Rev. Phys. **4**, 16 (2021)
18. Y. Aharonov, D.Z. Albert, L. Vaidman, Phys. Rev. Lett. **60**, 1351 (1988)
19. M.V. Berry, J. Opt. **15**, 044006 (2013)
20. M.V. Berry, M.R. Dennis, J. Phys. A Math. Theor. **42**, 022003 (2009)
21. G. Chen, Z.Q. Wen, C.W. Qiu, Light Sci. Appl. **8**, 56 (2019)
22. E.T.F. Rogers, N.I. Zheludev, J. Opt. **15**, 094008 (2013)
23. M. Abramowitz, I.A. Stegun, *Handbook of Mathematical Functions: With Formulas, Graphs, and Mathematical Tables* (Courier Corporation, 1965)
24. L. Allen, M.W. Beijersbergen, R.J.C. Spreeuw, J.P. Woerdman, Phys. Rev. A **45**, 8185 (1992)
25. M.V. Berry, N.L. Balazs, Am. J. Phys. **47**, 264 (1979)
26. G.A. Siviloglou, D.N. Christodoulides, Opt. Lett. **32**, 979 (2007)
27. G.A. Siviloglou, J. Broky, A. Dogariu, D.N. Christodoulides, Phys. Rev. Lett. **99**, 213901 (2007)
28. V.V. Kotlyar, A.A. Kovalev, Opt. Commun. **313**, 290 (2014)
29. J.A. Davis, M.J. Mitry, M.A. Bandres, I. Ruiz, K.P. McAuley, D.M. Cottrell, Appl. Optics **48**, 3170 (2009)
30. M.A. Bandres, J.C. Gutiérrez-Vega, Opt. Express **15**, 16719 (2007)
31. J.A. Davis, M.J. Mitry, M.A. Bandres, D.M. Cottrell, Opt. Express **16**, 12866 (2008)
32. J.E. Morris, M. Mazilu, J. Baumgartl, T. Čižmár, K. Dholakia, Opt. Express **17**, 13236 (2009)
33. M.I. Carvalho, M. Facão, Opt. Express **18**, 21938 (2010)
34. N.K. Efremidis, Z. Chen, M. Segev, D.N. Christodoulides, Optica **6**, 686 (2019)
35. P. Polynkin, M. Kolesik, J.V. Moloney, G.A. Siviloglou, D.N. Christodoulides, Science (80-) **324**, 229 (2009)
36. J. Baumgartl, M. Mazilu, K. Dholakia, Nat. Photonics **2**, 675 (2008)
37. P. Zhang, J. Prakash, Z. Zhang, M.S. Mills, N.K. Efremidis, D.N. Christodoulides, Z. Chen, Opt. Lett. **36**, 2883 (2011)
38. A. Mathis, F. Courvoisier, L. Froehly, L. Furfaro, M. Jacquot, P.A. Lacourt, J.M. Dudley, Appl. Phys. Lett. **101**, 71110 (2012)
39. A. Salandrino, D.N. Christodoulides, Opt. Lett. **35**, 2082 (2010)
40. P. Zhang, S. Wang, Y. Liu, X. Yin, C. Lu, Z. Chen, X. Zhang, Opt. Lett. **36**, 3191 (2011)
41. A. Minovich, A.E. Klein, N. Janunts, T. Pertsch, D.N. Neshev, Y.S. Kivshar, Phys. Rev. Lett. **107**, 116802 (2011)
42. F.G. Mitri, J. Opt. Soc. Am. A **33**, 1661 (2016)
43. P. Saari, I.M. Besieris, Photonics **11**, 1129 (2024)
44. V.V. Kotlyar, A.A. Kovalev, V.A. Soifer, J. Opt. Soc. Am. A **29**, 741 (2012)
45. Y.E. Geints, I.V. Minin, O.V. Minin, Opt. Lett. **47**, 1786 (2022)
46. M.V. Berry, S. Popescu, J. Phys. A Math. Gen. **39**, 6965 (2006)
47. A.G. Nalimov, V.V. Kotlyar, Y.V. Khanenko, Comput. Opt. **47**, 734 (2023)
48. A. Bekshaev, K.Y. Bliokh, M. Soskin, J. Opt. **13**, 053001 (2011)
49. K.Y. Bliokh, A.Y. Bekshaev, F. Nori, Nat. Commun. **5**, 3300 (2014)
50. A.J. Bracken, G.F. Melloy, J. Phys. A Math. Gen. **27**, 2197 (1994)
51. M.V. Berry, J. Phys. A Math. Theor. **43**, 415302 (2010)
52. A.P. Prudnikov, U.A. Brychkov, O.I. Marichev, *Integrals and Series. Special Functions* (Nauka, Moscow, 1983)

53. V.V. Kotlyar, A.A. Kovalev, A.G. Nalimov, Comput. Opt. **33**, 369 (2009)
54. V.V. Kotlyar, A.A. Kovalev, A.G. Nalimov, A.M. Telegin, Opt. Mem. Neural Netw. **34**, 476 (2025)
55. V. Kotlyar, A. Kovalev, A. Nalimov, J. Opt. Soc. Am. B **43**, 128 (2026)
56. V.V. Kotlyar, A.G. Nalimov, A.A. Kovalev, M. Arfan, S.S. Stafeev, J. Opt. **28**, 15605 (2026)

Chapter 2
Reverse Canonical Energy Flow Without Evanescent Waves

2.1 Non-paraxial Sinc Beam Without Evanescent Waves

The problem of the reverse (negative) flow of light energy has been known for a long time [1]. This is understood as the presence of negative values of the longitudinal component of the Poynting vector. Negative values of the Poynting vector were found in the tight focus of vector light beams [1–4], in the cross section of non-paraxial Bessel beams [5] and X-beams [6]. However the Poynting vector has one drawback: it is defined with an accuracy of up to the rotor of any continuous function. The Poynting vector is included in the differential law of conservation of energy in the form of divergence, and the divergence from the rotor is zero. Therefore, the Poynting vector consists of the sum of two vectors: the canonical energy flow and the spin flow [7, 8]. For example, in two-dimensional (2D) light fields, there is only one projection of the electric field strength, and therefore for such fields the Poynting vector coincides with the canonical energy flow. In [9–11], the reverse canonical energy flow was found for some 2D light fields. For the sake of completeness, we note that reverse flow was also found in quantum mechanics [12], in plasmonic lenses [13], in light fields with superoscillations [14], and at the focus of spiral laser beams [15].

Previously, we have shown [11] that 2D non-paraxial laser beams of three types with TE polarization have a canonical reverse energy flow in the near field. That is, the wave vector of the light field in the initial plane has a negative longitudinal projection. In [11], the authors associated this reverse energy flow with the presence of evanescent waves in the near field. In this section we show that the reverse canonical flow for a sinc-beam also occurs without evanescent waves. It is also shown that with a decrease in the degree of non-paraxiality of the beam, the magnitude of the reverse flow decreases, and a paraxial sinc beam has no reverse flow.

V. Kotlyar et al., *Inverse Canonical Energy Flow in Laser Beams*,
https://doi.org/10.1007/978-3-032-27289-8_2

2.1.1 Reverse Canonical Energy Flow for a Sinc-Beam with the Evanescent Wave

Consider a 2D TE-polarized non-paraxial sinc-beam; its plane wave spectrum amplitude is described by the rect-function (step function). Then the complex amplitude of such a beam in the initial plane ($z = 0$) will be:

$$E(x, z=0) = \int_{-\infty}^{\infty} A(\xi)\exp(ikx\xi)d\xi = \int_{-\infty}^{\infty} \mathrm{rect}\left(\frac{\xi}{a}\right)\exp(ikx\xi)d\xi = 2a\frac{\sin(kxa)}{kxa}. \tag{2.1}$$

In (2.1) x, z are the transverse and longitudinal coordinates, $k = 2\pi/\lambda$ is the wave number of light with wavelength λ, a is a dimensionless parameter that determines the half-width of the plane waves spectrum (if $a > 1$, then the spectrum also contains evanescent waves). $A(\xi) = rect(\xi/a) = \{1,\ |\xi| < a;\ 0\ |\xi| > a\}$ is the amplitude of the plane waves spectrum. The complex amplitude of the sinc-beam at any distance z will be:

$$E(x, z) = \int_{-\infty}^{\infty} A(\xi)\exp\left(ikx\xi + ikz\sqrt{1-\xi^2}\right)d\xi = \int_{-a}^{a} \exp\left(ikx\xi + ikz\sqrt{1-\xi^2}\right)d\xi. \tag{2.2}$$

The derivative of the field (2.2) with respect to the longitudinal coordinate z on the initial plane will be:

$$\frac{\partial}{\partial z}E(x, z)\bigg|_{z=0} = ik\int_{-a}^{a}\sqrt{1-\xi^2}\exp(ikx\xi)d\xi. \tag{2.3}$$

The Poynting vector, up to irrelevant constants, consists of the sum of two vectors [1–3] that describe the orbital energy flux (canonical energy flow) P_{or} and the spin flow P_{sp}:

$$\begin{aligned}&\mathbf{P}=\mathbf{P}_{or}+\mathbf{P}_{sp}=\mathrm{Im}(\mathbf{E}^{*}\cdot(\nabla)\mathbf{E})+\frac{1}{2}(\nabla\times\mathbf{S}),\\&\mathbf{S}=\mathrm{Im}(\mathbf{E}^{*}\times\mathbf{E}).\end{aligned}\tag{2.4}$$

The longitudinal projection of the orbital energy flow, according to (2.4), could be calculated using the formula:

$$P_{or,z}=\mathrm{Im}\left(E_x^{*}\frac{\partial}{\partial z}E_x+E_y^{*}\frac{\partial}{\partial z}E_y+E_z^{*}\frac{\partial}{\partial z}E_z\right).\tag{2.5}$$

Since in our 2D case of diffraction with TE polarization there remains only one projection of the electric field vector (2.2), we substitute it into (2.5), and obtain (a > 1):

$$\begin{aligned}&P_{or,z}=\mathrm{Im}\left(E_y^{*}\frac{\partial}{\partial z}E_y\right)\bigg|_{z=0}=\\&4ak\frac{\sin(kxa)}{kxa}\int\limits_0^1\sqrt{1-\xi^2}\cos(kx\xi)d\xi.\end{aligned}\tag{2.6}$$

The integral (2.6) is known [15]; therefore instead of (2.6) we get:

$$\begin{aligned}&P_{or,z}=4ak\frac{\sin(kxa)}{kxa}\int\limits_0^1\sqrt{1-\xi^2}\cos(kx\xi)d\xi=\\&4\pi ak\left(\frac{\sin(kxa)}{kxa}\right)\left(\frac{2J_1(kx)}{kx}\right).\end{aligned}\tag{2.7}$$

From (2.7) it follows that for a > 1, i.e. when evanescent waves are present in the near field, there is a reverse energy flow. In (2.7), $J_1(\xi)$ is the first-order Bessel function. Figure 2.1 shows the plots of the sinc-function and the Bessel function divided by the argument, from the coordinate x. The first zero of the Bessel function does not depend on the parameter a, and has the coordinate $x_2 = 1.2(\lambda/2)$. And the first zero of the sinc-function depends on a, and has the coordinate $x_1 = (\lambda/2a)$, less than the coordinate x_2, since $a > 1$. From Fig. 2.1 it is clear that the reverse energy flow in the initial plane will take place between these two zeros, that is, it will have coordinates x from the interval: $x_1 < x < x_2$. With an increase in the parameter $a > 1$, the distance between the zeros with coordinates x_1 and x_2 increases and the magnitude of the reverse flow also increases. In [11], based on this, it was concluded that the reverse flow takes place in the presence of evanescent waves. But even at $a \le 1$, when there are no evanescent waves, the reverse flow also takes place. Indeed, at $a = 1$ from (2.7) it follows:

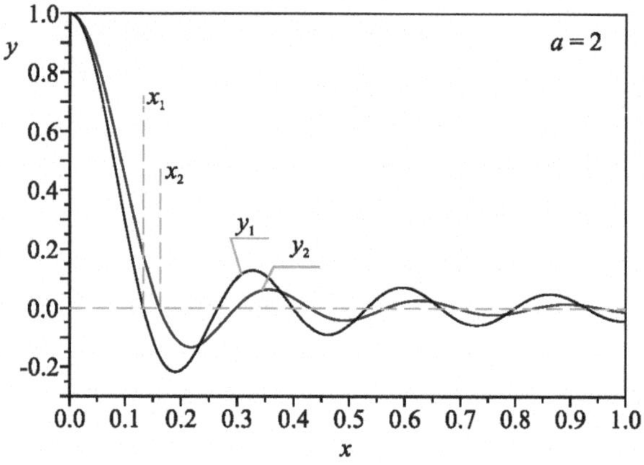

Fig. 2.1 Functions $y_1 = \sin(\xi)/\xi$ and $y_2 = 2J_1(\xi)/\xi$ from coordinates x, where $\xi = kxa$, $a = 2$, $\lambda = 532$ nm. Points with coordinates $x_1 = \lambda/(2a)$ and $x_2 = 1.2\lambda/2$ are the first zero of functions y_1 and y_2

$$P_{or,z}(a=1) = 4\pi k\left(\frac{\sin(kx)}{kx}\right)\left(\frac{2J_1(kx)}{kx}\right). \tag{2.8}$$

From (2.8) it is seen that even at $a = 1$, when there are no evanescent waves, the reverse flow will still exist between the first two zeros of the sinc-function and the Bessel function. Only the interval between the coordinates x_1 and x_2 will decrease and will be equal to: $\lambda/2 < x < 1.2(\lambda/2)$. In this case, the magnitude of the reverse flow will decrease. Will there be a reverse flow with further decrease of the parameter a, we will find out from the following sections.

2.1.2 *Transition from Non-paraxial to Paraxial Type of Sinc-Beam*

When choosing the parameter a from the interval $0 < a < 1$, it is no longer possible to use an explicit expression for the canonical energy flow in the initial plane (2.7), since the integral in (2.7) will no longer be in the range from 0 to 1, but it will be in the range from 0 to a:

$$P_{or,z} = 4ak\frac{\sin(kxa)}{kxa}\int\limits_0^a \sqrt{1-\xi^2}\cos(kx\xi)d\xi. \tag{2.9}$$

The integral in (2.9) is no longer known. If the parameter a decreases by a very small amount, $a = 1 - \delta$, $\delta << 1$, then instead of (2.9) we can write approximately:

$$P_{or,z} \simeq 4ak\left(\frac{\sin(kx)}{kx}\right)\left(\frac{2\pi J_1(kx)}{kx} - \delta\sqrt{1-\delta^2}\cos(kx\delta)\right),$$
$$\delta < < 1. \tag{2.10}$$

From (2.10) it follows that the interval of change of coordinates x, in which the reverse flow takes place, will decrease by the value δ: $\lambda/2 < x < 1.2(\lambda/2) - \delta$. And the magnitude of the reverse flow will also decrease slightly. For small values of $a < <1$ (paraxial beam), the square root under the integral can be expanded in a Taylor series and the first few terms can be left:

$$\sqrt{1-\xi^2} = 1 - \frac{\xi^2}{2} - \frac{\xi^4}{8} - \frac{\xi^6}{16} - \ldots \tag{2.11}$$

Let us select from (2.11) only the first two terms and write an approximate expression for the canonical energy flux in the initial plane for $a << 1$:

$$P_{or,z}(a < < 1) = 4ak\left(\frac{\sin(kxa)}{kxa}\right) \times$$
$$\left(\frac{2a\sin(kxa)}{kxa} - \frac{a^3}{2}\frac{\sin(kxa)}{kxa}\left[1 - \frac{2}{(kxa)^2}\right] - \frac{a^3\cos(kxa)}{(kxa)^2}\right). \tag{2.12}$$

From (2.12) it follows that if we leave only the first term in the second parentheses, then when the beam is paraxial, there will be no reverse flow, since instead of (2.12) we get:

$$P_{or,z}(a < < 1) = 8a^2k\left(\frac{\sin(kxa)}{kxa}\right)^2 \geq 0. \tag{2.13}$$

If we take into account all the terms in (2.12), then at x close (on the right) to the first zero of the sinc-function, in the first brackets the sinc-function will be equal to a small negative value, and in the second brackets the first two terms, also proportional to the sinc-function, can be neglected in comparison with the third term, which will be positive, since $kxa \sim \pi$. Therefore, the entire expression (2.12) at the point x close (and greater) to the point $kxa = \pi$ will have a negative value. That is, with an increase in the parameter a (when passing from a paraxial beam to a non-paraxial one), a reverse flow will take place near the initial plane, the value of which will continuously increase as it grows in the interval $0 < a < 1$. And at the same time, there will be no evanescent waves near the initial plane.

It is clear that what happens near the first zero of intensity (or amplitude) of the sinc-beam will also happen near all other zeros, only the magnitude of the return flow at subsequent zeros of intensity will be significantly smaller, since the intensity in the side lobes is also significantly less than in the main lobe.

2.1.3 *Simulation of the Reverse Energy Flow for a Sinc-Beam at $0 < a < 1$*

The simulation was performed using the Rayleigh-Sommerfeld integral [16]. The simulation parameters were: initial field 1500 × 1500 counts, 10 × 10 μm (count size approximately λ/80), wavelength λ = 532 nm, a = 0.1…3. The output field had a size of 1500 × 3 counts within the boundaries (−0.5 μm < x < 0.5 μm). The field was calculated twice, at a distance of 10 nm from the initial plane and at a distance of 20 nm from it to be able to take the derivative. Figures 2.2 and 2.3 show the cross-sections of the field intensity at a distance of 10 nm from the initial plane $I = |E(x)|^2$ (Figs. 2.2a and 2.3a) and the projections of the canonical flux $P_{or,z}$ onto the z axis for $a = 0.9$ (Fig. 2.2b) and $a = 2$ (Fig. 2.3b).

From Figs. 2.2 and 2.3 it follows that the maximum value of the reverse flow at $a = 2$ (Fig. 2.3b) is approximately 15% of the maximum value of the direct flow. Reverse flow appears near the first side lobe of intensity (Fig. 2.3a). With a decrease of $a = 0.9$ (Fig. 2.2b) the reverse flow decreases rapidly. In Fig. 2.2b it is almost invisible, but it is there and is approximately 1% of the direct flow. The magnitude of the reverse flow $P_{or,\,z}$ in Fig. 2.4 was calculated as the ratio of the reverse flow to the

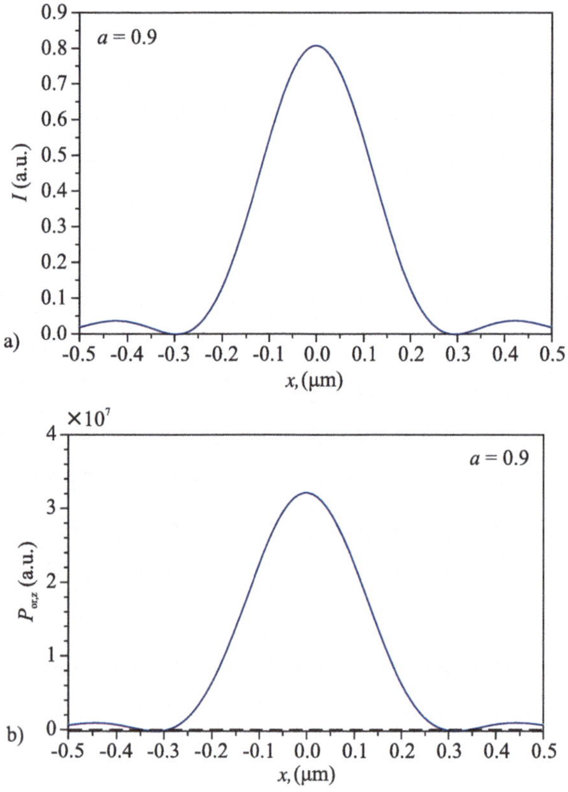

Fig. 2.2 Cross-sections along the x-axis of the projection of the intensity of the resulting field (**a**) and the canonical energy flux onto the z-axis (**b**) at a distance of 10 nm from the initial plane for $a = 0.9$

Fig. 2.3 Cross-sections along the x-axis of the projection of the intensity of the resulting field (**a**) and the canonical energy flux onto the z-axis (**b**) at a distance of 10 nm from the initial plane for $a = 2$

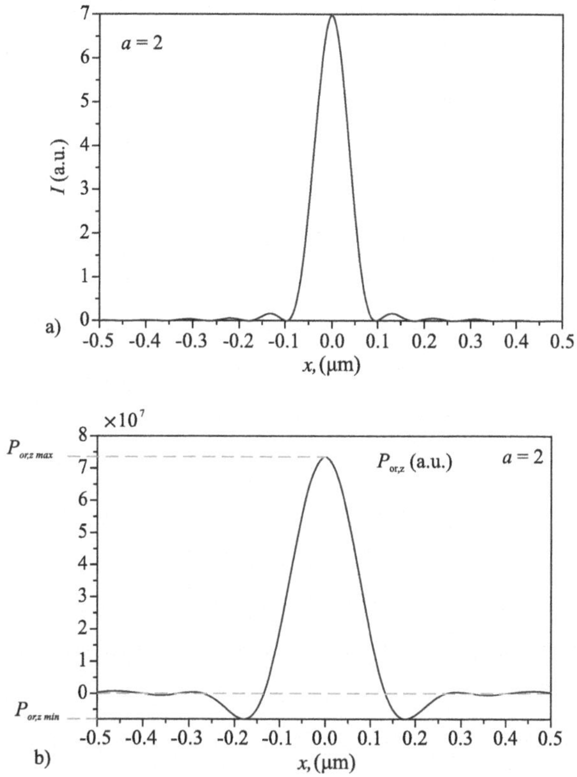

direct flow, which are found as the maximum and minimum values:—$P_{or,\ z\ min}/P_{or,\ z\ max}$. Figure 2.4 shows a graph of the dependence of the relations—$P_{or,\ z\ min}/P_{or,\ z\ Max}$. on the parameter a. To obtain this graph, the ratio—$P_{\mathrm{or,z\ min}}/P_{\mathrm{or,z\ max}}$ was calculated for a in the range $0.1 < a < 3$ with a step of 0.1. From Fig. 2.4 it follows that for $a > 1$ (there are evanescent waves) the reverse flow decreases slow with decreasing a (Fig. 2.4a), and for $a < 1$ (there are no evanescent waves) the reverse flow decreases quickly with decreasing a. For small values of $a < <1$ (the beam is paraxial) the reverse flow is equal to zero.

Figure 2.5 shows two graphs of the dependence of the transverse coordinate x on a in the initial plane for a sinc-beam, in which the intensity is zero (the nearest zero from the origin) (solid line) and in which the reverse flow has a maximum value (dashed curve). It is evident from Fig. 2.5 that both curves almost coincide. This means that the reverse flow in the near field of the sinc- beam is formed near the zeros of the intensity distribution.

Figure 2.6 shows how the reverse flow decreases when moving away from the initial plane. With the selected parameter $a = 0.9$, there are no evanescent waves. For each observation plane (at different z), the light field was also calculated twice, in this plane and at a distance of +2 nm from it to take the derivative with respect to z. The simulation step was 5 nm. That is, for the first observation plane, two

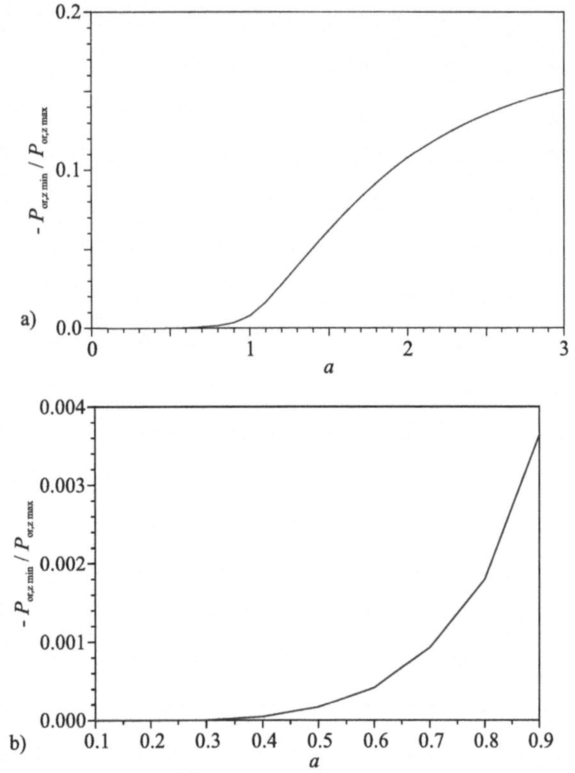

Fig. 2.4 Dependence of the ratio of the reverse and forward flows $-P_{\mathrm{or,z\,min}}/P_{\mathrm{or,z\,max}}$ on the parameter a in the ranges of values $0.1 < a < 3$ (**a**) and $0.1 < a < 0.9$ (**b**)

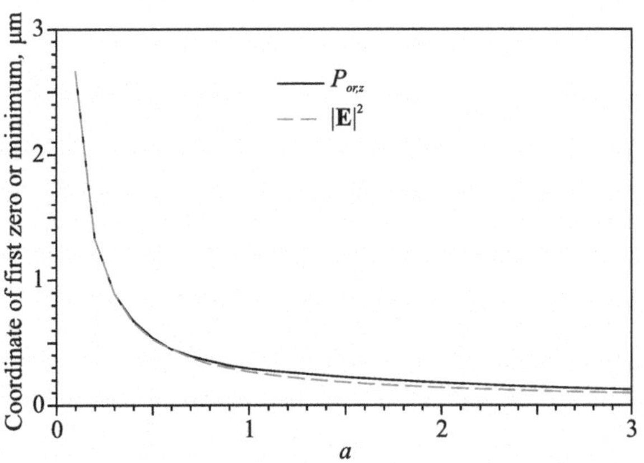

Fig. 2.5 Dependences of the x coordinates of the first zero of intensity $|\mathbf{E}|^2$ (dashed line) and the minimum of the reverse flow $P_{\mathrm{or,z}}$ on the paraxiality parameter a

calculations were performed, at a distance of 10 and 12 nm from the initial one, for the second plane—at distances of 15 and 17 nm, etc. Positive values of the graph correspond to the presence of a reverse flow in the nearest minimum $P_{\mathrm{or,z}}$. It is also evident from Fig. 2.6 that the reverse flow disappears at a distance of 50 nm (if there

Fig. 2.6 Dependence of the ratio $-P_{or,z\ min}/P_{or,z\ max}$ for $a = 0.9$ (**a**) and $a = 2$ (**b**) on the distance from the initial plane z

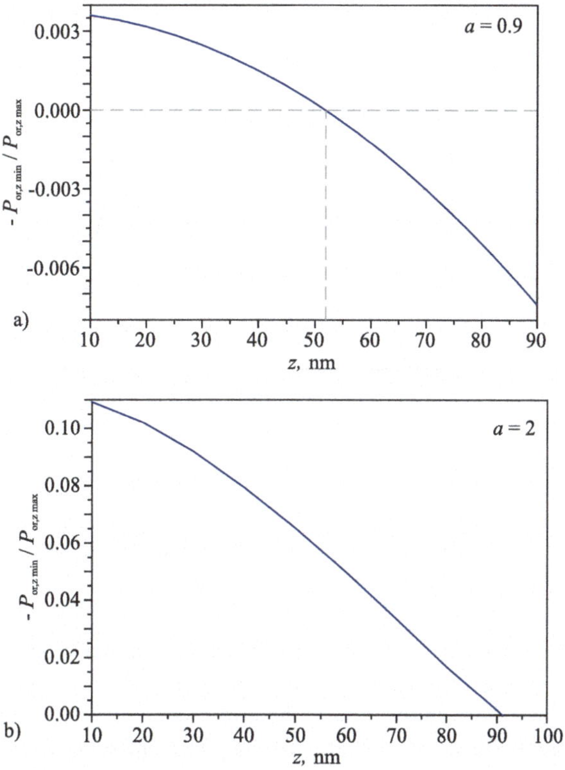

are no evanescent waves) and at a distance of 90 nm (in the presence of evanescent waves) from the initial plane.

2.1.4 Mechanism of Forming a Reverse Energy Flow

From Fig. 2.5 it follows that the coordinate of the maximum of the reverse flow almost coincides with the coordinate of zero intensity of the sinc-beam. This suggests that the reverse flow occurs where there is a sharp jump in the phase of the initial field. Indeed, when passing through the point of zero intensity of the 2D sinc-beam, the phase function of the amplitude changes abruptly from zero to π. Then, when the beam propagates from the initial plane, a smooth wave front should be formed without phase jumps by an amount less than 2π. In this case, at the point where there was zero intensity in the initial plane, there will already be a non-zero intensity during propagation, and near this point the phase will continuously change (without jumps) along the transverse coordinate. In order to "eliminate" the initial phase jump by π (that is, to make a smooth transition from zero to some value less than π, that is, to reduce the phase difference), locally the flow of light energy near

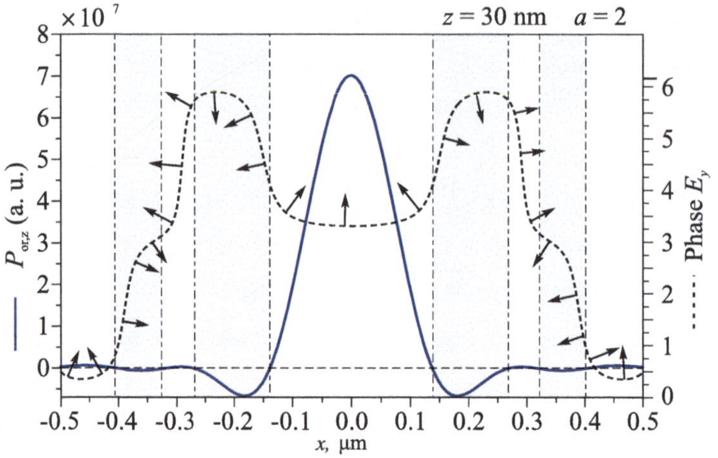

Fig. 2.7 Energy flow $P_{or,z}$ and field phase arg(E_y) at a distance of 30 nm from the initial plane. Arrows on the phase graph show the direction of the energy flow. The magnitude of the arrows is not equal to the phase gradient modulus for better clarity, but is perpendicular to the phase function. Gray areas highlight areas where the direction of the energy flow is reversed

zero intensity and near the initial plane should propagate in the opposite direction. Figure 2.7 shows the phase of the sinc-beam at a distance of 30 nm from the initial plane. Also in this figure, the arrows show the direction of local propagation of the energy flow (the direction of the phase gradient vector, without taking into account its magnitude). From Fig. 2.7 it is evident that the reverse energy flow is formed in places of rapid phase change near the intensity zeros.

A light subwavelength sinc-field in the initial plane can be formed in practice using a 2D gradient microlens, for example, a Mikaelian lens [17]. In this work, it is shown that if a point light source is placed at the input of a 2D gradient lens (in silicon) with a secant law of refractive index distribution along the transverse coordinate, then at the output (in the plane at the output end of the microlens) a subwavelength intensity distribution of the square sinc-function type with a central maximum width of approximately 0.12λ will be formed. Such a subwavelength intensity distribution can be created by surface evanescent waves that run along the output surface of the gradient microlens towards each other, forming a standing wave with a subwavelength period.

2.1.5 Backward Energy Flow for 2D Bessel Beam

Similarly to the previous one, it can be shown that a 2D TE-polarized Bessel beam with infinite energy and without evanescent waves in the near field also forms a reverse flow near the intensity zeros. Indeed, the amplitude of the spectrum of a plane Bessel beam of zero order has the form [18]:

$$A(\xi) = \left(1 - \xi^2\right)^{-1/2}, \quad |\xi| < 1. \tag{2.14}$$

In the initial plane the Bessel beam with spectrum (2.14) will have the form:

$$\begin{gathered} E(x, z = 0) = \int\limits_{-\infty}^{\infty} A(\xi) \exp(ikx\xi) d\xi = \\ 2 \int\limits_{0}^{1} \frac{\cos(kx\xi)}{\sqrt{1 - \xi^2}} d\xi = \pi J_0(kx). \end{gathered} \tag{2.15}$$

In (2.15) $J_0(t)$ is the Bessel function of the first kind of zero order. Next, similarly to (2.6), we find the longitudinal projection of the canonical energy flux vector in the initial plane for the Bessel beam:

$$P_{or,z} = \mathrm{Im}\left(E_y^* \frac{\partial}{\partial z} E_y \right)\bigg|_{z=0} = 2\pi k J_0(kx) \frac{\sin(kx)}{kx}. \tag{2.16}$$

From (2.16) it follows that the maximum reverse energy flow in the Bessel beam will be between the first zero of the Bessel function $J_0(kx)$, which is at the point $x_1 \sim 0.76$ $(\lambda/2)$, and the first zero of the sinc function: $\sin(kx)/(kx)$, which is at the point $x_2 = (\lambda/2)$. And the mechanism of formation of the reverse energy flow near the zeros of the Bessel function, in which there is a phase jump of π, will be similar to how it occurs in the sinc-beam.

2.1.6 Energy Backflow of 2D Beam at the Focus

Instead of Eq. (2.2), the expression for the amplitude of a non-paraxial 2D beam with autofocus can be written as:

$$E(x, z) = \int\limits_{-\infty}^{\infty} A(\xi) \exp\left(ikx\xi + ik(z - z_0)\sqrt{1 - \xi^2} \right) d\xi, \tag{2.17}$$

where z_0 is the autofocus distance. For the beam (2.17), the canonical reverse flow forms not in the near field but precisely in the focal plane at $z = z_0$. For a sinc beam in the focal plane, the amplitude (2.17) simplifies to:

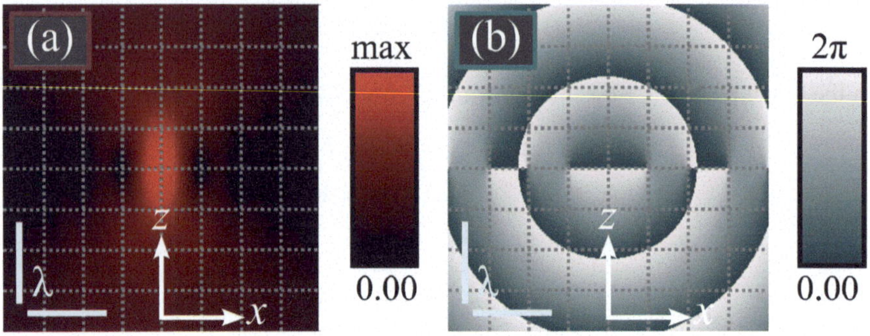

Fig. 2.8 Distribution of the amplitude modulus (2.17) (**a**) and its argument (phase) (**b**) in coordinates (x, z) at $z_0 = 2\lambda$, $\lambda = 532$ nm, frame size $4\lambda \times 4\lambda$ (**a, b**)

$$E(x, z = z_0) = 2\frac{\sin(kx)}{kx}, \tag{2.18}$$

and the reverse flow is described by Eq. (2.8). Figure 2.8 shows the amplitude modulus (a) and phase (b) in the longitudinal xz plane for a sinc beam with autofocusing. It is clearly seen that the phase in the focal plane exhibits a vortex structure. To the right of the center with a topological charge of −1, and to the left with a topological charge of +1. The centers of these optical vortices coincide with the zeros of function (2.18). In the vicinity of the intensity zeros, a circulation of the energy flow is observed. For a topological charge of −1, the energy flow circulates clockwise, while for a topological charge of +1, the energy flow circulates counterclockwise. Moreover, on certain segments of the closed trajectories, the energy flow is directed opposite to the main propagation direction of the beam.

2.1.7 Energy Backflow of the Beam with Zero Intensity on the Optical Axis

In this section we consider another simple 2D non-paraxial beam that always has zero intensity on the optical axis and contains no evanescent waves. To this end, we choose the amplitude of the plane wave spectrum as follows:

$$A(\xi) = \xi, \quad |\xi| < 1. \tag{2.19}$$

Substituting Eq. (2.19) into Eq. (2.2) and using the tabulated integrals from Ref. [18], we arrive at the following expression:

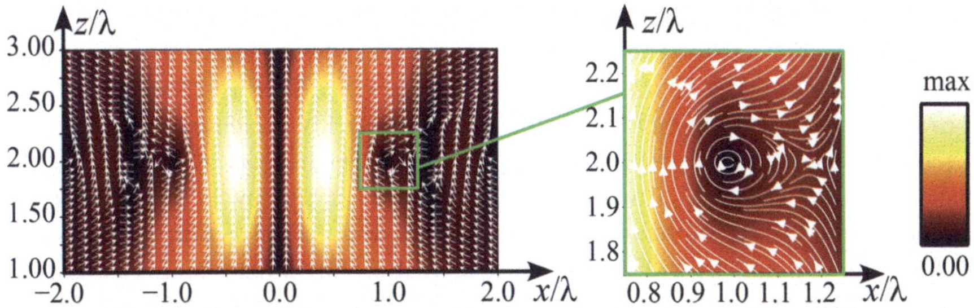

Fig. 2.9 Intensity (color) and energy flow (arrows) in the xz plane for beam (2.20) with autofocus at $z_0 = 2\lambda$, $\lambda = 532$ nm. The inset shows an enlarged fragment

$$
\begin{aligned}
E(x, z=0) &= \int_{-1}^{1} \xi \exp(ikx\xi)d\xi = 2i\sqrt{\frac{\pi}{2kx}} J_{3/2}(kx), \\
\frac{\partial}{\partial z} E(x, z=0) &= ik \int_{-1}^{1} \xi\sqrt{1-\xi^2} \exp(ikx\xi)d\xi = -\frac{\pi}{kx} J_2(kx).
\end{aligned}
\tag{2.20}
$$

The beam described by Eq. (2.20) has zero intensity on the optical axis ($x = 0$). Considering Eq. (2.20), we can write an exact expression for the longitudinal component of the canonical energy flow vector in the near field of this beam:

$$
P_{or,z} = \sqrt{2}k\left(\frac{\pi}{kx}\right)^{3/2} J_{3/2}(kx)J_2(kx). \tag{2.21}
$$

The Bessel function of order 3/2 in Eq. (2.21) has its first zero on the right at $x_1 = 0.72\lambda$, while the first zero on the right for the Bessel function of order 2 in Eq. (2.21) is at $x_2 = 0.81\lambda$. Between these zeros, both Bessel functions have opposite signs, resulting in a negative energy flow. If the light field in Eq. (2.20) is shifted along the optical axis, analogous to Eq. (2.17), it will autofocus at the distance $z = z_0$. Figure 2.9 shows the intensity (color) and the energy flow (arrows) in the xz plane for beam (2.20) with a focal length $z_0 = 2\lambda$. The inset in Fig. 2.9 demonstrates clockwise circulation of the energy flow around the first intensity zero on the right.

The following results were obtained in this section. It was shown theoretically and numerically that for a 2D TE-polarized sinc-beam, in which the amplitude of the plane wave spectrum has the form of a step, a reverse canonical energy flow takes place near the initial plane in some local regions in the beam cross section (Fig. 2.3b). That is, the derivative of the phase with respect to the longitudinal coordinate has a negative value at some points, and the wave vector at these points has a negative projection onto the optical axis. Note that the longitudinal projection of the Poynting vector for a 2D TE-polarized field coincides with the longitudinal projection of the canonical energy flow vector. The formation of the reverse energy flow in the near field is not necessarily associated with the presence of evanescent waves, as was previously believed [9]. We showed that the reverse energy flow takes place in a non-paraxial field and without evanescent waves (Fig. 2.4). It was also shown that the reverse energy flow in the near field quickly decreases (Fig. 2.6). The reverse energy flow in the sinc-beam is formed near the intensity zeros in the initial plane (Fig. 2.5). It follows from this that the reverse energy flow is formed near the points of rapid change of the phase function in the beam cross-section, for example, near the phase jumps by π (Fig. 2.8). Similarly, it is shown that the reverse energy flow takes place near the intensity zeros in the 2D TE-polarized Bessel beam of zero order (2.16).

2.2 2D Optical Vortices and a Reverse Energy Flow Occurs near the Intensity Zeros

In modern optics, optical effects that are difficult to detect because they are associated with small energy fluxes are of great interest. Such effects include superoscillations [18–21], superresolution, the effect of giant wave vectors [22–26], and the reverse energy flux [1–6]. The slow light effect is associated with large values of wave vectors [27, 28]. Note that optical vortices are formed at intensity zeros [29–32], and when passing the intensity zero in the cross section of the vortex beam, the azimuthal projection of the wave vector k_φ tends to infinity as n/r, as r tends to zero. Here n is the topological charge of the optical vortex (e.g., a Laguerre-Gauss beam), (r, φ) are the polar coordinates in the beam cross section [7].

Let us briefly dwell on the indicated effects. Superoscillations arise when a complex amplitude with a limited Fourier spectrum exhibits oscillations with a spatial frequency greater than the maximum frequency in the Fourier spectrum [19]. Regions with the superoscillation effect are formed in a superposition that consists of a large number of plane waves, where the modulus of the phase gradient of the complex field amplitude is greater than the wave number of each plane wave [20]. In [13, 26], reviews of modern works on the use of superoscillations for obtaining superresolution in optics are given. The reverse energy flow, as negative values of the longitudinal component of the Poynting vector, was previously studied in a sharp focus [1, 2, 14], in non-paraxial Bessel beams [5] and in X-waves [6]. However, using the Poynting vector to find the reverse energy flux is difficult, since the Poynting vector, in addition to the canonical energy flux vector, also includes the spin flux vector [7, 8]. For example, the reverse energy flux was found for a 2D non-paraxial Airy beam in [9], and in [10], the reverse energy flux for a 2D non-paraxial beam was also investigated. In two-dimensional light fields with TE-polarization, there is only one projection of the electric vector, and therefore the Poynting vector has no spin component. For 2D light fields, the longitudinal projection of the Poynting vector coincides with the longitudinal projection of the canonical energy flux.

We consider a 2D nonparaxial light field with TE-polarization in this paper. We have previously considered such a sinc-beam [11] and shown that in the near field, in the presence of evanescent waves, a canonical reverse energy flow occurs. In this paper, we show that the reverse energy flow is formed even without evanescent waves. The reason for the reverse energy flow is that the 2D light field has intensity zeros, around which an optical vortex is formed and the energy flow circulates around them. The longitudinal projection of the wave vector becomes gigantic near the intensity zeros, and the phase velocity of light decreases, right down to zero at the very point of zero intensity. All of these effects occur in the subwavelength region around the zero intensity, demonstrating the presence of superoscillations near the center of the 2D optical vortex.

2.2.1 *The Longitudinal Component of the Wave Vector in the Near Field of a Sinc Beam*

In his work [7], M. Berry introduced the concept of a local wave vector **k** as the phase gradient of the complex amplitude of a scalar wave field

$$\begin{aligned}\mathbf{J} &= \mathrm{Im}(E^*(\mathbf{r})\nabla E(\mathbf{r})) = \\ &|E(\mathbf{r})|^2\nabla(\arg E(\mathbf{r})) = |E(\mathbf{r})|^2\mathbf{k}(\mathbf{r}).\end{aligned} \tag{2.22}$$

In (2.22) **J** is the [7] optical current vector of a scalar light field with complex amplitude $E(\mathbf{r}) = A(\mathbf{r})\exp(i\Phi(\mathbf{r}))$. In this paper, we use more modern terminology and, instead of optical current, speak of the canonical energy flux. The Poynting vector, up to insignificant constants, is equal to the sum of two vectors [7, 8], which describe the orbital energy flux (canonical energy flux) P_{or} and the spin flux P_{sp}:

$$\begin{aligned}&\mathbf{P} = \mathbf{P}_{or} + \mathbf{P}_{sp} = \mathrm{Im}(\mathbf{E}^*\cdot(\nabla)\mathbf{E}) + \frac{1}{2}(\nabla \times \mathbf{S}),\\ &\mathbf{S} = \mathrm{Im}(\mathbf{E}^* \times \mathbf{E}).\end{aligned} \tag{2.23}$$

The canonical energy flux is seen from comparison of (2.22) and (2.23) to be the optical current. In this paper, we consider the features of the change in the longitudinal component of the local wave vector for a 2D non-paraxial sinc-field of light [11]. The complex amplitude of any 2D light field satisfying the Helmholtz equation

$$\left(\frac{\partial^2}{\partial x^2} + \frac{\partial^2}{\partial z^2} + k^2\right)E(x,z) = 0, \tag{2.24}$$

$k = 2\pi/\lambda$ is the wave number of light with wavelength λ, can be represented as the result of expansion over the plane waves spectrum:

$$E(x,z)=\int_{-\infty}^{\infty} A(\xi)\exp\left(ikx\xi+ikz\sqrt{1-\xi^2}\right)d\xi. \tag{2.25}$$

In (2.25), x and z are the transverse and longitudinal Cartesian coordinates. By a sinc beam we mean a beam for which the plane wave spectrum amplitude in (2.25) is equal to the step:

$$A(\xi)=rect(\xi/a)=\{1,\ \ |\xi|<a;\ \ 0\ \ |\xi|>a\}, \tag{2.26}$$

where a is a dimensionless parameter that determines the half-width of the plane wave spectrum (if a > 1, then the spectrum also contains evanescent waves). For a sinc beam, instead of (4), taking into account (2.26), we write the complex amplitude:

$$\begin{gathered} E(x,z)=\int_{-a}^{a} \exp\left(ikx\xi+ikz\sqrt{1-\xi^2}\right)d\xi, \\ E(x,z=0)=\int_{-a}^{a} \exp(ikx\xi)d\xi=2a\frac{\sin(kxa)}{(kxa)}. \end{gathered} \tag{2.27}$$

It is shown in [28] that for beam (2.27) in the initial plane ($z = 0$) the longitudinal projection of the canonical energy flux vector can be written as:

$$\begin{gathered} P_{or,z}=\mathrm{Im}\left(E^*(x,z)\frac{\partial E(x,z)}{\partial z}\right)|_{z=0}= \\ 4\pi ak\left(\frac{\sin(kxa)}{kxa}\right)\left(\frac{2J_1(kx)}{kx}\right). \end{gathered} \tag{2.28}$$

In (2.28), the function $J_1(\xi)$ is a first-order Bessel function. By explicitly separating the sinc-beam intensity in the initial plane and the z-derivative of the phase in (2.28), as shown in (2.22), we obtain:

$$\begin{gathered} P_{or,z}=|E(x)|^2\frac{\partial}{\partial z}(\arg E(x))= \\ \pi k\left(\frac{2a\sin(kxa)}{kxa}\right)^2\left(\frac{2J_1(kx)}{\sin(kxa)}\right). \end{gathered} \tag{2.29}$$

The first parentheses in (2.29) show the intensity of the sinc beam, and the expression in the second parentheses, multiplied by $2k\pi$, is the derivative of the phase of the field (2.27) with respect to z, or according to (2.22) the longitudinal projection of the local wave vector:

Table 2.1 Coordinates of the first few zeros of the Bessel and sine functions in (2.30)

x	x_1	x_2	x_3	x_4	x_5	x_6	x_7
$J_1(kx)$		$1.2\lambda/2$		$2.2\ \lambda/2$		$3.2\ \lambda/2$	
$\sin(kx)$	$\lambda/2$		$2\ \lambda/2$		$3\ \lambda/2$		$4\ \lambda/2$

$$k_z = 2\pi k\left(\frac{J_1(kx)}{\sin(kxa)}\right). \tag{2.30}$$

Next, we consider expression (2.30) in more detail. The behavior of the ratio of two functions in the numerator and denominator in (2.30) is determined by the zeros of these functions. Approximate values of the first few zeros of these functions for $a = 1$ (no evanescent waves) are shown in Table 2.1.

In the interval between the zeros $x_{2n+1} < x < x_{2n+2}$, $n = 0, 1, 2, \ldots$, the numerators and denominators of the functions in (2.30) is seen from Table 2.1 to have different signs, and therefore the projection of the wave vector onto the longitudinal axis is negative. That is, in these intervals of the transverse axis, a reverse canonical energy flow occurs. Moreover, at special points where the sine function is zero, the projection of the wave vector onto the z-axis tends to infinity: for $x < x_{2n+1}$, $n = 0, 1, 2,\ldots k_z$ tends to positive infinity, and for $x > x_{2n+1}$ k_z tends to negative infinity. Near the zeros of the sine function (zeros of intensity), light slows down, since a large (giant) wave number means that the light at this location has a very short wavelength (slow light). Regions in the cross section of a light beam where the modulus of the wave vector (it is k_z in our case) is greater than the wave number of light: $k_z > k = 2\pi/\lambda$, are called regions with the superoscillation effect [19]. The distance between the extreme points of the intervals, where the reverse flow takes place, is subwavelength and is approximately 0.1λ. At the extreme points of the intervals, the longitudinal energy flux is zero. At odd points, the energy flux is zero because the intensity is zero. And at even points, the longitudinal energy flux is zero because the longitudinal projection of the wave vector is zero: $k_z = 0$. When passing through the extreme points of the specified intervals, the longitudinal projection of the wave vector k_z changes sign. At the even zero points of the Bessel function x_{2n+2} in (2.30), the longitudinal energy flux (2.28) is zero, and the longitudinal projection of the wave vector is zero, $k_z = 0$. But since the intensity of the sinc beam at the points x_{2n+2} is not zero, this means that the light energy at these points is directed along the transverse x-axis. Note that all the described metamorphoses of light (slow light, transverse propagation of light, reverse energy flow) occur at a distance from the initial plane much smaller than the wavelength ($z < < \lambda$), when the wave process has not yet "settled down". Interestingly, the intensity distribution of the sinc beam (2.27) at $a = 1$ has a size at half-maximum of intensity equal to the scalar diffraction limit for 2D beams, FWHM $= 0.44\lambda$. That is, the intensity distribution is not subwavelength, and the distribution of the longitudinal energy flow (2.29) has significantly subwavelength regions.

The mechanism for generating the reverse flow is as follows. The initial sinc-field of light (2.27) has intensity zeros at the points x_{2n+1}. At these intensity zeros, the phase changes abruptly by π. As the light propagates along the z-axis, the locations where the intensity zero was experiences an intensity minimum, and instead of a phase jump of π, there is a smooth phase change by an amount less than π. That is, as the light propagates from the initial plane, the phase difference between two adjacent points decreases. This is only possible when the light (energy flow) locally propagates in the opposite direction. Note also that the derivative of the complex amplitude of the sinc-beam (2.27) with respect to the transverse coordinate is a real function, and therefore the transverse energy flow in the initial plane is zero:

$$\begin{gathered} P_{or,x} = \operatorname{Im}\left(E^*(x,z) \frac{\partial E(x,z)}{\partial x} \right)\Big|_{z=0} = \\ \operatorname{Im}\left(\frac{4\sin(kxa)}{kx} \frac{\partial}{\partial x} \left(\frac{\sin(kxa)}{kx} \right) \right) = 0. \end{gathered} \tag{2.31}$$

When light propagates, due to diffraction, the wave vector of course has a transverse component at each point in the beam cross-section.

2.2.2 *Numerical Simulation Results*

Figure 2.10 depicts the curves of the Bessel function (curve 1) and sine function (curve 2) from Eq. (2.29) for $a = 1$. Microns are plotted along the horizontal axis. The odd coordinates x_{2n+1} in Fig. 2.10 are the zeros of the sine function (intensity zeros), and their values are given in Table 2.1. The even coordinates x_{2n+2} are the zeros of the Bessel function, whose values are also given in Table 2.1.

The sine and Bessel functions is seen from Fig. 2.10 to have opposite signs between adjacent odd and even zeros. This means that a reverse energy flow is formed in these intervals, meaning the longitudinal projection of the energy flow (2.29) is negative.

Figure 2.11 shows the distribution of the longitudinal projection of the wave vector (2.30) in the initial plane for a sinc beam. In the vicinity of the intensity zeros (in the vicinity of the points x_{2n+1}), the values of the wave vector projection are seen from Fig. 2.11 to tend to plus and minus infinity, as predicted by expression (2.30). At even points x_{2n+2}, the longitudinal projection is zero, meaning that the energy at these points propagates along the initial plane. Further numerical simulation is performed using the Rayleigh-Sommerfeld integral [15]. The simulation parameters are: the initial field size is 1000 × 1000 samples, 10 × 10 μm (sample size is approximately λ/53), wavelength is λ = 532 nm, $E_y = 2a \ \sin c(kxa)$, $a = 1$. The output field has a size of 1000 × 10 samples within the boundaries (−0.5 μm < X < 0.5 μm), (−2.5 nm < Y < 2.5 nm). The field is calculated twice, at a distance of 17 nm (approximately λ/30) from the initial plane and at a distance of

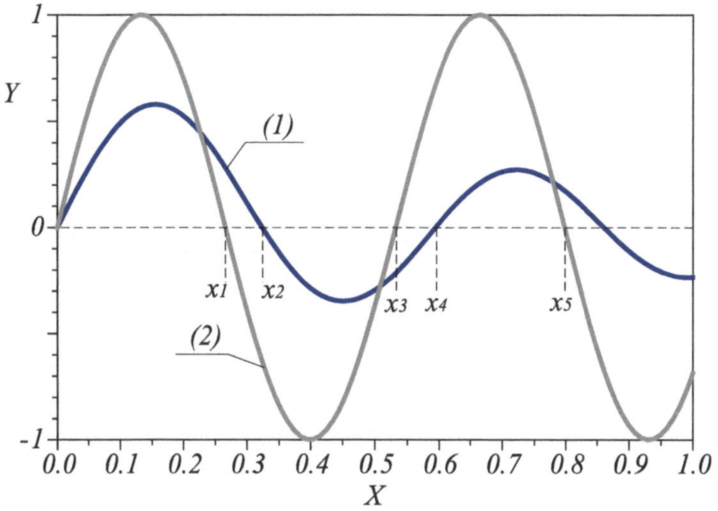

Fig. 2.10 Graphs of the dependence of $J_1(kx)$ (curve 1) and $\sin(kx)$ (curve 2) on the x coordinate, $k = 2\pi/\lambda$, $\lambda = 532$ nm

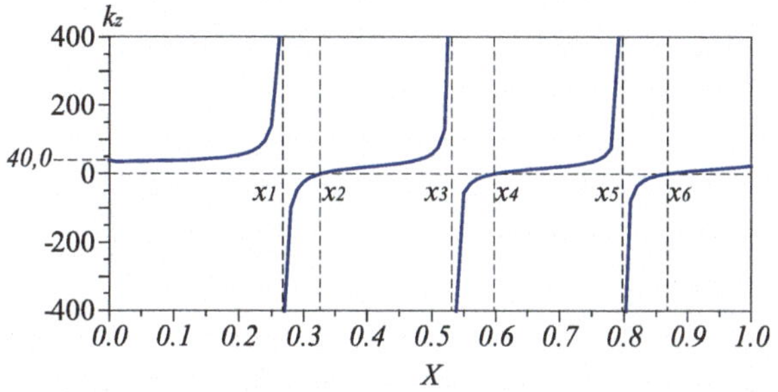

Fig. 2.11 Graph of the dependence of the wave vector longitudinal component (2.30) on the transverse coordinate x. Microns are plotted along the horizontal axis

34 nm from it to be able to take the derivative with respect to z. The intensity (a) and longitudinal energy flux (b) distributions near the initial plane for the sinc beam are shown in Fig. 2.12.

The longitudinal energy flux is seen from Fig. 2.12 to be negative in the region of the first side lobe of the intensity distribution, which is consistent with Fig. 2.11. Shown in Fig. 2.13 is the phase distributions along the transverse coordinate x at two close distances from the initial plane.

The grey curve is clear from Fig. 2.13 to lie above the blue curve everywhere except for the dotted areas, where, on the contrary, the blue curve lies above the grey curve. That is, the phase in the regions of the reverse flow during the propagation of the sinc-beam does not increase, as usual, but decreases. This shows that the light in these regions propagates in the opposite direction. To confirm this result, we repeat a similar calculation using a more accurate difference method for solving Maxwell's equations. The simulation is performed using the FDTD method using the RSoft FullWAVE software package. The simulation parameters are as follows: the entire

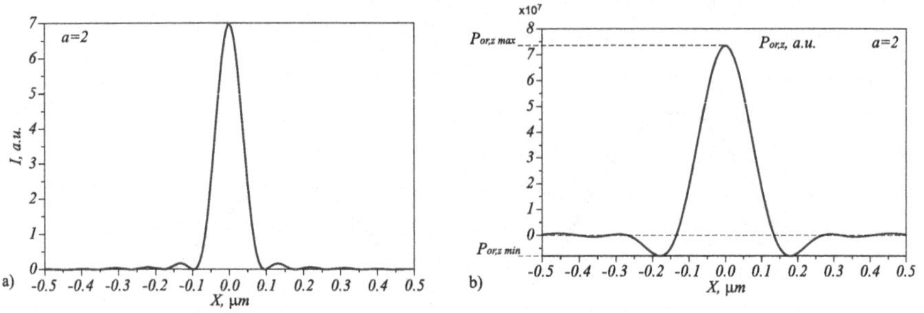

Fig. 2.12 Profiles along the x-axis of the intensity projection of the sinc beam (**a**) and the longitudinal canonical energy flux (**b**) at a distance of 17 nm from the initial plane at $a = 1$

Fig. 2.13 Phase distributions of the field E_y (2.27) at a distance of $z_1 = 17$ nm (blue curve) and $z_2 = 34$ nm (gray curve) from the initial plane. The dotted lines indicate regions where the derivative of the phase with respect to z is negative, i.e., where the value of $P_{or,z}$ (2.29) is negative

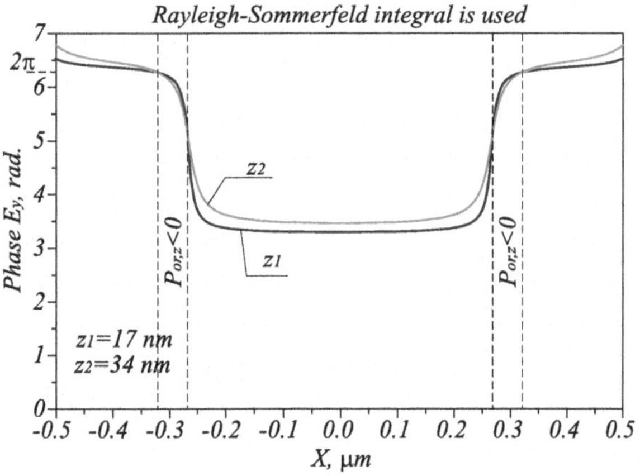

simulation field is 10x10x1 μm, the initial field is the same as in Fig. 2.12 ($E_y = 2a \quad \sin c(kxa)$, $a = 1$). The initial field is specified in the 10 × 10 μm region. The partition along all three coordinates is λ/30. The observation planes are also spaced from each other by λ/30 (approximately 17 nm), the distance to the first observation plane is also 17 nm. Phase profiles similar to those in Fig. 2.13 are shown in Fig. 2.14.

It is clear from a comparison of Figs. 2.13 and 2.14 that they almost coincide.

2.2.3 Generalizations of Results

In the previous sections, we demonstrate that several interesting optical effects occur in the near field of a non-paraxial 2D sinc beam. Near zero-intensity points, regions with superoscillation effects form, where the longitudinal projection of the wave vector takes on large (giant) values. At locations with large negative values of the wave vector longitudinal component, a reverse energy flow occurs. In the region of

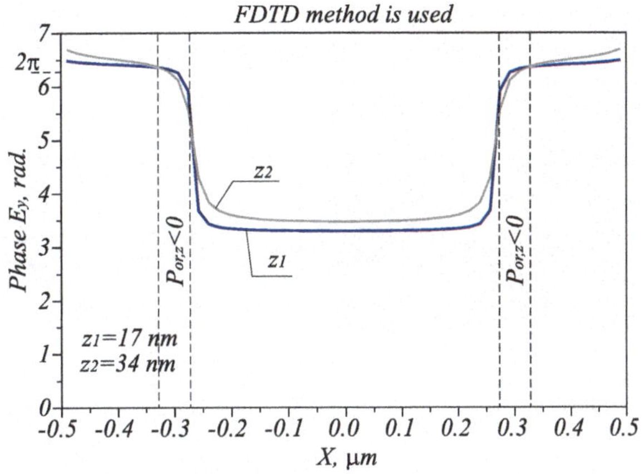

Fig. 2.14 The same as in Fig. 2.13, but the simulation is carried out using the FDTD method

giant wave vector values, the light has a phase velocity much smaller than the speed of light (slow light). However, these effects are not unique to a sinc beam, but are inherent in any 2D non-paraxial light field that has first-order amplitude zeros or intensity zeros. Indeed, similar to how we obtained (2.30) from (2.29), we can generally obtain the following expression for the longitudinal projection of the energy flow:

$$\begin{aligned} P_{or,z} &= \mathrm{Im}\left(E^*(x,z)\frac{\partial E(x,z)}{\partial z}\right) = \\ &|E(x,z)|^2\mathrm{Im}\left(E^{-1}(x,z)\frac{\partial E(x,z)}{\partial z}\right), \\ &\frac{\partial}{\partial z}(\arg E(x,z)) = \mathrm{Im}\left(E^{-1}(x,z)\frac{\partial E(x,z)}{\partial z}\right). \end{aligned} \quad (2.32)$$

In (2.32), the second expression for the phase derivative with respect to z has the complex amplitude $E(x, y)$ in the denominator. This means that if the complex amplitude has an isolated zero (not necessarily at $z = 0$) of the first order, then in the vicinity of this zero the field amplitude can be written as:

$$E(x,z) \approx (x + x_0) \pm i\alpha(z - z_0), \quad (2.33)$$

where (x_0, z_0) are the coordinates of the intensity zero, α is the ellipticity parameter of the 2D optical vortex. Follows from (2.33) that if it is known that in the initial field ($z = 0$) the light field has intensity zeros, for example, as in the sinc-beam (2.27) at the points $x_p = \lambda p/(2a)$, $p = \pm 1, \pm 2, \pm 3, \ldots$, then intensity zeros of light field with complex amplitude

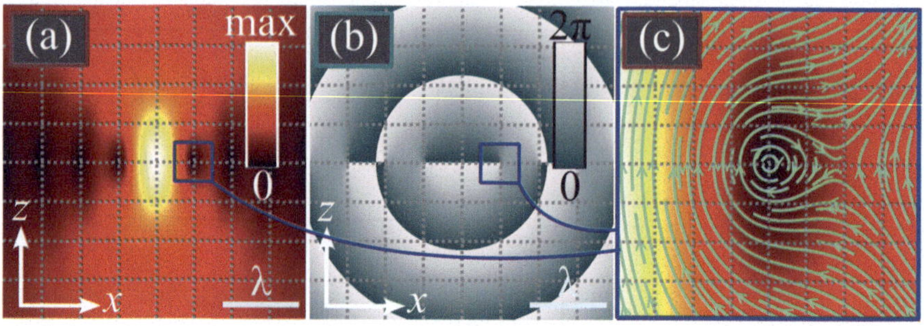

Fig. 2.15 Distribution of the amplitude modulus (2.34) at $a = 1$ (**a**) and its argument (phase) (**b**) in coordinates (x, z) at $z_0 = 2\lambda$, $\lambda = 532$ nm, frame sizes are $4\lambda \times 4\lambda$. The vertical axis is along z, the horizontal axis is along x. Lines with directions of energy flow (**c**) are in the blue square

$$E(x, z) = \int_{-a}^{a} \exp\left(ikx\xi + ik(z - z_0)\sqrt{1 - \xi^2}\right) d\xi, \tag{2.34}$$

are at points with coordinates (x_p, z_0). And around these intensity zeros there are 2D optical vortices with topological charges of +1 and −1. And the energy flow circulates around these optical vortices clockwise if the topological charge is −1, or counterclockwise if the topological charge is +1. For example, in the sinc beam (2.27), all vortices at $x > 0$ have a topological charge of −1, and all vortices at $x < 0$ have a topological charge of +1. Part of the energy flow that circulates around the intensity zeros of the 2D non-paraxial light field is directed in the opposite direction to the optical axis. For example, in the sinc beam (2.27), the energy flow around the vortices at $x > 0$ circulates clockwise, and around the vortices at $x < 0$, the energy flow circulates counterclockwise. That is, it can be concluded that if a 2D non-paraxial light field has an isolated intensity zero, then a reverse energy flow takes place in the vicinity of this intensity zero. Shown in Fig. 2.15 are the amplitude (a) and phase (b) distributions for a sinc beam (2.34) shifted along the z-axis at $a = 1$. 2D optical vortices are depicted in Fig. 2.15b to be visible on the horizontal axis at $z_0 = 2\lambda$ (3 vortices with a charge of −1 on the right and 3 vortices with a charge of +1 on the left). That is, the phase changes from zero to 2π when going around the centers of the vortices.

Our theoretical and numerical analysis reveals a superoscillation effect in the near-field region surrounding points of zero light intensity. Using a nonparaxial 2D sinc beam that does not contain evanescent waves, we show that this effect arises where the longitudinal component of the wave vector significantly exceeds the total light wave number. As the wave passes a zero-intensity point, this large positive longitudinal wave vector component inverts to a negative value. The immense magnitude of the wave vector near these zeros implies a drastically reduced local wavelength and phase velocity, characteristic of "slow light." Furthermore, the negative longitudinal component signifies the onset of a reversed energy flow immediately beyond the zero-intensity point. This backward flow persists until the wave vector becomes purely transverse (its longitudinal component reaches zero), at which point the energy is redirected laterally. All these phenomena—slow light and reverse flow—are confined to a region extremely close to the source, at a distance substantially less than the light wavelength.

2.3 Light Obstacle

In a recent work [33], a 2D nonparaxial light field has been investigated. An explicit analytical expression has been derived for the complex amplitude of this field at an arbitrary distance from the initial plane. Besides, it was found in [33] that this nonparaxial beam has a countable number of local subwavelength areas with the energy flow directed backward. Energy backflow has also been found in several 2D nonparaxial light beams [9–11, 34–36] and, perhaps, it will be possible to found the reverse flow in some other known 2D nonparaxial beams: Weber and Mathieu beams [37]. In this work, we demonstrate both theoretically and numerically that the beams [33] have along the optical axis a set of pairs of close intensity zeros, symmetric with respect to the optical axis, and the backward energy flow between these zeros does not allow passing of the forward energy flow. Thus, the forward energy flow has to bypass this "obstacle". We call this phenomenon a light obstacle.

2.3.1 2D Fractional Order Bessel Beam

We consider a 2D nonparaxial light field with the following complex amplitude [33]:

$$\begin{aligned}
E_1(x,z) &= \int_{-\infty}^{\infty} A_1(\xi)\exp\left(ikx\xi + ik(z-z_0)\sqrt{1-\xi^2}\right)d\xi \\
&= \int_{-1}^{1} \frac{\exp\left(ikx\xi + ik(z-z_0)\sqrt{1-\xi^2}\right)}{\left(1-\xi^2\right)^{3/4}}d\xi \\
&= \int_{-1}^{1} \frac{\cos(kx\xi)\exp\left(ik(z-z_0)\sqrt{1-\xi^2}\right)}{\left(1-\xi^2\right)^{3/4}}d\xi \\
&= 2\left(\frac{\pi}{2}\right)^{3/2}\sqrt{k|z-z_0|}\left[J_{-\frac{1}{4}}\left(k\frac{\sqrt{x^2+(z-z_0)^2}-x}{2}\right)J_{-\frac{1}{4}}\left(k\frac{\sqrt{x^2+(z-z_0)^2}+x}{2}\right)\right. \\
&\left. +i\,\mathrm{sgn}(z-z_0)J_{\frac{1}{4}}\left(k\frac{\sqrt{x^2+(z-z_0)^2}-x}{2}\right)J_{\frac{1}{4}}\left(k\frac{\sqrt{x^2+(z-z_0)^2}+x}{2}\right)\right],
\end{aligned} \tag{2.35}$$

with $J_\nu(t)$ being the νth-order Bessel function. In contrast to the work [33], we shift the beam along the optical axis by a distance of z_0, to make the beam autofocusing. The integrals in Eq. (2.35) indicate that the amplitude of the light field is expanded into the plane waves spectrum. In Eq. (2.35), x and z are transverse and longitudinal coordinates, respectively, z_0 is the autofocusing distance. Amplitude of the plane waves spectrum in Eq. (2.35) is given by

$$A_1(\xi) = \left(1 - \xi^2\right)^{-3/4}, \quad |\xi| < 1. \tag{2.36}$$

The integrals in Eq. (2.35) are evaluated analytically since they are reference integrals given in [38]. The light field (2.35) propagates in positive direction of the axis z and has infinite energy. Further, we show that the beam in its focus plane $z = z_0$ has intensity zeros, determined by the zeros of the Bessel function of the order $-1/4$. According to Eq. (2.35), the amplitude of the light field in the focus plane $z = z_0$ is equal to:

$$\begin{aligned} E_1(x, z=0) &= \int_{-\infty}^{\infty} A_1(\xi) \exp(ikx\xi) d\xi = 2 \int_0^1 \frac{\cos(kx\xi)}{\left(1 - \xi^2\right)^{3/4}} d\xi \\ &= 4\left(\frac{\pi}{2}\right)^{3/2} \Gamma^{-1}\left(\frac{3}{4}\right)\left(\frac{kx}{2}\right)^{1/4} J_{-1/4}(kx). \end{aligned} \tag{2.37}$$

with $\Gamma(t)$ being the Gamma function. The first zero of the Bessel function at $x > 0$ is nearly equal to 2, i.e. $J_{-1/4}(2) = 0$. The transverse coordinate of this zero is $x_1 = 0.32\lambda$. At $x < 0$, the first zero of the Bessel function is equal to -2, i.e. $J_{-1/4}(-2) = 0$, and its coordinate is $-x_1 = -0.32\lambda$. Around these intensity zeros of the beam (1), energy flow circulates. We note that around the zero with the coordinate $x_1 = 0.32\lambda$ energy rotates clockwise, whereas around the intensity zero with the coordinate $-x_1 = -0.32\lambda$ energy rotates counterclockwise.

To confirm this, we should write the longitudinal component of the canonical energy flow vector [7, 39] of the TE-polarized beam (2.35) in an explicit form. We suppose that the polarization vector of the beam (2.35) is directed along the coordinate y. Then, at $z = z_0$, we get:

$$\begin{aligned} P_{or,z} &= \operatorname{Im}\left(E_y^* \frac{\partial}{\partial z} E_y\right)\Big|_{z=0} \\ &= 4k \int_0^1 \frac{\cos(kx\xi)}{\left(1 - \xi^2\right)^{3/4}} d\xi \int_0^1 \frac{\cos(kx\xi)}{\left(1 - \xi^2\right)^{1/4}} d\xi. \end{aligned} \tag{2.38}$$

The integrals in Eq. (2.38) are reference integrals [38] and thus Eq. (2.38) can be rewritten as:

$$\begin{aligned} P_{or,z}(x,z=0) &= 4k\left(\frac{\pi}{2}\right)^3\Gamma^{-1}\left(\frac{3}{4}\right) \\ &\times\left(\frac{1}{2}\Gamma^{-1}\left(\frac{5}{4}\right)+2\Gamma^{-1}\left(\frac{1}{4}\right)\right)J_{-1/4}(kx)J_{1/4}(kx). \end{aligned} \tag{2.39}$$

The first zero of the Bessel function of the order 1/4 is nearly 2.8, i.e. $J_{1/4}(2.8) = 0$, and the coordinate of this zero at $x > 0$ is equal to $x_2 = 0.44\lambda$. The zero of the Bessel function of the order 1/4, symmetrical with respect to the optical axis, has a coordinate $-x_2 = -0.44\lambda$. Between the root $x_1 = 0.32\lambda$ and the root $x_2 = 0.44\lambda$, the Bessel function in Eq. (2.39) has different signs and the energy flow is thus negative. The energy flow is negative in the interval $x_2—x_1 = 0.44\lambda—0.32\lambda = 0.12\lambda$. Before the first zero, in the interval $0 < x < x_1$, energy flow (5) is positive. Thus, around the first intensity zero of the beam (1) in the focus plane $z = z_0$, energy flow circulates clockwise. The same way, around the first intensity zero with the coordinate $-x_1 = -0.32\lambda$, energy flow circulates counterclockwise. In the interval between the first two zeros, with a distance between them being equal to $2x_1 = 0.64\lambda$, light propagates in the forward direction along the optical axis z. Light, propagating in forward direction, encounters intensity zeros and bypasses them from both sides. Therefore, the area around the intensity zero with the size of nearly 2 $(x_2—x_1) = 0.24\lambda$, where the energy circulates along a closed contour, is an obstacle for the light, which bypasses this obstacle from both sides.

2.3.2 *Longitudinal Projection of the Canonical Energy Flux Vector for a 2D Fractional Bessel Beam*

Now we consider the expression for the complex amplitude of the beam (2.35) at a certain distance from the focus plane $z = z_0$. Since the first roots of the Bessel functions in Eq. (2.35) are nearly equal to 2 and 2.8, i.e. $J_{-1/4}(2) = 0$, $J_{1/4}(2.8) = 0$, then, making the arguments equal to these values, we obtain a system of equations for finding the transverse and longitudinal coordinates of these zeros:

$$\begin{cases} \frac{\pi}{\lambda}\left(\sqrt{x^2+z^2}-x\right)=2, \\ \frac{\pi}{\lambda}\left(\sqrt{x^2+z^2}+x\right)=2.8. \end{cases} \tag{2.40}$$

Solving the system (2.40), we obtain approximate coordinates of the intensity zeros: $x_3 = 0.13\lambda$, $-x_3 = -0.13\lambda$, $z = z_0 + 0.75\lambda$. Thus, at the distance $z_1 = 0.75\lambda$ from the focus plane, two symmetrical close intensity zeros appear near the optical axis at a distance of 0.26λ from each other. Around these intensity zeros (as around any isolated intensity zero of a 2D nonparaxial beam), circulates energy flow. For determining the direction of energy rotation around these zeros, we consider a general expression for the longitudinal component of the canonical energy flow vector, similar to Eq. (2.39), but with an arbitrary z:

$$\begin{aligned}P_{or,z}(x,z)=&\left(\frac{\pi^3k^2(z-z_0)^2}{8\sqrt{x^2+(z-z_0)^2}^2}\right)\{J_{-1/4}(u)J_{-1/4}(v)\\&\times\left[J_{-3/4}(u)J_{1/4}(v)-J_{5/4}(u)J_{1/4}(v)\right.\\&\left.+J_{1/4}(u)J_{-3/4}(v)-J_{1/4}(u)J_{5/4}(v)\right]\\&-J_{1/4}(u)J_{1/4}(v)\\&\times\left[J_{-5/4}(u)J_{-1/4}(v)-J_{3/4}(u)J_{-1/4}(v)\right.\\&\left.+J_{-1/4}(u)J_{-5/4}(v)-J_{-1/4}(u)J_{3/4}(v)\right],\end{aligned}\tag{2.41}$$

with

$$\begin{aligned}u&=\frac{k}{2}\left(\sqrt{x^2+(z-z_0)^2}-x\right),\\v&=\frac{k}{2}\left(\sqrt{x^2+(z-z_0)^2}+x\right).\end{aligned}\tag{2.42}$$

In Eq. (2.41), an expression was used for the z-derivative of the amplitude (2.35):

$$\begin{aligned}&\frac{\partial}{k\partial z}\int_{-1}^{1}\frac{\cos(kx\xi)\exp\left(ik(z-z_0)\sqrt{1-\xi^2}\right)}{\left(1-\xi^2\right)^{3/4}}d\xi\\&=i\int_{-1}^{1}\frac{\cos(kx\xi)\exp\left(ik(z-z_0)\sqrt{1-\xi^2}\right)}{\left(1-\xi^2\right)^{1/4}}d\xi\\&=2i\left(\frac{\pi}{2}\right)^{3/2}\sqrt{\frac{1}{k|z-z_0|}}\{J_{1/4}(u)J_{1/4}(v)+kw(z)\left(J_{-3/4}(u)J_{1/4}(v)\right.\\&\quad\left.-J_{5/4}(u)J_{1/4}(v)+J_{1/4}(u)J_{-3/4}(v)-J_{1/4}(u)J_{5/4}(v)\right)\\&\quad-i\,\mathrm{sgn}(z-z_0)\left[J_{-1/4}(u)J_{-1/4}(v)+kw(z)\left(J_{-5/4}(u)J_{-1/4}(v)\right.\right.\\&\quad\left.\left.-J_{3/4}(u)J_{-1/4}(v)+J_{-1/4}(u)J_{-5/4}(v)-J_{-1/4}(u)J_{3/4}(v)\right)\}\right],\end{aligned}\tag{2.43}$$

Although the expression (2.41) looks complicated, it allows drawing some conclusions. Since the transverse coordinate of the first zero at the distance of 0.75λ from the focus equals $x_3=0.13\lambda$, we choose a point with a little smaller coordinate, for instance, $x=0.1\lambda$. Then, the arguments of the Bessel functions in Eq. (2.41) are equal to $u = 2.06$ and $v = 2.69$. Thus, the expression in the curly brackets in Eq. (2.41) reads as

$$\begin{aligned}
&\{J_{-1/4}(u)J_{-1/4}(v)\left[J_{-3/4}(u)J_{1/4}(v) - J_{5/4}(u)J_{1/4}(v)\right.\\
&\quad \left.+J_{1/4}(u)J_{-3/4}(v) - J_{1/4}(u)J_{5/4}(v)\right]\\
&- J_{1/4}(u)J_{1/4}(v)\left[J_{-5/4}(u)J_{-1/4}(v) - J_{3/4}(u)J_{-1/4}(v)\right.\\
&\quad \left.+J_{-1/4}(u)J_{-5/4}(v) - J_{-1/4}(u)J_{3/4}(v)\right]\}\\
&= \{J_{-1/4}(2.06)J_{-1/4}(2.69)\left[J_{-3/4}(2.06)J_{1/4}(2.69)\right.\\
&\quad - J_{5/4}(2.06)J_{1/4}(2.69) + J_{1/4}(2.06)J_{-3/4}(2.69)\\
&\quad \left.- J_{1/4}(2.06)J_{5/4}(2.69)\right]\\
&- J_{1/4}(2.06)J_{1/4}(2.69)\left[J_{-5/4}(2.06)J_{-1/4}(2.69)\right.\\
&\quad - J_{3/4}(2.06)J_{-1/4}(2.69) + J_{-1/4}(2.06)J_{-5/4}(2.69)\\
&\quad \left.- J_{-1/4}(2.06)J_{3/4}(2.69)\right]\}\\
&= \{(-0.03)(-0.31)[(-0.46)(0.044)\\
&\quad - (0.549)(0.044) + (0.368)(-0.491) - (0.368)(0.549)]\\
&- (0.368)(0.044)[(-0.549)(-0.31)\\
&\quad - (0.557)(-0.31) + (-0.03)(-0.291) - (-0.03)(0.349)]\}\\
&< 0.
\end{aligned} \tag{2.44}$$

Equation (2.44) indicates that the energy flow at $x < x_3$ is negative. Thus, the energy around the intensity zero with the coordinate x_3 circulates counterclockwise (topological charge of the longitudinal optical vortex is +1). The same way, it can be shown that the energy flow (2.41) to the right from the intensity zero with the coordinate $-x_3$, i.e. in the area $-x_3 < x < 0$, is negative. The equations system (2.36) can be generalized. For arbitrary values $b > a > 0$, the arguments u and v of the Bessel functions in Eq. (2.42) can be made equal to $u = a$, $v = b$ and then the corresponding transverse and longitudinal coordinates can be found:

$$\begin{cases} x = \dfrac{\lambda}{2\pi}(b - a), \\ z = \dfrac{\lambda}{\pi}\sqrt{ab}. \end{cases} \tag{2.45}$$

Using Eq. (2.45) and zeros of the Bessel functions of the orders −1/4 and 1/4, we can obtain the coordinates of all pairs of intensity zeros near the optical axis, residing at a certain distance from the focus plane. The next several zeros of the Bessel functions, $J_{-1/4}(\gamma_n) = 0$, $J_{1/4}(\delta_n) = 0$ with $n = 1, 2, 3, 4$, are approximately equal to $\gamma_1 = 2$, $\gamma_2 = 5.1$, $\gamma_3 = 8.2$, $\gamma_4 = 11.4$; $\delta_1 = 2.8$, $\delta_2 = 5.9$, $\delta_3 = 9.0$, $\delta_4 = 12.2$. It is seen that the difference between the roots remains almost constant: δ_n—$\gamma_n = 0.8$. Therefore, pairs of intensity zeros near the optical axis at different distances from the focus are separated by almost equal distance:

$$\begin{aligned} x_4 &= \pm(\delta_2-\gamma_2)\lambda/(2\pi) \approx \pm 0.13\lambda, \quad z_2 = 1.75\lambda, \\ x_5 &= \pm 2\pi(\delta_3-\gamma_3)\lambda/(2\pi) \approx \pm 0.13\lambda, \quad z_3 = 2.73\lambda, \\ x_6 &= \pm 2\pi(\delta_4-\gamma_4)\lambda/(2\pi) \approx \pm 0.13\lambda, \quad z_4 = 3.75\lambda. \end{aligned} \tag{2.46}$$

According to Eq. (2.46), only the first pair of the intensity zeros with the coordinates $x_3 = (\delta_1 - \gamma_1)\lambda/(2\pi) \approx \pm 0.13\lambda$ resides at the distance of $z_1 = 0.75\lambda$ from the focus plane, whereas other pairs of zeros repeated after each wavelength along the optical axis. The distance between the intensity zeros is conserved since the beam (2.35) is diffraction-free and has infinite energy, similarly to the well known Bessel beam.

2.3.3 Numerical Simulation

In contrast to the two nearest intensity zeros in the focus plane, the energy flow within the pairs of zeros (2.46), close to the optical axis, is backward, as is demonstrated by an example in Eq. (2.44). Thus, the energy flow around the intensity zero $x_4 > 0$ circulates counterclockwise, while the energy flow around the intensity zero $-x_4 < 0$ circulates clockwise. The forward energy flow that reaches these intensity zeros, does not pass between them. Rather, it passes around them. The same way, the forward energy flow does not pass between other pairs of intensity zeros with the coordinates from Eq. (2.46). Thus, these intensity zeros generate a light obstacle, forcing the light to bypass it. Illustrated in Fig. 2.16 are simulation results, confirming the above given theoretical conclusions. Figure 2.16 shows intensity (Fig. 2.16a) and phase (Fig. 2.16b) distributions of the 2D beam (2.35) with the focus distance $z_0 = 3\lambda$, where $\lambda = 532$ nm. The horizontal axis shows the coordinate x in wavelengths, while the vertical axis show the coordinate z, also in wavelengths. Figure 2.16b reveals that there is a set of longitudinal optical vortices with their centers defined by the roots of the Bessel functions in Eq. (2.35), i.e. when either $u = \gamma_n$ and $v = \delta_m$, or, vice versa, $u = \delta_m$ and $v = \gamma_n$, where $n = 1,2,3,\ldots$, $m = 1,2,3,\ldots$ As seen in Fig. 2.16c, around the two intensity zeros in the focus ($z = z_0 = 3\lambda$), energy circulates counterclockwise on the left and clockwise on the right. Therefore, the forward flow passes between the intensity zeros. As seen in Fig. 2.16d, between two intensity zeros with the coordinates $x_3 = \pm 0.13\lambda$, $z_1 = 0.75\lambda$, close to the optical axis, a reverse flow is generated, which does not allow the forward flow passing along the optical axis. Therefore, the forward flow is forced to bypass both intensity zeros. The picture of energy flow lines in Fig. 2.16d remind a bull's face with two nostrils and wide-set eyes. The topological charges of the left and of the right eye (+1 and −1) are opposite to the topological charges of the vortices in the nostrils (−1 on the left and +1 on the right). So, we demonstrated that the light in a subwavelength scale, at the level of tenths of a wavelength, can generate

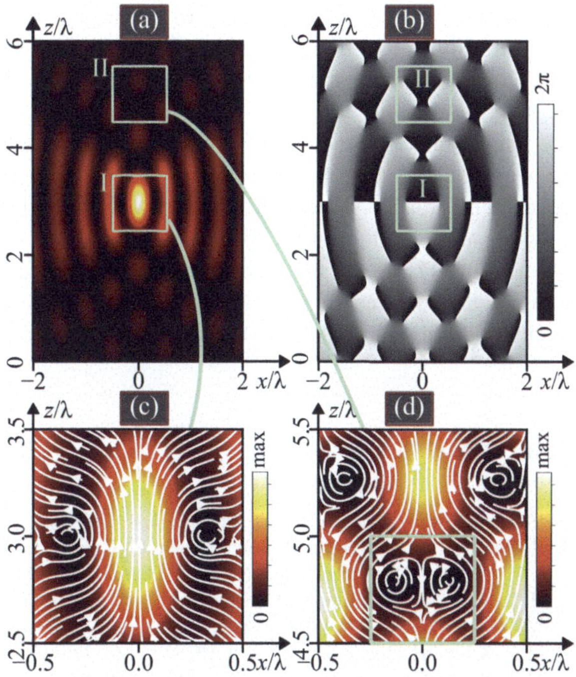

Fig. 2.16 Intensity (**a**) and phase (**b**) distributions of the beam (2.35) with $z_0 = 3\lambda$, $\lambda = 532$ nm, as well as magnified intensity distributions with energy flow directions, shown by arrows, in the highlighted squares I (**c**) and II (**d**)

in free space such a circulation of light energy that can become an obstacle to the forward energy flow, forcing it to bypass this obstacle [40].

At the end of this Chapter, the following conclusions can be made. In the work [41], M.V. Berry has demonstrated for the first time that in the waist of a bounded 2D nonparaxial Gaussian beam without evanescent waves, two phase singularity points (screw dislocations) exist, symmetric with respect to the optical axis and having the topological charges of +1 and −1, and two phase saddle points exist nearby [42]. Around the phase singularity points (intensity zeros), energy flow circulates, proportional to the wave vector. Reverse energy flow is generated in the interval between a point of phase screw dislocation and a phase saddle point. In [41], Fig. 2.3 illustrates that the energy flow makes a curve with a shape of digit 'eight', embracing both points (phase dislocation and saddle point). However, this is not the case. In [43], by using an exact solution, it was shown for a nonparaxial 2D sinc-beam that the energy flow circulates only around the intensity zero and does not go beyond the saddle point (Fig. 2.15c). We note that in 3D paraxial beams, phase singularity points also can be generated near the focus in longitudinal section, as was shown by G.P. Karman et al. [44, 45]. However, due to paraxiality, there is no saddle phase point near the phase singularity point, whereas the reverse flow can be generated in an interval between the phase singularity point and phase saddle point. Paraxial light fields cannot have a reverse energy flow. In [12], several examples of 1D complex amplitudes were investigated theoretically, that can have a negative phase gradient (reverse energy flow), including in quantum mechanics.

Experimentally, reverse energy flow can be registered near the first dark ring of zero intensity in a subwavelength focus, generated, for instance, by a zone plate [46]. For determining the reverse energy flow, two phase distributions should be measured in two close planes, shifted relative to each other along the optical axis. The longitudinal phase derivative is proportional to the longitudinal component of the wave vector. For measuring the phase with a resolution, smaller than the wavelength, transverse section of the 3D beam in the focus can be scanned by using a nanosized fluorescent sphere at the tip of the cantilever, as it was done, for instance, in [47]. Since the reverse energy flow arises near the intensity zero and occupies a region nearly tenth fraction of the wavelength, Weak Measurements method, described in [48, 49], should be employed for its detection.

References

1. V.S. Ignatowsky, Trans. Opt. Inst. Petrogr. **1**, 4 (1920)
2. B. Richards, E. Wolf, Proc. R. Soc. A Math. Phys. Eng. Sci. **253**, 358 (1959)
3. V.V. Kotlyar, S.S. Stafeev, A.G. Nalimov, Phys. Rev. A **99**, 033840 (2019)
4. H. Li, C. Wang, M. Tang, X. Li, Opt. Express **28**, 18607 (2020)
5. A.V. Novitsky, D.V. Novitsky, J. Opt. Soc. Am. A **24**, 2844 (2007)
6. M.A. Salem, H. Bağcı, Opt. Express **19**, 8526 (2011)
7. M.V. Berry, J. Opt. A Pure Appl. Opt. **11**, 094001 (2009)
8. A. Bekshaev, K.Y. Bliokh, M. Soskin, J. Opt. **13**, 053001 (2011)
9. P. Vaveliuk, O. Martinez-Matos, Opt. Express **20**, 26913 (2012)
10. I. Besieris, P. Saari, Phys. Rev. A **107**, 033502 (2023)
11. V.V. Kotlyar, A.A. Kovalev, A.G. Nalimov, A.M. Telegin, J. Opt. Soc. Am. A **42**, 1396 (2025)
12. M.V. Berry, J. Phys. A Math. Theor. **43**, 415302 (2010)
13. G. Chen, Z.Q. Wen, C.W. Qiu, Light Sci. Appl. **8**, 56 (2019)
14. W. Lu, J. Zhou, Y. Zhang, P. Shi, X. Yuan, C. Min, Opt. Lett. **49**, 6009 (2024)
15. A.G. Nalimov, V.V. Kotlyar, Y.V. Khanenko, Comput. Opt. **47**, 734 (2023)
16. V.V. Kotlyar, A.A. Kovalev, A.G. Nalimov, Comput. Opt. **33**, 369 (2009)
17. A.P. Prudnikov, U.A. Brychkov, O.I. Marichev, *Integrals and Series. Special Functions* (Nauka, Moscow, 1983)
18. M.V. Berry, J. Phys. A Math. Gen. **27**, L391 (1994)
19. M.V. Berry, S. Popescu, J. Phys. A Math. Gen. **39**, 6965 (2006)
20. M.V. Berry, M.R. Dennis, J. Phys. A Math. Theor. **42**, 022003 (2009)
21. M.V. Berry, J. Opt. **15**, 044006 (2013)
22. G. Yuan, E.T.F. Rogers, N.I. Zheludev, Light Sci. Appl. **8**, 2 (2019)
23. F.M. Huang, N.I. Zheludev, Nano Lett. **9**, 1249 (2009)
24. E.T.F. Rogers, N.I. Zheludev, J. Opt. **15**, 094008 (2013)
25. E.T.F. Rogers, J. Lindberg, T. Roy, S. Savo, J.E. Chad, M.R. Dennis, N.I. Zheludev, Nat. Mater. **11**, 432 (2012)
26. G. Gbur, Nano **8**, 205 (2019)
27. B. Corcoran, C. Monat, M. Pelusi, C. Grillet, T.P. White, L. O'Faolain, T.F. Krauss, B.J. Eggleton, D.J. Moss, Opt. Express **18**, 7770 (2010)
28. C. Monat, B. Corcoran, M. Ebnali-Heidari, C. Grillet, B.J. Eggleton, T.P. White, L. O'Faolain, T.F. Krauss, Opt. Express **17**, 2944 (2009)
29. J.F. Nye, M.V. Berry, Proc. R. Soc. A Math. Phys. Eng. Sci. **336**, 165 (1974)

30. M.S. Soskin, V.N. Gorshkov, M.V. Vasnetsov, J.T. Malos, N.R. Heckenberg, Phys. Rev. A—At. Mol. Opt. Phys. **56**, 4064 (1997)
31. E. Abramochkin, V. Volostnikov, Opt. Commun. **83**, 123 (1991)
32. V.V. Kotlyar, H. Elfstrom, J. Turunen, A.A. Almazov, S.N. Khonina, V.A. Soifer, J. Opt. Soc. Am. A **22**, 849 (2005)
33. P. Saari, I.M. Besieris, Photonics **11**, 1129 (2024)
34. V. Kotlyar, A. Kovalev, A. Nalimov, J. Opt. Soc. Am. B **43**, 128 (2026)
35. P. Saari, I. Besieris, Eur. J. Phys. **42**, 55301 (2021)
36. V.V. Kotlyar, A.A. Kovalev, A.G. Nalimov, A.M. Telegin, Opt. Mem. Neural Networks **34**, 476 (2025)
37. Y. Zhang, Z. Mo, D. Xu, S. He, Y. Ding, Q. Huang, Z. Lu, D. Deng, Opt. Lett. **47**, 3059 (2022)
38. A.P. Prudnikov, Y.A. Brychkov, O.I. Marichev, *Integrals and Series, Special Functions* (Gordon and Breach, New York, 1981)
39. B. Ghosh, A. Daniel, B. Gorzkowski, A.Y. Bekshaev, R. Lapkiewicz, K.Y. Bliokh, J. Opt. Soc. Am. B **41**, 1276 (2024)
40. V.V. Kotlyar, A.G. Nalimov, A.A. Kovalev, M. Arfan, S.S. Stafeev, J. Opt. **28**, 15605 (2026)
41. M.V. Berry, J. Mod. Opt. **45**, 1845 (1998)
42. J.F. Nye, J.V. Hajnal, J.H. Hannay, Proc. R. Soc. London. Ser. A Math. Phys. Eng. Sci. **417**, 7 (1988)
43. V. Kotlyar, A. Nalimov, A. Kovalev, A. Telegin, Opt. Lett. **51**, 973 (2026)
44. G.P. Karman, M.W. Beijersbergen, A. van Duijl, J.P. Woerdman, Opt. Lett. **22**, 1503 (1997)
45. G.P. Karman, A. Van Duijl, J.P. Woerdman, Opt. Lett. **23**, 403 (1998)
46. V.V. Kotlyar, S.S. Stafeev, Y. Liu, L. O'Faolain, A.A. Kovalev, Appl. Opt. **52**, 330 (2013)
47. T. Bauer, P. Banzer, E. Karimi, S. Orlov, A. Rubano, L. Marrucci, E. Santamato, R.W. Boyd, G. Leuchs, Science (80-). **347**, 964 (2015)
48. R. Dhara, N. Modak, S. Guchhait, N. Ghosh, Adv. Phys. Res. **4**, 2400136 (2025)
49. Z. Zhu, D. Hay, Y. Zhou, A. Fyffe, B. Kantor, G.S. Agarwal, R.W. Boyd, Z. Shi, Phys. Rev. Appl. **12**, 34036 (2019)

Chapter 3
Reverse Canonical Energy Flow in 3D Laser Beams

3.1 Reverse Canonical Energy Flux in a Nonparaxial Linearly Polarized Beam near the Initial Plane

The canonical energy flux was introduced in optics in works [1–5]. For a scalar light field, the canonical energy flux is equal to a product of the light intensity (power density) and of the gradient of phase distribution in the beam cross-section. It is also called the orbital energy flux [6]. For a light field propagating in the positive direction of the optical axis, orbital energy flux is always positive. However, nonparaxial light fields, that have evanescent waves in the near field, can have reverse canonical energy flux. For instance, reverse canonical energy flux was discovered in the near field of a nonparaxial 2D Airy beam [7]. Since the orbital energy flux vector is one of two terms, comprising the Poynting vector [8], which also has a second term—the spin flux, then the negative values of the longitudinal component of the Poynting vector, previously found in a number of works [9–11], should be interpreted correctly. It has been shown earlier [12] that negative values of the longitudinal component of the Poynting vector arise when the positive value of the longitudinal component of the orbital energy flux is less than the negative value of the longitudinal component of the spin flux vector. However, if the light field has no spin flux, then the longitudinal component of the Poynting vector coincides with the longitudinal component of the canonical energy flux vector. The canonical reverse flux was also found near interfaces between two mediums [13, 14, 21], where surface (evanescent) waves arise. The reverse energy flux was also studied in spatiotemporal localized beams [15, 16]. Usually, the reverse canonical flux arises in the presence of evanescent waves [7], but it was shown later [17] that the canonical reverse flux is generated far from the initial plane, where there are no evanescent waves, due to the superoscillation effect [18, 19]. The superoscillation effect arises when a paraxial light field with a limited spatial spectrum can have local amplitude modulations with a period smaller than the boundary values of the Fourier spectrum [18]. It is also known [20] that near the center of the optical vortex $(x + iy)^n$, the

V. Kotlyar et al., *Inverse Canonical Energy Flow in Laser Beams*,
https://doi.org/10.1007/978-3-032-27289-8_3

modulus of the wave vector K, equal to the phase gradient $K = |\nabla(n\arctan(y/x))| = n/\sqrt{x^2+y^2}$, can be greater than the wave number of the light wave $k = 2\pi/\lambda$. The objective of this section is to show theoretically and numerically, using a simple example of a non-paraxial light beam with linear polarization, that in the near field near the intensity zeros, a reverse canonical energy flow is formed. In this section, we employ the plane waves spectrum expansion [21] and study propagation of a nonparaxial linearly polarized light field with the amplitude of the plane waves spectrum being equal to a circle function circl(ρ/a), where ρ is the radial polar coordinate in the spectral plane and a is a scaling parameter (radius of circle in the spectral plane), which guarantees at $a > 1$ presence of evanescent waves in the near field. We obtain an explicit expression for the longitudinal component of the canonical energy flux. This expression indicates that near the initial plane, a reverse canonical flux arises, whose coincides with the first sidelobe of the light field intensity. The numerical simulation is carried out in two independent ways: using the plane wave spectrum decomposition and the Rayleigh-Sommerfeld integrals.

3.1.1 Canonical Energy Flux in the Near Field of a Nonparaxial Circl-Beam

We consider here a vector light field with the electric strength vector, having near the initial plane only one transverse component along the Cartesian axis x and the longitudinal component along the axis z [21]:

$$\begin{aligned} E_x(r,\varphi,z) &= 2\pi\int_0^\infty A(\xi)\exp\left(ikz\sqrt{1-\xi^2}\right)J_0(kr\xi)\xi d\xi, \\ E_z(r,\varphi,z) &= -2\pi i\cos\varphi\int_0^\infty A(\xi)\exp\left(ikz\sqrt{1-\xi^2}\right)J_1(kr\xi)\frac{\xi^2}{\sqrt{1-\xi^2}}d\xi. \end{aligned} \tag{3.1}$$

In Eq. (3.1), $A(\xi)$ stands for the amplitude of the plane waves spectrum, depending only on the radial polar coordinate $\xi = \sqrt{\xi_x^2+\xi_y^2}$, (r, φ) are the polar coordinates in the beam cross section, k is the wavenumber of monochromatic light, $J_0(x)$ and $J_1(x)$ are the Bessel functions of the zero and first orders. Longitudinal derivatives of both field components (3.1) in the initial plane ($z = 0$) are derived as

$$\frac{\partial E_x(r,\varphi,z)}{\partial z}|_{z=0} = 2\pi ik \int\limits_0^1 \sqrt{1-\xi^2}A(\xi)J_0(kr\xi)\xi d\xi -$$
$$2\pi k \int\limits_1^\infty \sqrt{\xi^2-1}A(\xi)J_0(kr\xi)\xi d\xi, \tag{3.2}$$
$$\frac{\partial E_z(r,\varphi,z)}{\partial z}|_{z=0} = 2\pi k\cos\varphi \int\limits_0^\infty A(\xi)J_1(kr\xi)\xi^2 d\xi.$$

The Poynting vector **P**, after omitting insignificant constants, is the sum of two vectors [8]: the orbital energy flux vector (canonical energy flux) $\mathbf{P}_{\mathrm{or}}$ and the spin flux vector $\mathbf{P}_{\mathrm{sp}}$:

$$\begin{aligned} &\mathbf{P} = \mathbf{P}_{\mathrm{or}} + \mathbf{P}_{\mathrm{sp}} = \mathrm{Im}(\mathbf{E}^* \cdot (\nabla)\mathbf{E}) + \frac{1}{2}(\nabla \times \mathbf{S}), \\ &\mathbf{S} = \mathrm{Im}(\mathbf{E}^* \times \mathbf{E}). \end{aligned} \tag{3.3}$$

The vector **S** in Eq. (3.3) is the vector of spin angular momentum (up to a constant). Longitudinal component of the canonical energy flux, as follows from Eq. (3.3) is given by

$$P_{\mathrm{or},z} = \mathrm{Im}\left(E_x^*\frac{\partial}{\partial z}E_x + E_y^*\frac{\partial}{\partial z}E_y + E_z^*\frac{\partial}{\partial z}E_z\right). \tag{3.4}$$

Substitution of expressions for the field components (3.1) and for their derivatives (3.2) into the expression for the energy flux (3.4) yields:

$$\begin{aligned} P_{\mathrm{or},z}|_{z=0} = k(2\pi)^2 \Bigg[&\int\limits_0^\infty A(\xi)J_0(kr\xi)\xi d\xi \int\limits_0^1 \sqrt{1-\xi^2}A(\xi)J_0(kr\xi)\xi d\xi \\ &+\cos^2\varphi \int\limits_0^1 A(\xi)J_1(kr\xi)\frac{\xi^2 d\xi}{\sqrt{1-\xi^2}} \int\limits_0^\infty A(\xi)J_1(kr\xi)\xi^2 d\xi \Bigg]. \end{aligned} \tag{3.5}$$

Longitudinal component of the canonical energy flux in the initial plane (3.5) consists of two terms. The first term is circularly symmetric, whereas the second one depends on the azimuthal angle φ. At some values of this azimuthal angle, the second term can turn to zero. Then, for $\varphi = \pi/2$ and $\varphi = 3\pi/2$, expression (3.5) contains only the first term:

$$P_{\text{or},z}\Big|_{z=0}^{\varphi=\pm\pi/2} = k(2\pi)^2 \int_0^\infty A(\xi)J_0(kr\xi)\xi d\xi \times \int_0^1 \sqrt{1-\xi^2}A(\xi)J_0(kr\xi)\xi d\xi. \tag{3.6}$$

The first integral in Eq. (3.6) describes the amplitude of the transverse component of the electric field strength vector along the axis x in the initial plane. Since this integral is evaluated from zero to infinity, the beam amplitude is contributed also by inhomogeneous evanescent waves. The second integral in Eq. (3.6) describes only the propagating waves, since the integration is done from zero to unity. Equation (3.6) indicates that at certain values r, the integrals can have different signs, and then the reverse flux arises at these radii r. Equation (3.6) is written in a general form and for its analytical treatment, we now consider a partial case when the amplitude of plane waves spectrum has the form of a circl-function ($a > 1$):

$$A(\xi) = \text{circl}\left(\frac{\xi}{a}\right) = \begin{cases} 1, & \xi < a, \\ 0, & \xi > a. \end{cases} \tag{3.7}$$

In this partial case, the integrals from Eq. (3.6) can be evaluated by using reference integrals:

$$\begin{aligned} &\int_0^1 \sqrt{1-\xi^2}J_0(kr\xi)\xi d\xi = \frac{1}{(kr)^2}\left(\frac{\sin(kr)}{kr} - \cos(kr)\right), \\ &\int_0^a J_0(kr\xi)\xi d\xi = a^2\frac{J_1(kra)}{kra}. \end{aligned} \tag{3.8}$$

Using Eq. (3.8), we obtain instead of Eq. (3.6):

$$P_{\text{or},z}\Big|_{z=0}^{\varphi=\pm\pi/2} = 2\pi k\left(\pi a^2\frac{2J_1(kra)}{kra}\right)\frac{1}{(kr)^2}\left(\frac{\sin(kr)}{kr} - \cos(kr)\right). \tag{3.9}$$

For certain values of the radial variable r, both factors in Eq. (3.9) can have different signs. Therefore, on the vertical axis y, reverse canonical energy flux arises at these distances r. For instance, the first zero of the function $J_1(kra)$ is at $r = 3.8\lambda/(2\pi a)$, while the first zero of the second factor in Eq. (3.9) (this is the spherical Bessel function of the order 1 or the usual Bessel function of the order 3/2) is at $r = 4.2\lambda/(2\pi)$. Therefore, the reverse flux on the vertical axis arises between these two zeros. Using two other reference integrals:

$$\int_0^1 J_1(kr\xi)\frac{\xi^2}{\sqrt{1-\xi^2}}d\xi = \sqrt{\frac{\pi}{2kr}}J_{3/2}(kr) = \frac{1}{kr}\left(\frac{\sin(kr)}{kr} - \cos(kr)\right), \tag{3.10}$$

$$\int_0^a J_1(kr\xi)\xi^2 d\xi = a^2\frac{J_2(kra)}{kr},$$

we obtain an explicit expression for the longitudinal canonical energy flux in the initial plane of linearly polarized circl-beam:

$$P_{\mathrm{or},z}\Big|_{z=0} = \frac{4\pi^2 ka^2}{(kr)^2}\left(\frac{\sin(kr)}{kr} - \cos(kr)\right)\times \left[\frac{J_1(kra)}{kra} + \cos^2\varphi J_2(kra)\right]. \tag{3.11}$$

At $\varphi = \pi/2$ and at $\varphi = 3\pi/2$, expression (3.11) coincides with (3.9). Expression (3.11) indicates that the energy flux density distribution near the initial plane has an asymmetrical shape. In a transverse cross section, maximal flux values reside on the horizontal axis, since the linear polarization vector is directed along the x-axis, and since the square of cosine on this axis is equal to 1. The reverse flux occurs where the multipliers in round and square brackets in Eq. (3.11) are of different signs. The multiplier in round brackets in Eq. (3.11) describes the amplitude of propagating waves, whereas the multiplier in square brackets describes propagating and evanescent ($a > 1$) waves near the initial plane. Since the first (off-axis) zero of the multiplier in round brackets in Eq. (11), $J_{3/2}(x)$, is located at an approximate distance of $4.2\lambda/(2\pi)$ from the optical axis, and the first zeros of the Bessel functions $J_1(x)$ and $J_2(x)$ from the multiplier in square brackets reside at distances $3.8\lambda/(2\pi a)$ and $5.1\lambda/(2\pi a)$ from the optical axis, then, for $a > 1.5$, the reverse flux occurs at a distance r in the interval $5.1\lambda/(2\pi a) < r < 4.2\lambda/(2\pi)$. On the vertical axis ($x = 0$), as follows from Eq. (3.9), the reverse energy flux in the initial plane arises after the first zero of the first-order Bessel function, i.e. $x > 3.8\lambda/(2\pi a) = 0.2\lambda$ at $a = 3$. Simulation confirms this conclusion. The greater is the value of the parameter $a > 1$, the greater is contribution of the evanescent waves to the energy flux (3.11) and the greater is the magnitude of the reverse canonical energy flux near the initial plane. If we choose the plane waves spectrum in the form of a delta function $A(\xi) \sim \delta(\xi—a)$, with $a > 1$, then the evanescent waves propagate along the initial plane and form standing waves with amplitudes proportional to $E_x \sim J_0(kra)$ and $E_z \sim J_1(kra)$. Although the evanescent waves do not transmit energy along the z axis they participate in generating the reverse energy flux, since they participate in generating the phase function of the total field, whose z-derivative determines the sign of the longitudinal component of the canonical energy flux vector.

3.1.2 Reverse Canonical Energy Flux in the Presence of Evanescent Waves

We consider here a simple example illustrating how evanescent waves participate in generating the reverse canonical flux near the initial plane of a light beam. We suggest that a light field is a superposition of a plane wave with the amplitude A propagating along the axis z and of an evanescent wave with the amplitude B propagating along the horizontal axis x:

$$E(x,z) = A\exp(ikz) + B\exp\left(-kz\sqrt{a^2-1} + ikxa\right). \tag{3.12}$$

In Eq. (3.12), a is a dimensionless parameter that defines the wavenumber of the evanescent wave, $k_x = ka$, $a > 1$, k is the wavenumber of the plane wave. The phase function of the superposition (3.12) is equal to

$$\begin{aligned} &\Phi = \arg(E(x,z)) = \arctan\left(\frac{P}{Q}\right), \\ &P(x.z) = A\sin(kz) + B\exp\left(-kz\sqrt{a^2-1}\right)\sin(kxa), \\ &Q(x.z) = A\cos(kz) + B\exp\left(-kz\sqrt{a^2-1}\right)\cos(kxa). \end{aligned} \tag{3.13}$$

Following Eq. (3.4), we can now derive the longitudinal component of the canonical reverse flux for the field (3.12) at an arbitrary z:

$$\begin{aligned} &P_{\text{or},z} = Q\frac{\partial}{\partial z}P - P\frac{\partial}{\partial z}Q = \\ &kA\Big[A + B\exp\left(-kz\sqrt{a^2-1}\right)\times \\ &\left(\cos(kz-kxa) + \sqrt{a^2-1}\sin(kz-kxa)\right)\Big]. \end{aligned} \tag{3.14}$$

For more convenient analysis, expression (3.14) can be rewritten for the initial plane ($z = 0$):

$$P_{\text{or},z}|_{z=0} = kA\left[A + B\left(\cos(kxa) - \sqrt{a^2-1}\sin(kxa)\right)\right]. \tag{3.15}$$

As seen from Eq. (3.15), energy flux is negative if the following inequality is satisfied:

$$\begin{aligned}&\cos\left(kxa + \arctan\left(\sqrt{a^2-1}\right)\right) < -\frac{A}{aB}, \quad A < aB,\\&M - N < kxa < 2\pi - M - N,\\&M = \arccos\left(\frac{-A}{aB}\right), \quad N = \arctan\left(\sqrt{a^2-1}\right).\end{aligned} \tag{3.16}$$

For instance, if $A = B$ and $a = 3$, then the area of the reverse energy flux in the initial plane, nearest to the optical axis, resides in the subwavelength interval: $0.03\lambda < x < 0.16\lambda$.

3.1.3 Numerical Simulation

We conducted numerical simulation by using the Rayleigh-Sommerfeld integral. For the field component E_y, the Rayleigh-Sommerfeld integral is given by [22]:

$$\begin{aligned}&E_x(x, y, z > 0) =\\&-\frac{z}{2\pi}\int\int_{u,v} E_x(u, v, z = 0)\frac{\exp(ikL)}{L^2}\left(ik - \frac{1}{L}\right)dudv,\end{aligned} \tag{3.17}$$

where $L = \sqrt{(u-x)^2 + (v-y)^2 + z^2}$, z is the longitudinal coordinate, k is the wavenumber of light with the wavelength of λ. The field component E_z can be computed based upon the initial transverse field component E_y:

$$\begin{aligned}&E_z(x, y, z > 0) =\\&\frac{1}{2\pi}\int\int_{u,v} E_x(u, v, z = 0)(u - x)\frac{\exp(ikL)}{L^2}\left(ik - \frac{1}{L}\right)dudv.\end{aligned} \tag{3.18}$$

For simulation, the following parameters were used: initial field 1000×1000 pixels, $4 \times 4\ \lambda$ (one pixel is nearly $\lambda/250$), amplitude distribution is $E_x = J_1(kra)/(kra)$, $E_y = 0$, parameter a is $a = 3$, r is the radial polar coordinate (in μm), wavenumber $k = 2\pi/\lambda$, wavelength $\lambda = 532$ nm. The output field has the size of 500×500 pixels in the domain $(-\lambda < x < \lambda, -\lambda < y < \lambda)$. The field was computed twice, at a distance of $z = 0.01\lambda$ from the initial plane and at a distance of $z + dz = 0.02\lambda$ from it for subsequent computation of the derivative: $dE(z)/dz \sim [E(z + dz)\text{-}E(z)]/dz$. Shown in Fig. 3.1 are the intensity of the component E_x and its cross-section in the initial plane.

Figure 3.2 illustrates intensity distribution $|\mathbf{E}|^2$ (Fig. 3.2a), its cross section (Fig. 3.2b), as well as distribution of the longitudinal z-component of the canonical flux $P_{or,z}$ (Fig. 3.2c) and its cross section (Fig. 3.2d) at a distance of 10 nm from the initial plane.

Shown in Fig. 3.3 are the intensity distributions (a) and the longitudinal projection of the canonical energy flux (b), calculated using integrals (3.17)–(3.18). The

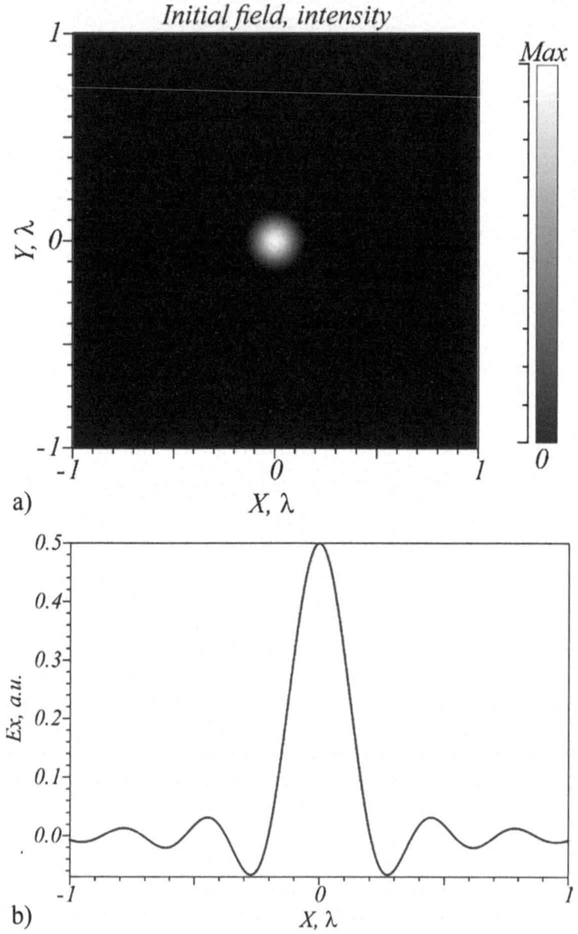

Fig. 3.1 Intensity (**a**) and the field component E_x (**b**) in the initial plane of the field (3.7) with $a = 3$

intensity in Fig. 3.3a is calculated in the initial plane (z = 0). To calculate the derivative with respect to the z coordinate in (3.4), two complex amplitudes are calculated at distances $z = 0$ and $z + dz = 10$ nm. The number of samples for the entire frame of length 2λ is chosen to be 1000.

Comparison of the energy flux cross-sections in Figs. 3.2d and 3.3b reveals that both curves almost coincide and that negative values of the energy flux in both curves begin at the first sidelobe of the amplitude (Fig. 3.1b) or intensity (Figs. 3.2b and 3.3a) at $x = 0.2\lambda$. This is consistent with the theoretical value obtained using the Eq. (3.9) at $a = 3$: $x > 3.8\lambda/(2\pi a) = 0.2\lambda$.

In this section, we have obtained the following results. Using the plane waves expansion (3.1) for a linearly polarized nonparaxial light field with a constant circular-shape plane waves spectrum (3.7), an explicit expression was obtained for the longitudinal component of the canonical energy flux in the initial plane (3.11). This expression reveals that, at first, longitudinal component of the energy flux does

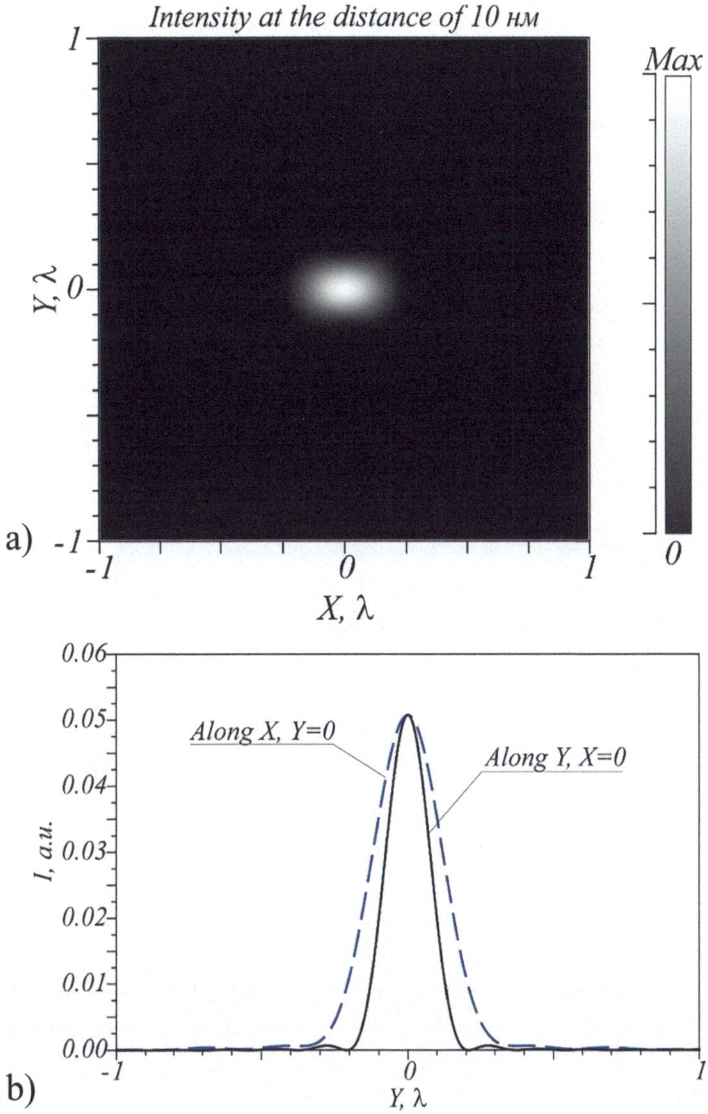

Fig. 3.2 Intensity $|\mathbf{E}|^2$ (**a**) and its cross sections (**b**) along the Cartesian axes at a distance of 10 nm from the initial plane (at $a = 3$), as well as longitudinal component of the canonical energy flux $P_{or,z}$ (**c**) and its cross sections along the Cartesian axes (**d**) at the same distance

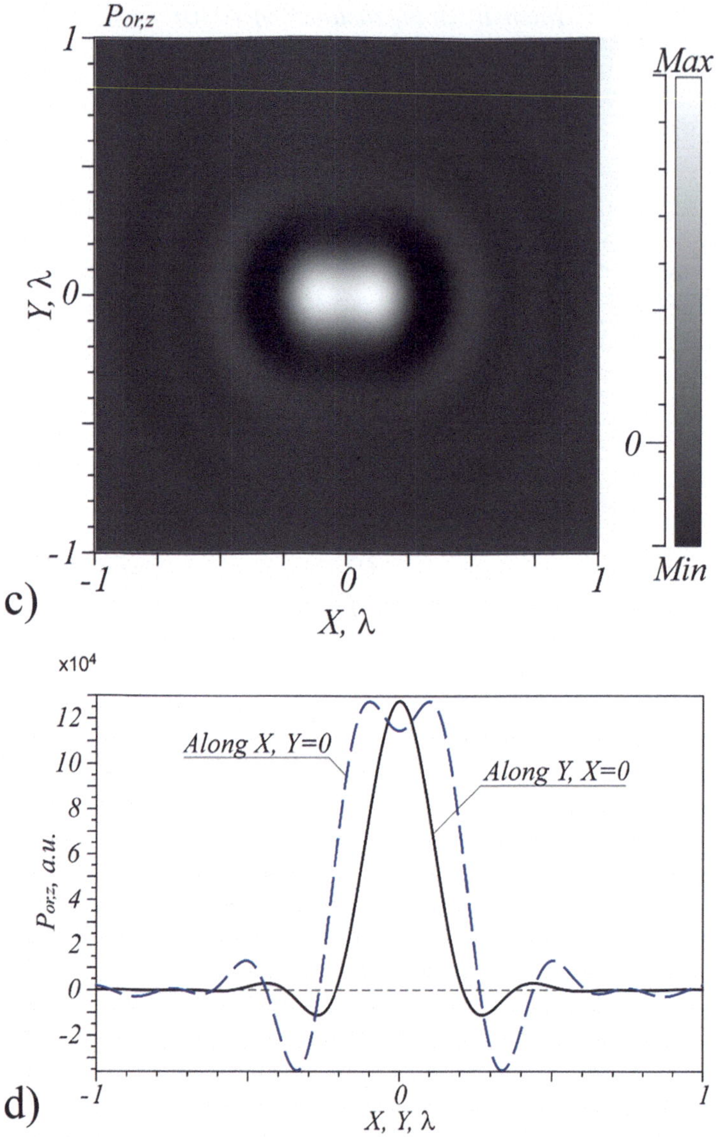

Fig. 3.2 (continued)

not have a radially symmetric distribution (due to nonzero longitudinal component in the light field). We note that the longitudinal component of the Poynting vector for the considered case would be radially symmetric, since it does not include the longitudinal field component. At second, Eq. (3.11) indicates that the canonical flux can take negative values, starting from the first zero of the first-order Bessel

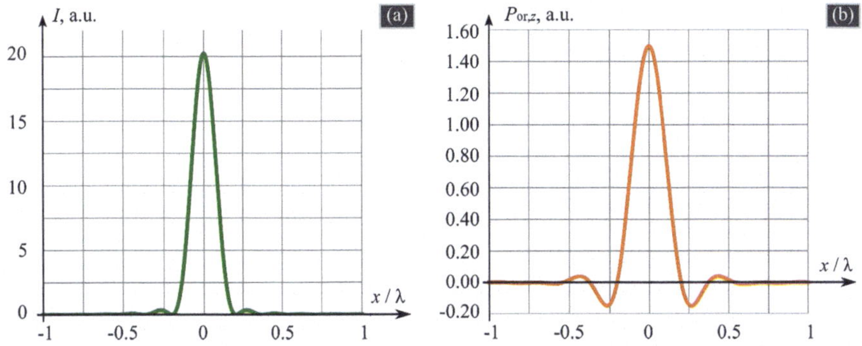

Fig. 3.3 Distributions of intensity (**a**) and of the canonical energy flux (**b**) along the vertical axis ($x = 0$) in the initial plane ($z = 0$) for the circl-beam (3.7) with $a = 3$, computed by using the plane waves spectrum expansion (3.1)

function. Moreover, the wider is the plane waves spectrum, including evanescent waves, the closer to the optical axis are the areas with the reverse flux. The magnitude of the reverse flux is approximately 7 times smaller than that of the direct flux. Numerical simulation, conducted by two methods, using the Rayleigh-Sommerfeld integrals (3.17), (3.18) and using the plane waves expansion (3.1), confirm the theoretical conclusions.

The fact that the reverse flow is 7 times smaller than the forward flow and is formed in the near field makes its experimental detection difficult. The difficulties are associated with the formation of a light field with a complex amplitude of $2J_1(kr)/(kr)$ in the initial plane, as follows from (3.8) for $a = 1$. This requires to use an optical element with a sharp change in surface relief of approximately $\lambda/2(n\text{-}1)$ at a radius of $r = 3.8\lambda/(2\pi)$, where n is the refractive index of the material, to obtain a phase delay of π. While the light passes this step in the material, the conditions of the initial plane change. Another problem is the formation of an interference pattern in the near field with an external reference beam to obtain the phase distribution. The phase distributions in the near field have to be obtained on two nearby planes to calculate the derivative with respect to the longitudinal coordinate z (3.4). Another way to generate the amplitude $2J_1(kr)/(kr)$ is to focus the initial light field, limited by a circular aperture, using a spherical lens. Focusing have to be sharp to form an almost diffraction-limited focal spot with a radius of $r = 3.8\lambda/(2\pi) = 0.6\lambda$ or less. In this case, all three components of the light field — E_x, E_y, and E_z — are present in the focal plane and should be measured separately. We plan to conduct such an experiment. However, there are currently no experiments detecting the inverse canonical energy flux.

Previously, the reverse energy flow was investigated in non-paraxial scalar beams [23–25], in the sharp focus of vector light fields [26–28]. The reverse energy flow was also considered in the cross section of light beams [29, 30] and near microparticles placed in a light beam [31, 32].

3.2 Reverse Canonical Energy Flow at the Sharp Focus of Vector Laser Beams

Reverse energy flow or negative values of the longitudinal component of the Poynting vector was theoretically discovered rather long time ago [33, 34]. In addition, reverse flow was investigated not only at the sharp focus of a coherent light [8, 35, 36], but also in vector laser beams [9, 10], in highly diverging light fields [37]. As was shown in [1, 2, 7], the Poynting vector is a sum of two vectors: orbital energy flow vector (canonical energy flow) and spin flow vector (curl of the spin angular momentum vector). Thus, it turned out that the spin component of a light field contributes to the Poynting vector. Therefore, the Poynting vector not always shows flow direction of light energy. Reverse canonical energy flow was studied in 2D light fields [6, 15, 24]. Two-dimensional light fields have only one component of the electric vector and detecting the reverse canonical flow is thus relatively simple. The reverse canonical flow was also found in quantum mechanics [38], in plasmonic lenses [39], in light fields with superoscillations [40].

In this section, employing the Richards-Wolf formulae, we both theoretically and numerically found a reverse canonical flow at the sharp focus of certain initial light fields, such as linearly polarized optical vortex with an arbitrary integer topological charge or cylindrical radially and azimuthally polarized light fields of an arbitrary order. We also shown that circularly polarized optical vortex does not have the reverse canonical flow in the focus plane.

3.2.1 Longitudinal Canonical Energy Flow at the Focus of a Linearly Polarized Optical Vortex

For a linearly polarized optical vortex, initial Jones vector is given by

$$\mathbf{E}_n(\theta,\varphi) = A(\theta)\exp(in\varphi)\begin{pmatrix} 1 \\ 0 \end{pmatrix}, \tag{3.19}$$

where (r, φ) are the polar coordinates in the beam transverse section, n is an integer topological charge of the vortex, vector of linear polarization is directed along the horizontal axis x. Function $A(\theta)$ is a real-vale function that defines circularly symmetric amplitude of the initial field and depends on the tilt angle θ of a ray outgoing from a point on the initial spherical wavefront and converging to the center of the focus plane. For the initial field (3.19), electric and magnetic field vectors in the plane of sharp focus have the following components [41]:

$$\begin{aligned} E_x &= \frac{i^{n-1}}{2} e^{in\varphi} \left(2I_{0,n} + e^{2i\varphi} I_{2,n+2} + e^{-2i\varphi} I_{2,n-2} \right), \\ E_y &= \frac{i^n}{2} e^{in\varphi} \left(e^{-2i\varphi} I_{2,n-2} - e^{2i\varphi} I_{2,n+2} \right), \\ E_z &= i^n e^{in\varphi} \left(e^{-i\varphi} I_{1,n-1} - e^{i\varphi} I_{1,n+1} \right). \end{aligned} \tag{3.20}$$

The functions $I_{\mu\nu}$ in Eq. (3.20) depend only on the radial polar coordinate r:

$$I_{\nu,\mu} = 2kf \int\limits_0^\alpha \sin^{\nu+1}\left(\frac{\theta}{2}\right) \cos^{3-\nu}\left(\frac{\theta}{2}\right) \cos^{1/2}(\theta) A(\theta) e^{ikz\cos\theta} J_\mu(kr\sin\theta) d\theta, \tag{3.21}$$

where $k = 2\pi/\lambda$ is the wavenumber of monochromatic light with the wavelength λ, f is the focal distance of the focusing lens, α is the maximal tilt angle of rays to the optical axis, that defines numerical aperture of the aplanatic lens NA $= \sin\alpha$, $J_\nu(\xi)$ is the νth-order Bessel function of the first kind. Description of a light field at the focus by using the functions (3.21) was derived for the first time in the classical work by B. Richards and E. Wolf [34].

Further, using expressions (3.20) and (3.21), we obtain an expression for the longitudinal component of the canonical energy flow vector at the sharp focus of the initial light field (3.19). Up to insignificant multipliers, the Poynting vector $\mathbf{P}$ is a sum of two vectors [8]: orbital energy flow vector (canonical energy flow) $\mathbf{P}_{\text{or}}$ and spin flow P_{sp}:

$$\begin{aligned} \mathbf{P} &= \mathbf{P}_{\text{or}} + \mathbf{P}_{\text{sp}} = \operatorname{Im}(\mathbf{E}^* \cdot (\nabla)\mathbf{E}) + \frac{1}{2}(\nabla \times \mathbf{S}), \\ \mathbf{S} &= \operatorname{Im}(\mathbf{E}^* \times \mathbf{E}). \end{aligned} \tag{3.22}$$

In Eq. (3.22), $\mathbf{S}$ is the spin angular momentum vector (up to a constant multiplier). As follows from Eq. (3.22), the longitudinal component of the canonical energy flow, parallel to the optical axis, is given by

$$P_{\text{or},z} = \operatorname{Im}\left(E_x^* \frac{\partial}{\partial z} E_x + E_y^* \frac{\partial}{\partial z} E_y + E_z^* \frac{\partial}{\partial z} E_z \right). \tag{3.23}$$

Substitution of Eq. (3.23) into Eqs. (3.20) and (3.21) yields:

$$\begin{aligned} P_{\text{or},z}(r,\varphi) &= \frac{k}{2} \left[2I_{0,n}\bar{I}_{0,n} + I_{2,n+2}\bar{I}_{2,n+2} + I_{2,n-2}\bar{I}_{2,n-2} + \right. \\ & 2I_{1,n+1}\bar{I}_{1,n+1} + 2I_{1,n-1}\bar{I}_{1,n-1} + \cos(2\varphi)\left(I_{0,n}\bar{I}_{2,n+2} + I_{0,n}\bar{I}_{2,n-2} + \right. \\ & \left. \left. I_{2,n+2}\bar{I}_{0,n} + I_{2,n-2}\bar{I}_{0,n} - 2I_{1,n-1}\bar{I}_{1,n+1} - 2I_{1,n+1}\bar{I}_{1,n-1} \right) \right], \end{aligned} \tag{3.24}$$

where the integrals $\bar{I}_{\nu,\mu}$ are slightly different than the integrals (3.21) and are equal to

$$\bar{I}_{\nu,\mu} = 2kf \int_0^{\alpha} \sin^{\nu+1}\left(\frac{\theta}{2}\right) \cos^{3-\nu}\left(\frac{\theta}{2}\right) \cos^{3/2}(\theta) A(\theta) e^{ikz\cos\theta} J_\mu(kr\sin\theta) d\theta. \quad (3.25)$$

The integrals (3.21) and (3.25) depend only on the radial coordinate r in the focus plane and differ from each other by the different power of the multiplier cos θ. Therefore, their products with equal indices in Eq. (3.24) can yield not only positive values, but negative values as well. Besides, Eq. (3.24) indicates that the energy flow distribution depends also on the azimuthal coordinate φ. Since cos 2φ equals -1 on the vertical axis y (at $\varphi = \pi/2$), we can expect that negative values of the energy flow on this axis. In addition, the reverse flow should arise both in the presence of an optical vortex and in its absence ($n = 0$). At $n = 0$, instead of Eq. (3.24), we get:

$$\begin{aligned} P_{\mathrm{or},z}(r,\varphi) = k\big[&I_{0,0}\bar{I}_{0,0} + I_{2,2}\bar{I}_{2,2} + 2I_{1,1}\bar{I}_{1,1} + \\ &\cos(2\varphi)\left(I_{0,0}\bar{I}_{2,2} + I_{2,2}\bar{I}_{0,0} + 2I_{1,1}\bar{I}_{1,1}\right)\big]. \end{aligned} \quad (3.26)$$

3.2.2 *Longitudinal Canonical Energy Flow at the Focus of a Cylindrically Polarized Initial Field*

If an initial light field is an mth-order cylindrical vector beam, the Jones vector for the electric field is given by

$$\mathbf{E}_m(\theta,\varphi) = A(\theta)\begin{pmatrix} \cos m\varphi \\ \sin m\varphi \end{pmatrix}. \quad (3.27)$$

In [42], using the Richards-Wolf formalism, explicit expressions were derived for all components of the electric field strength vector in the sharp focus plane of a cylindrical vector field (3.27):

$$\begin{aligned} E_x(r,\varphi) &= i^{m-1}\left(\cos(m\varphi)I_{0,m} + \cos((m-2)\varphi)I_{2,m-2}\right), \\ E_y(r,\varphi) &= i^{m-1}\left(\sin(m\varphi)I_{0,m} - \sin((m-2)\varphi)I_{2,m-2}\right), \\ E_z(r,\varphi) &= 2i^m\cos((m-1)\varphi)I_{1,m-1}. \end{aligned} \quad (3.28)$$

The integrals $I_{\mu\nu}$ in Eq. (3.28) are the same as in Eq. (3.21). Further, substituting Eq. (3.28) into Eq. (3.23), we obtain the longitudinal component of the canonical energy flow vector at the sharp focus of a cylindrical vector field:

$$P_{\text{or},z}(r,\varphi) = k\big[I_{0,m}\bar{I}_{0,m} + I_{2,m-2}\bar{I}_{2,m-2} + 4I_{1,m-1}\bar{I}_{1,m-1}\cos^2((m-1)\varphi) + \cos(2(m-1)\varphi)\big(I_{0,m}\bar{I}_{2,m-2} + I_{2,m-2}\bar{I}_{0,m}\big)\big]. \tag{3.29}$$

For $m = 0$ (linearly polarized initial field), Eq. (3.29) reduces to an expression for the canonical energy flow, coinciding with Eq. (3.26):

$$\begin{aligned} P_{\text{or},z}(r,\varphi) &= k\big[I_{0,0}\bar{I}_{0,0} + I_{2,2}\bar{I}_{2,2} + 4I_{1,1}\bar{I}_{1,1}\cos^2\varphi + \\ &\quad + \cos(2\varphi)\big(I_{0,0}\bar{I}_{2,2} + I_{2,2}\bar{I}_{0,0}\big)\big] \\ &= k\big[I_{0,0}\bar{I}_{0,0} + I_{2,2}\bar{I}_{2,2} + 2I_{1,1}\bar{I}_{1,1} + \\ &\quad + \cos(2\varphi)\big(I_{0,0}\bar{I}_{2,2} + I_{2,2}\bar{I}_{0,0} + 2I_{1,1}\bar{I}_{1,1}\big)\big]. \end{aligned} \tag{3.30}$$

Equation (3.30), similarly to Eq. (3.26), indicates that on the vertical axis at certain distances from the center, small values of the reverse energy flow can arise. At $m = 1$ (radially polarized initial field), Eq. (3.29) is independent of the azimuthal coordinate φ, and the reverse flow thus has the shape of narrow rings of a certain radius. These rings are close to the rings of zero intensity. At $m = 2$, canonical energy flow (3.29) depends on the azimuthal coordinate φ and the areas with the reverse flow are distributed near the vertical axis:

$$\begin{aligned} P_{\text{or},z}(r,\varphi) &= k\big[I_{0,2}\bar{I}_{0,2} + I_{2,0}\bar{I}_{2,0} + 4I_{1,1}\bar{I}_{1,1}\cos^2(\varphi) + \\ &\quad + 2\cos(2\varphi)\big(I_{0,2}\bar{I}_{2,0} + I_{2,0}\bar{I}_{0,2}\big)\big]. \end{aligned} \tag{3.31}$$

As seen from Eq. (3.31), cosine of the double angle is equal to -1 at $\varphi = \pi/2$, and the reverse energy flow is generated near the vertical axis y.

Now we consider an initial field with cylindrical polarization of another type, but related with the field (3.27):

$$\mathbf{E}_m(\theta,\varphi) = A(\theta)\begin{pmatrix} -\sin m\varphi \\ \cos m\varphi \end{pmatrix}. \tag{3.32}$$

The initial light field (3.27) is the field with mth-order radial polarization, whereas the light field (3.32) is the field with mth-order azimuthal polarization. At $m = 0$, the field (3.27) has horizontal linear polarization, while the field (3.32) has vertical linear polarization. At $m = 1$, the field (3.27) has conventional radial polarization, while the field (3.32) has conventional azimuthal polarization. For a cylindrical field (3.32), the strength vector of the electric field at the sharp focus has the following components [43]:

$$
\begin{aligned}
E_x(r,\varphi) &= i^{m+1}[\sin(m\varphi)I_{0,m} + \sin((m-2)\varphi)I_{2,m-2}],\\
E_y(r,\varphi) &= i^{m+1}[-\cos(m\varphi)I_{0,m} + \cos((m-2)\varphi)I_{2,m-2}],\\
E_z(r,\varphi) &= -2i^{m}\sin((m-1)\varphi)I_{1,m-1}.
\end{aligned}
\tag{3.33}
$$

Comparison of Eqs. (3.33) and (3.28) reveals that, up to signs, Eq. (3.33) can be obtained from Eq. (3.28) by replacing the sines by cosines and vice versa. Longitudinal component of the canonical energy flow vector (3.23) at the focus of the initial field (3.33) is given by

$$
\begin{aligned}
&P_{\mathrm{or},z}(r,\varphi) = k\big[I_{0,m}\bar{I}_{0,m} + I_{2,m-2}\bar{I}_{2,m-2}+\\
&4I_{1,m-1}\bar{I}_{1,m-1}\sin^2((m-1)\varphi)-\\
&-\cos(2(m-1)\varphi)\big(I_{0,m}\bar{I}_{2,m-2} + I_{2,m-2}\bar{I}_{0,m}\big)\big].
\end{aligned}
\tag{3.34}
$$

Equation (3.34) illustrates that the energy flow distribution at the focus of light with mth-order azimuthal polarization looks similar to the energy flow distribution of light with mth-order radial polarization (3.29), but the maximal reverse flow, according to Eq. (3.34), is generated near the horizontal axis, rather than near vertical axis, as indicated by Eq. (3.29). At $m = 1$, the reverse energy flow (3.34), similarly to the reverse energy flow (3.29), has radial symmetry and resides on rings.

3.2.3 *Longitudinal Canonical Energy Flow at the Focus of a Circularly Polarized Optical Vortex*

If the initial light field is an optical vortex with the topological charge of n and with left circular polarization, the Jones vector of the electric field is equal to

$$
\mathbf{E}_n(\theta,\varphi) = \frac{A(\theta)}{\sqrt{2}}\exp(in\varphi)\begin{pmatrix}1\\ -i\end{pmatrix}. \tag{3.35}
$$

At the sharp focus of the initial field (3.35), the strength vector of the electric field has the following components [41]:

$$
\begin{aligned}
E_x &= \frac{i^{n-1}}{\sqrt{2}}e^{in\varphi}\big(I_{0,n} + e^{-2i\varphi}I_{2,n-2}\big),\\
E_y &= \frac{i^{n}}{\sqrt{2}}e^{in\varphi}\big(-I_{0,n} + e^{-2i\varphi}I_{2,n-2}\big),\\
E_z &= \sqrt{2}i^{n}e^{i(n-1)\varphi}I_{1,n-1}.
\end{aligned}
\tag{3.36}
$$

Substituting Eq. (3.36) into Eq. (3.23), we obtain the longitudinal component of the canonical energy flow vector at the sharp focus of a circularly polarized optical vortex:

$$P_{\mathrm{or},z}(r,\varphi) = k\left[I_{0,n}\bar{I}_{0,n} + I_{2,n-2}\bar{I}_{2,n-2} + 2I_{1,n-1}\bar{I}_{1,n-1}\right]. \tag{3.37}$$

Energy flow distribution at the focus (3.37) looks similar to the intensity distribution at the focus of a circularly polarized optical vortex:

$$I = |E_x|^2 + \left|E_y\right|^2 + |E_z|^2 = I_{0,n}^2 + I_{2,n-2}^2 + 2I_{1,n-1}^2. \tag{3.38}$$

As seen from Eq. (3.37), the longitudinal component of the energy flow is almost equal to the intensity (3.38), multiplied by the wavenumber. Both distributions (3.37) and (3.38) have circularly symmetric shape, i.e. depend only on the radial coordinate r, and both distributions do not have negative values. It is obvious for the intensity (3.38), since it is a sum of squares, whereas each term in the expression for the energy flow (3.37) is a product of almost equal integrals (3.21) and (3.25) with equal indices. Numerical simulation confirms this conclusion: circularly polarized optical vortex does not have the reverse canonical flow in the focus plane.

3.2.4 Numerical Simulation

Illustrated below are results of a numerical simulation, performed by employing the vectorial Richards-Wolf integrals [34]. For several initial vector fields, we applied the Richards-Wolf integrals to compute the amplitude distributions of components of the electric field strength vector near the sharp focus of an ideal spherical lens. Using these amplitude distributions, we computed the longitudinal component of the canonical energy flow vector. The following parameters were used for computations: wavelength $\lambda = 532$ nm, focal distance $f = 20\lambda$, numerical aperture $NA = 0.95$, computation area $-1\ \mu\mathrm{m} \le x, y \le 1\ \mu\mathrm{m}$, radial distribution of the initial amplitude $A(\theta) = 1$.

Figure 3.4 depicts distributions of the intensity (Fig. 3.4a, d) and of the energy flow (Fig. 3.4b, e), as well as of its sign (Fig. 3.4c, f) at the focus of the initial field (3.19) with $n = 0$ (Fig. 3.4a–c) and with $n = 3$ (Fig. 3.4d–f). As seen in Fig. 3.4, the energy flow distribution looks like the intensity distribution, and near the vertical axis, areas with reverse (negative) energy flow are arise, as is predicted by Eq. (3.24). Maximal magnitude (by modulus) of the reverse flow amounts a fraction of a percent of the maximal magnitude of the direct flow.

Shown in Fig. 3.5 are distributions of the intensity (Fig. 3.5a, d), energy flow (Fig. 3.5b, e), and of its sign (Fig. 3.5c, f) at the focus of the initial field (3.28) with $m = 1$ (Fig. 3.5a–c) and with $m = 2$ (Fig. 3.5d–f). As seen from Fig. 3.5, the energy flow distribution looks similar to the intensity distribution, and that areas with the

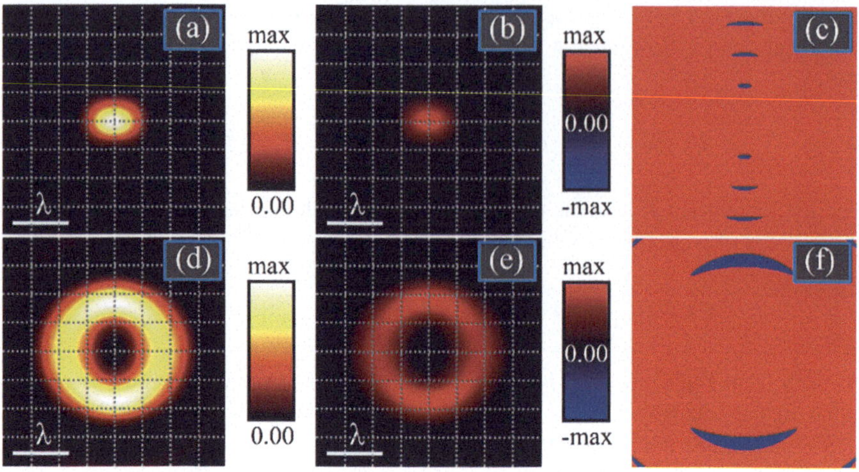

Fig. 3.4 Intensity distribution at the focus of an optical vortex (3.19) with linear horizontal polarization (**a**, **d**), as well as distribution of the canonical energy flow (3.23) (**b**, **e**) and of its sign (**c**, **f**). Red and blue color denotes positive and negative values, respectively. Topological charge is $n = 0$ (**a–c**) and $n = 3$ (**d–f**)

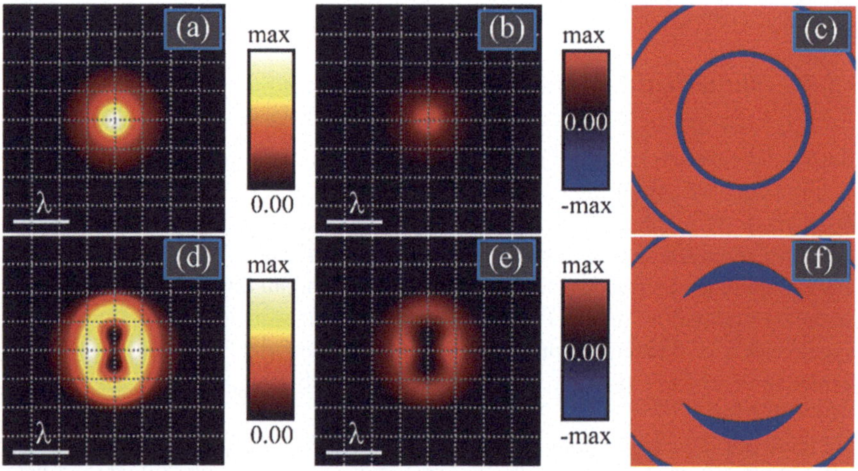

Fig. 3.5 Intensity distribution at the focus of a cylindrical vector field (3.27) (**a**, **d**), as well as distribution of the canonical energy flow (3.23) (**b**, **e**) and of its sign (**c**, **f**). Red and blue color denotes positive and negative values, respectively. The order of the cylindrical beam is $m = 1$ (**a–c**) and $m = 2$ (**d–f**)

reverse flow exist. At $m = 1$, energy flow has radial symmetry (Fig. 3.5b), and the reverse flow distribution has a shape of concentric rings (Fig. 3.5c), whose radii are close to the radii of zero intensity. At $m = 2$, radial symmetry of the energy flow disappears, and the areas with the reverse flow reside mostly along the vertical axis

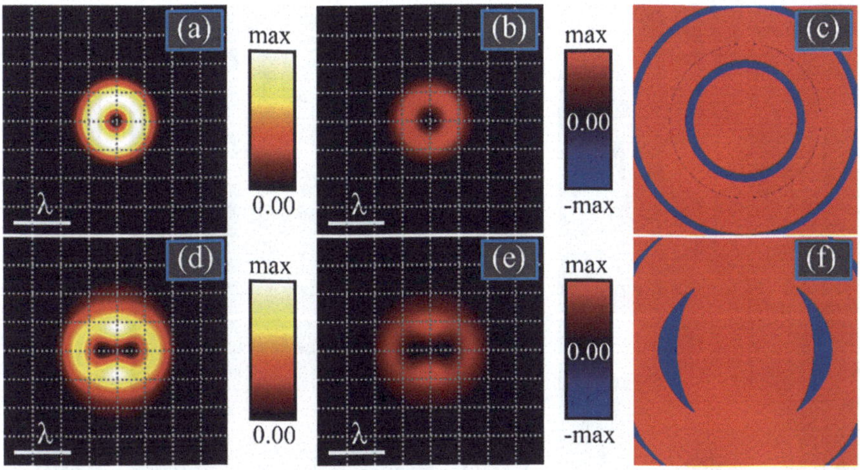

Fig. 3.6 Intensity distribution at the focus of a cylindrical vector field (3.32) (**a**, **d**), as well as distribution of the canonical energy flow (3.23) (**b**, **e**) and of its sign (**c**, **f**). Red and blue color denotes positive and negative values, respectively. The order of the cylindrical beam is $m = 1$ (**a–c**) and $m = 2$ (**d–f**)

(Fig. 3.5f). At $m = 0$, cylindrical polarization reduces to linear, and intensity and energy flow distributions coincide with Fig. 3.4a–c.

Figure 3.6 illustrates distributions of the intensity (Fig. 3.6a, d), energy flow (Fig. 3.6b, e), and of its sign (Fig. 3.6c, f) at the focus of the initial field (3.32) with $m = 1$ (Fig. 3.3a–c) and with $m = 2$ (Fig. 3.3d–f). Figure 3.6 reveals that the energy flow distribution looks similar to the intensity distribution and that areas with the reverse flow arise in the transverse section. At $m = 1$, energy flow is circularly symmetric (Fig. 3.6b), and generated reverse flow distribution has a shape of concentric rings (Fig. 3.6c), whose radii are close to the radii of the intensity zeros. At $m = 2$, circular symmetry of the energy flow disappears, and the areas of the reverse flow reside mostly along the horizontal axis (Fig. 3.6f). At $m = 0$, cylindrical polarization reduces to linear, and the intensity and the energy flow are the same as in Fig. 3.4, but rotated by the angle of $\pi/2$.

Depicted in Fig. 3.7 are intensity distributions of a circularly polarized optical vortex at the focus (Fig. 3.7a, c), as well as the energy flow distributions (Fig. 3.7b, d) for two values of the topological charge n: 0 (Fig. 3.7a, b) and 3 (Fig. 3.7c, d). As seen in Fig. 3.7b, d, reverse flow at the focus of the field (3.35) is not generated. This is consistent with Eq. (3.37).

3.2.5 Discussion of the Results

Equation (3.22) indicates that the Poynting vector is different from the canonical energy flow vector, if the curl of the spin angular momentum (SAM) is nonzero.

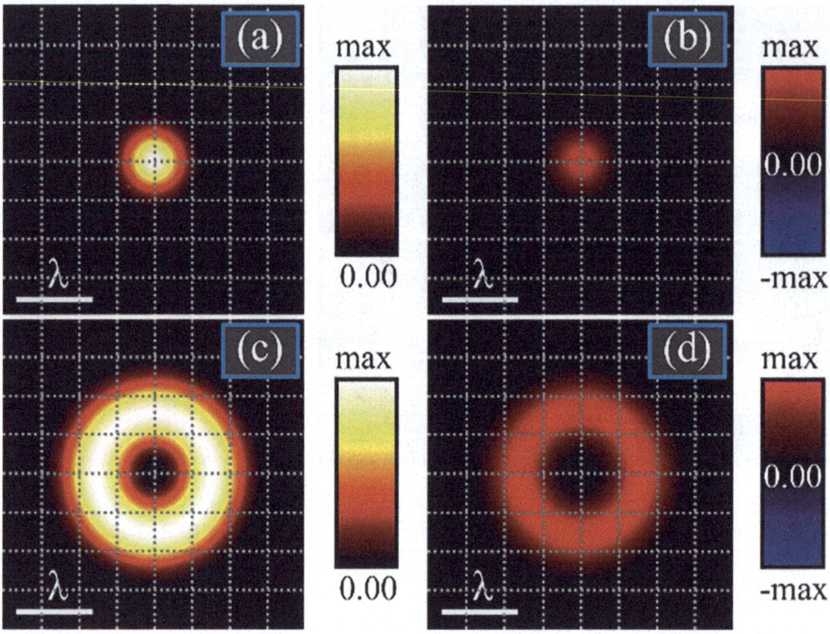

Fig. 3.7 Intensity distribution at the focus of an optical vortex (3.35) with left circular polarization (**a**, **c**), as well as distribution of the canonical energy flow (3.23) (**b**, **d**). Topological charge is $n = 0$ (**a**, **b**) and $n = 3$ (**c**, **d**)

There is no such a light field that all its SAM components are identically zero. However, there are light fields with one or two zero components of the SAM vector. Instances of such light fields are light fields with azimuthal polarization. Indeed, as seen from Eq. (3.33) at $m = 1$, the longitudinal component of the strength vector of the electric field E_z is zero, and thus at the focus only transverse field components remain:

$$\begin{aligned} E_x(r,\varphi) &= i \sin\varphi(I_{0,1}+I_{2,1}), \\ E_y(r,\varphi) &= -i \cos\varphi(I_{0,1}+I_{2,1}). \end{aligned} \tag{3.39}$$

Now we will show that the transverse components of the SAM vector at the focus plane are zero for the initial field with azimuthal polarization. The spin angular momentum vector is defined by the expression [44]:

$$\mathbf{S} = \frac{1}{16\pi\omega} \mathrm{Im}(\mathbf{E}^* \times \mathbf{E}), \tag{3.40}$$

with ω being the angular frequency of light. Below, the constant $1/(16\pi\omega)$ is omitted for brevity. The transverse components of the SAM vector (3.39) are given by:

$$S_x = 2\mathrm{Im}\left(E_y^* E_z\right), \quad S_y = 2\mathrm{Im}\left(E_z^* E_x\right). \tag{3.41}$$

As follows from Eq. (3.40), the transverse components of the SAM vector at the focus of the field (3.33) are zero since the longitudinal component E_z is also zero. Thus, the longitudinal component of the curl of the SAM vector is zero as well:

$$(\nabla \times \mathbf{S})_z = \frac{\partial S_y}{\partial x} - \frac{\partial S_x}{\partial y} = 0. \tag{3.42}$$

Therefore, for the initial light field with azimuthal polarization (3.32) with $m = 1$, the longitudinal component of the Poynting vector in the sharp focus plane coincides with the longitudinal component of the canonical energy flow:

$$P_z = P_{\mathrm{or},z}(r, \varphi) = k\left[I_{0,1}\bar{I}_{0,1} + I_{2,1}\bar{I}_{2,1} + I_{0,1}\bar{I}_{2,1} + I_{2,1}\bar{I}_{0,1}\right]. \tag{3.43}$$

The last two terms in the expression (3.43), containing products of the integrals (3.21) and (3.25) with different indices, can have negative values, in contrast to the first two terms. Therefore, the canonical reverse energy flow in the sharp focus plane of azimuthally polarized light, illustrated in Fig. 3.6c, is at the same time the reverse energy flow for the summary Poynting vector (3.22).

We have derived explicit analytical expressions for distribution of the longitudinal component of the canonical energy flow vector in the plane of the sharp focus of a linearly polarized optical vortex (3.24), of a vector field with radial (3.29) and azimuthal (3.34) polarizations of an arbitrary integer order, and of a circularly polarized optical vortex (3.37). It is demonstrated that initial fields with linear and cylindrical polarizations have at the focus plane areas with reverse energy flow, i.e. with negative values of the longitudinal component of the canonical energy flow vector (Figs. 3.4c, f, 3.5c, f and 3.6c, f). For a circularly polarized optical vortex, the reverse energy flow was not detected at the focus (3.37). It is shown that for an initial field with first-order azimuthal polarization, longitudinal components of the canonical energy flow and of the Poynting vector at the focus coincide (3.43).

3.3 Coincidence of the Longitudinal Components of the Poynting Vectors and the Canonical Energy Flow in the Tight Focus of an Azimuthally Polarized Optical Vortex

In modern optics, certain phenomena lack a complete and unambiguous interpretation. A prime example is the physical meaning of negative values of the longitudinal component of the Poynting vector. In their seminal 1959 work, Richards and Wolf [34] showed that in the plane of a tight focus for a linearly polarized input field, the longitudinal component of the Poynting vector can attain negative values, reaching approximately 1% of its maximum positive value. The authors [34] termed this behavior a 'reverse energy flow'. However, the interpretation of negative Poynting vector values as a reverse energy flow was challenged by the work of Bekshaev and Soskin in 2007 [4]. They demonstrated that the Poynting vector can be decomposed into the sum of two components: the orbital energy flow (canonical energy flow) and the spin flow. The orbital energy flow for a scalar light field is proportional to the intensity multiplied by the phase gradient, while the spin flow is proportional to the curl of the spin angular momentum. Crucially, the spin component does not transfer energy, as follows from the differential law of energy conservation: the divergence of the Poynting vector appears in this law, and the divergence of the curl of any vector is identically zero. Consequently, in regions with circular or elliptical polarization (where spin is present), negative Poynting vector values cannot be automatically interpreted as a reverse energy flow. For light with initial linear polarization, as in [34], all three components of the electric field arise at the tight focus, contributing to the longitudinal spin flow. Thus, in this case, the Poynting vector cannot be identified with the canonical energy flow. Subsequent studies have reported negative values of the Poynting vector at the tight focus of other vector fields [9, 35, 36] and for some non-paraxial vector beams [10, 11]. However, since all these fields possessed all three electric field components, the observed negative Poynting vector values in these cases cannot be unequivocally attributed to a reverse energy flow.

The situation is different for non-paraxial 2D light fields. Such TE-polarized fields have only one transverse electric component during propagation, and therefore, the longitudinal component of the Poynting vector coincides with the longitudinal component of the canonical energy flow. As a result, reverse canonical energy flows have been unambiguously identified for 2D non-paraxial light fields [7, 15, 24]. Furthermore, a reverse flow near the focus, associated with the superoscillation effect, was demonstrated in [19] for a 2D light field transmitted through a superlens. However, for some 3D vector fields, the Poynting vector may also coincide with the canonical energy flux vector. This occurs in light fields that lack a longitudinal electric field component. One type of such a vector field is discussed in this paper.

In this section, we theoretically demonstrate that at the tight focus of an optical vortex with an arbitrary integer topological charge and azimuthal polarization, the longitudinal components of the Poynting vector and the canonical energy flux coincide. Therefore, the regions of negative Poynting vector values identified in this work—which take the form of rings centered on the optical axis—can be identified as regions of reverse energy flow. In these regions, the longitudinal component of the local wave vector is negative, confirming that the energy flow is directed opposite to the propagation direction.

3.3.1 Poynting Vector and Canonical Energy Flow Vector at a Tight Focus

We consider the tight focusing of an optical vortex with a topological charge n and azimuthal polarization using an aplanatic lens (an ideal spherical lens). The Jones vector of such a light field in the initial plane is given by:

$$\mathbf{E} = A(\theta)e^{in\varphi}\begin{pmatrix} -\sin\varphi \\ \cos\varphi \end{pmatrix}. \tag{3.44}$$

Here φ is the azimuthal angle in the beam cross-section, $A(\theta)$ is the amplitude, which depends only on the radial variable r. The radial variable is related to the polar angle θ by the expression $r = f\sin\theta$, f is the focal length of a spherical lens. The light field at the tight focus of the aplanatic lens is described within the framework of the Richards -Wolf theory [34]. Expressions for the projections of the electric field vector at the tight focus of an azimuthally polarized optical vortex were obtained in [42]:

$$\begin{aligned}
&E_x = -\frac{1}{2}i^{n-1}\left[e^{i(n+1)\varphi}\left(I_{0,n+1} + e^{-i2\varphi}I_{2,n-1}\right) + \right.\\
&\left. e^{i(n-1)\varphi}\left(I_{0,n-1} + e^{i2\varphi}I_{2,n+1}\right)\right],\\
&E_y = \frac{1}{2}i^{n}\left[e^{i(n+1)\varphi}\left(I_{0,n+1} - e^{-i2\varphi}I_{2,n-1}\right) - \right.\\
&e^{i(n-1)\varphi}\left(I_{0,n-1} - e^{i2\varphi}I_{2,n+1}\right),\\
&E_z = 0.
\end{aligned} \tag{3.45}$$

From (3.45) it is evident that an azimuthally polarized optical vortex at the focus has only transverse components of the electric field and no longitudinal component. This property will be used in the following analysis. The Poynting vector

$$\mathbf{P} = \frac{1}{c}\,\mathrm{Re}(\mathbf{E}^* \times \mathbf{H}), \tag{3.46}$$

where **E** and **H** are the vectors of the electric and magnetic light fields, c is the speed of light in a vacuum, × denotes the vector product, Re() denotes the real part, and the asterisk (*) denotes complex conjugation, is the sum of two vectors [8]: the orbital energy flow (canonical energy flux) $\mathbf{P}_{or}$ and the spin flow $\mathbf{P}_{sp}$:

$$\begin{aligned}
&\mathbf{P} = \mathbf{P}_{or} + \mathbf{P}_{sp} = \frac{1}{\omega}\mathrm{Im}(\mathbf{E}^* \cdot (\nabla)\mathbf{E}) + \frac{1}{2}(\nabla \times \mathbf{S}),\\
&\mathbf{S} = \frac{1}{\omega}\mathrm{Im}(\mathbf{E}^* \times \mathbf{E}).
\end{aligned} \tag{3.47}$$

In (3.47), **S** is the spin angular momentum (SAM) vector, and ω is the cyclic frequency of light. The transverse projections of the SAM vector in (3.47) are:

$$S_x = \frac{2}{\omega}\mathrm{Im}\left(E_y^* E_z\right), \quad S_y = \frac{2}{\omega}\mathrm{Im}\left(E_z^* E_x\right). \tag{3.48}$$

From (3.48) it is clear that the transverse projections of the SAM vector at the focus contain the longitudinal projection E_z, which is equal to zero (3.45) for the light field (3.44). Therefore, the transverse projections of the SAM themselves are equal to zero. Consequently, the longitudinal projection of the curl of the SAM vector will also be zero:

$$(\nabla \times \mathbf{S})_z = \frac{\partial S_y}{\partial x} - \frac{\partial S_x}{\partial y} = 0. \tag{3.49}$$

This means that for vector light fields lacking a longitudinal electric field component at the focus, the longitudinal projection of the spin flow is also zero $\mathbf{P}_{sp,z} = 0$. Therefore, for such fields, the longitudinal components of the Poynting vector P_z (3.46) and the canonical energy flux $\mathbf{P}_{or,z}$ coincide. Let us show this in more detail. The projection of the canonical energy flux onto the longitudinal axis, as follows from Eq. (3.47), is:

$$P_{or,z} = \mathrm{Im}\left(E_x^* \frac{\partial}{\partial z} E_x + E_y^* \frac{\partial}{\partial z} E_y + E_z^* \frac{\partial}{\partial z} E_z\right). \tag{3.50}$$

Substituting the expressions from (3.45) into (3.50) we obtain:

$$\begin{gathered} P_{or,z}(r) = \frac{k}{2\omega}\left[(I_{0,n-1} + I_{2,n-1})(\bar{I}_{0,n-1} + \bar{I}_{2,n-1}) + \right. \\ \left. (I_{0,n+1} + I_{2,n+1})(\bar{I}_{0,n+1} + \bar{I}_{2,n+1})\right], \end{gathered} \tag{3.51}$$

where the integrals $I_{\nu,\mu}$ and $\bar{I}_{\nu,\mu}$ differ slightly from each other and have the form:

$$I_{\nu,\mu} = 2kf \int_0^\alpha \sin^{\nu+1}\left(\frac{\theta}{2}\right) \cos^{3-\nu}\left(\frac{\theta}{2}\right) \cos^{1/2}(\theta) A(\theta) e^{ikz\cos\theta} J_\mu(kr\sin\theta) d\theta, \tag{3.52}$$

$$\bar{I}_{\nu,\mu} = 2kf \int_0^\alpha \sin^{\nu+1}\left(\frac{\theta}{2}\right) \cos^{3-\nu}\left(\frac{\theta}{2}\right) \cos^{3/2}(\theta) A(\theta) e^{ikz\cos\theta} J_\mu(kr\sin\theta) d\theta. \tag{3.53}$$

In (3.52) and (3.53) $k = 2\pi/\lambda$ is the wave number of monochromatic light with wavelength λ, α is the maximum angle of inclination θ of the rays to the optical axis, which determines the numerical aperture of the aplanatic lens NA = sin α, $J_\mu(\xi)$ is the Bessel function of the first kind of order μ. From (3.52) and (3.53) it is clear that the integrals depend only on the radial variable r. Therefore, the distribution of the canonical energy flow (3.51) has circular symmetry. The longitudinal projection of the Poynting vector (3.46) at the focus of the vector field (3.44) has the form [42]:

$$P_z(r) = \frac{1}{2c}\left[(I_{0,n-1} + I_{2,n-1})(I_{0,n-1} - I_{2,n-1}) + (I_{0,n+1} + I_{2,n+1})(I_{0,n+1} - I_{2,n+1})\right]. \tag{3.54}$$

Eq. (3.54) includes only integrals (3.52). Comparing Eqs. (3.51) and (3.54), it is evident that their equality requires the following relations to hold:

$$\begin{aligned} I_{0,n-1} - I_{2,n-1} &= \bar{I}_{0,n-1} + \bar{I}_{2,n-1}, \\ I_{0,n+1} - I_{2,n+1} &= \bar{I}_{0,n+1} + \bar{I}_{2,n+1}. \end{aligned} \tag{3.55}$$

The second index in integrals (3.52) and (3.53) is equal to the order of the Bessel function. In the first expression in (3.55), the second index is the same for all integrals and equals $n - 1$, and in the second expression in (3.55), the second index is also the same for all integrals. Therefore, instead of proving the validity of the two expressions in (3.55), we prove the validity of a more general expression with an arbitrary second index:

$$I_{0,\mu} - I_{2,\mu} = \bar{I}_{0,\mu} + \bar{I}_{2,\mu}. \tag{3.56}$$

Using the explicit form of integrals (3.52) and (3.53), instead of (3.56) we obtain:

$$\begin{aligned} &2kf\int_0^{\alpha}\left(\sin\left(\frac{\theta}{2}\right)\cos^3\left(\frac{\theta}{2}\right) - \sin^3\left(\frac{\theta}{2}\right)\cos\left(\frac{\theta}{2}\right)\right)\cos^{1/2}(\theta)A(\theta)e^{ikz\cos\theta}J_\mu(kr\sin\theta)d\theta = \\ &= kf\int_0^{\alpha}(\sin\theta\cos\theta)\cos^{3/2}(\theta)A(\theta)e^{ikz\cos\theta}J_\mu(kr\sin\theta)d\theta = \\ &= 2kf\int_0^{\alpha}\left(\sin^3\left(\frac{\theta}{2}\right)\cos\left(\frac{\theta}{2}\right) + \sin\left(\frac{\theta}{2}\right)\cos^3\left(\frac{\theta}{2}\right)\right)\cos^{3/2}(\theta)A(\theta)e^{ikz\cos\theta}J_\mu(kr\sin\theta)d\theta. \end{aligned} \tag{3.57}$$

The longitudinal component of the canonical energy flow (3.51) for a light field with only transverse components (3.45) can be written in a more intuitive form:

$$P_{or,z} = |E_x|^2\frac{\partial}{\partial z}(\arg E_x) + |E_y|^2\frac{\partial}{\partial z}(\arg E_y). \tag{3.58}$$

That is, the longitudinal energy flow is proportional to the intensity distributions and phase derivatives along the longitudinal z-coordinate of each component of the electric vector. At those points in the focal plane where the phase derivatives are negative, a reverse energy flow occurs. The longitudinal phase derivatives are equal to the projections of the wave vectors [2]:

$$k_{x,z} = \frac{\partial}{\partial z}(\arg E_x), \quad k_{y,z} = \frac{\partial}{\partial z}(\arg E_y). \tag{3.59}$$

Therefore, the reverse energy flow at the focus means that in this part of the focal plane the wave vector is directed opposite to the direction of propagation, as it has a negative projection onto the optical axis.

3.3.2 Numerical Simulation

The focusing of an azimuthally polarized vector beam with a wavelength of $\lambda = 532$ nm was simulated using the Richards -Wolf integral. The incident wavefront was assumed to be plane. The numerical aperture was NA = 0.95. Figures 3.8 and 3.9 show the results of focusing a plane wave ($n = 0$) with azimuthal polarization using an aplanatic lens.

From Fig. 3.8 it is evident that the intensity and energy flux patterns are similar and have circular symmetry. From Fig. 3.9 it follows that the reverse flow is not large (approximately 1% of the maximum forward flux) and occurs near zero intensity. It also follows from Fig. 3.9b that the longitudinal projection of the wave vector is negative in the region of the reverse flow.

Figure 3.10a shows the distribution of the longitudinal Poynting vector component P_z in the yz-plane near the focus (approximately $-2\lambda < z < 2\lambda$) for the initial field with $n = 0$. It can be seen that the reverse flow (yellow) occurs not only in the

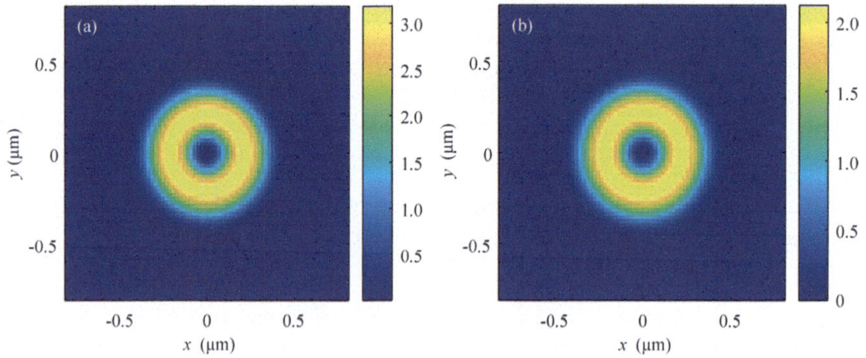

Fig. 3.8 Distribution of intensity $I = |E_x|^2 + |E_y|^2$ (**a**) and longitudinal component of the Poynting vector P_z(b) in the focal plane of the beam (3.44) with azimuthal polarization ($n = 0$)

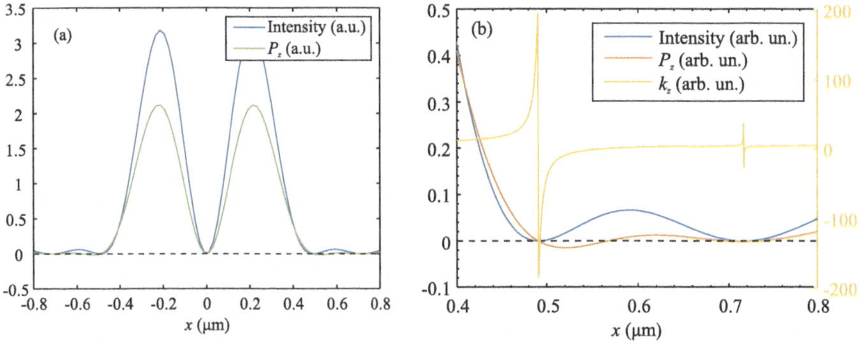

Fig. 3.9 Radial profiles of the intensity (blue curve) and longitudinal component of the Poynting vector (green curve) from Fig. 3.8 at the focus (**a**). An enlarged view of these sections near the zone of reverse energy flow (near the first two zeros of intensity) in the focal plane (**b**)

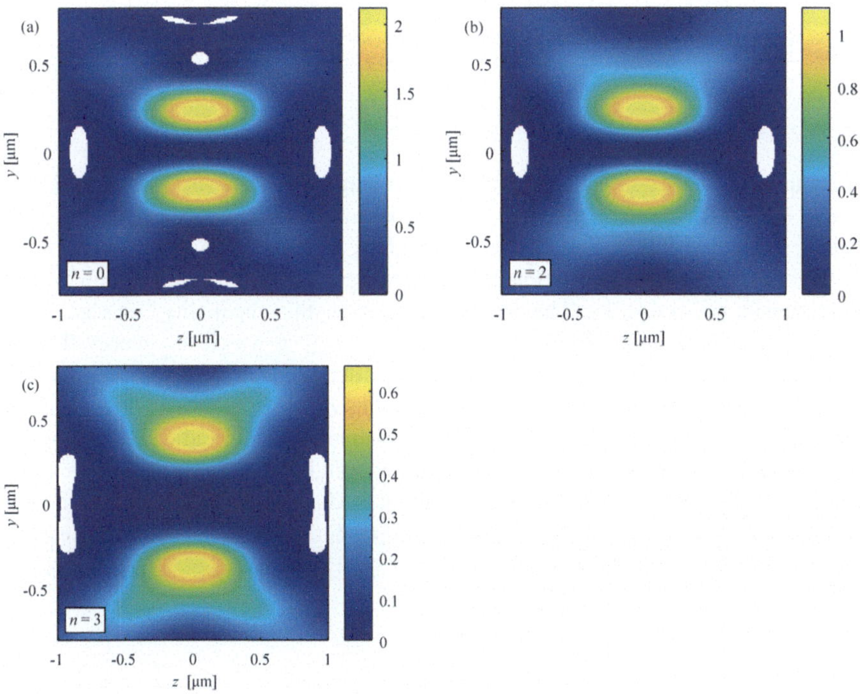

Fig. 3.10 Longitudinal projection of the Poynting vector in the *yz* plane for the initial field (3.44) at $n = 0$ (**a**), $n = 2$ (**b**) and $n = 3$ (**c**). Area with negative sign of the longitudinal projection of the Poynting vector is marked as white

focal plane ($z = 0$) but also along the optical axis, extending to almost two wavelengths both before and after the focus. Analogous distributions of P_z in the yz-plane are presented in Fig. 3.10b for $n = 2$ and in Fig. 3.3c for $n = 3$. For these vortex beams, the reverse flow is almost absent in the focal plane itself, but the region of reverse energy flow in the paraxial area expands significantly along the propagation direction, both before and after the focus.

As can be seen from Fig. 3.9b, the reverse flow in the focal plane for an azimuthally polarized field without a vortex occurs at a radius where the intensity is zero. This is consistent with theory. In [42], an expression was obtained for the intensity distribution in the focal plane for an azimuthally polarized optical vortex:

$$I(r) = \frac{1}{2}\left[\left(I_{0,n-1} + I_{2,n-1}\right)^2 + \left(I_{0,n+1} + I_{2,n+1}\right)^2\right]. \tag{3.60}$$

In the absence of an optical vortex ($n = 0$) the expression for intensity (3.60) simplifies to:

$$I(r) = \left(I_{0,1} + I_{2,1}\right).2 \tag{3.61}$$

The expression for the energy flow (3.54) also simplifies for $n = 0$:

$$P_z(r) = \frac{1}{c}\left(I_{0,1}^2 - I_{2,1}^2\right). \tag{3.62}$$

A comparison of Eqs. (3.61) and (3.62) reveals that the intensity is zero at a radius r_1 where the two integrals in Eq. (3.61) are equal in magnitude but opposite in sign: $I_{0,1} = - I_{2,1}$. At this radius r_1, the energy flow in Eq. (3.62) will also be zero. Furthermore, Eq. (3.62) shows that a reverse energy flow occurs within rings whose radii satisfy $r_1 < r < r_2$. The larger radius r_2 in this inequality, as seen from Eq. (3.62), is determined by the condition $I_{0,1} = I_{2,1}$, at which the energy flow is also zero: $P_z(r_2) = 0$. It is interesting that at a smaller radius r_1, both the intensity and the energy flow are zero, whereas at a larger radius r_2, only the energy flow (3.62) is zero, while the intensity (3.61) is non-zero. This means that at radius r_2, the energy flow is purely transverse (the wave vector has no longitudinal component). Next, we express the energy flow (3.62) for the azimuthal field in terms of the intensity and the z-derivative of the phase of the azimuthal component of the field (3.45). For $n = 0$, instead of (3.45), we obtain:

$$\begin{aligned} E_x &= -\sin\varphi\left(I_{0,1} + I_{2,1}\right), \\ E_y &= \cos\varphi\left(I_{0,1} + I_{2,1}\right). \end{aligned} \tag{3.63}$$

In polar coordinates, Eq. (3.63) can be written as:

$$\begin{aligned} E_r &= 0, \\ E_\varphi &= I_{0,1} + I_{2,1}. \end{aligned} \tag{3.64}$$

From Eq. (3.64), it is clear that at the focus of the azimuthally polarized field without a vortex, only the azimuthal component of the electric field is present. Therefore, the energy flow in Eq. (3.19) can be written as:

$$P_z(r) = \frac{1}{c}(I_{0,1} + I_{2,1})(I_{0,1} - I_{2,1}) = \\ \frac{1}{c}(I_{0,1} + I_{2,1})^2 \frac{(I_{0,1} - I_{2,1})}{(I_{0,1} + I_{2,1})} = I \frac{\partial}{\partial z} \arg E_\varphi. \tag{3.65}$$

From Eq. (3.65), it follows that the z-component of the wave vector in the focal plane for the azimuthally polarized field is given by:

$$k_z = \frac{\partial}{\partial z} \arg E_\varphi = \frac{(I_{0,1} - I_{2,1})}{c(I_{0,1} + I_{2,1})}. \tag{3.66}$$

From Eq. (3.66) it follows that under the condition $I_{0.1} = - I_{2.1}$ the denominator tends to infinity at radius r_1; moreover, at $r < r_1$ the projection of the wave vector on the z-axis tends to plus infinity, and for $r > r_1$ the projection of the wave vector tends to minus infinity (Fig. 3.9b). However, since at the radius r_1 the intensity is zero, the wave vector k_z is large (giant), but finite. When the wave vector of the light field becomes greater in absolute value than the wave number of light $k = 2\pi/\lambda$, then in this region in the focal plane the effect of superoscillations is realized [17]. At the radius r_2, where $I_{0,1} = I_{2,1}$, the numerator of the fraction in (3.66) is zero, and thus the longitudinal component of the wave vector is zero. At this radius the reverse flow ends, and the longitudinal projection of the wave vector becomes positive. It should be noted that for the reverse flow to arise outside the focal plane for optical vortices ($n > 0$), the presence of rings with zero intensity is necessary. However, as follows from Eq. (3.60), this requires the simultaneous satisfaction of two conditions: $I_{0,n-1} = -I_{2,n-1}$, $I_{0,n+1} = -I_{2,n+1}$. However, it is impossible to simultaneously satisfy these two conditions on a single ring. Therefore, in Fig. 3.11b and Fig. 3.12b, there are no reverse flows in the focal plane. There are only reverse energy flows near the optical axis before and after the focus, with a magnitude of approximately 1% or less of the forward energy flow.

Figure 3.11 shows the intensity (a) and longitudinal energy flow (dashed line in Fig. 3.11b) at a distance of 0.87 μm from the focal plane when focusing a second-order optical vortex with azimuthal polarization (3.44). It follows that the reverse flow is obtained near the optical axis and its magnitude is approximately 1% of the forward energy flux.

The mechanism for the formation of a reverse flow at a tight focus of azimuthally polarized light is illustrated in Fig. 3.12. The reverse flow is formed near the zero intensity points. In this case, the longitudinal projection of the wave vector assumes gigantic values and has a negative sign (i.e., the normal to the local wavefront is directed toward the source).

Fig. 3.11 Intensity (**a**) in the plane at a distance $z = 0.87$ μm from the focal plane of an azimuthally polarized second-order vortex ($n = 2$). The enlarged fragment of the longitudinal component of intensity and the Poynting vector (**b**)

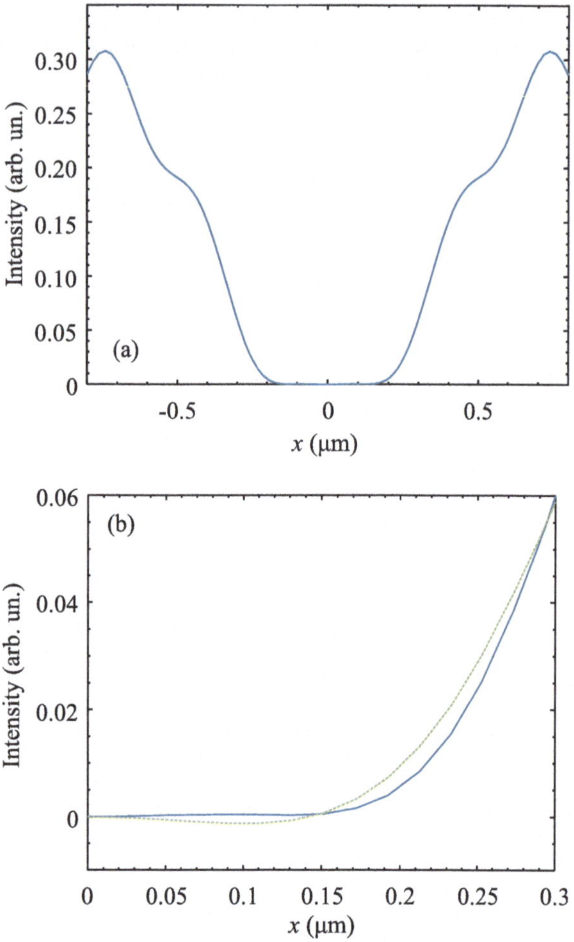

Figure 3.13 shows the phase distributions near the first intensity zero (blue curve) in the focal plane itself (green curve with a phase jump of π) at some distance from the focal plane $dz = 8$ nm (dashed curve). Figure 3.6 shows that before the first intensity zero, the longitudinal phase difference is positive (dashed curve above the green curve), and after the first intensity zero, where the reverse energy flow occurs, the longitudinal phase difference is negative (dashed curve below the green curve). That is, in the region of the reverse energy flow, the longitudinal phase derivative (3.66) is negative. Therefore, the reverse energy flow occurs because the light field compensates (smooths and reduces the phase difference) for the phase jump of π.

In this section, we obtained the following results. Using the Richards-Wolf formalism, which adequately describes light near the focus of a high-numerical-aperture system, we derived analytical expressions for the longitudinal components of the Poynting vectors and the canonical energy flow at the focus of an optical vortex with azimuthal polarization. It was shown theoretically that they are equal. It

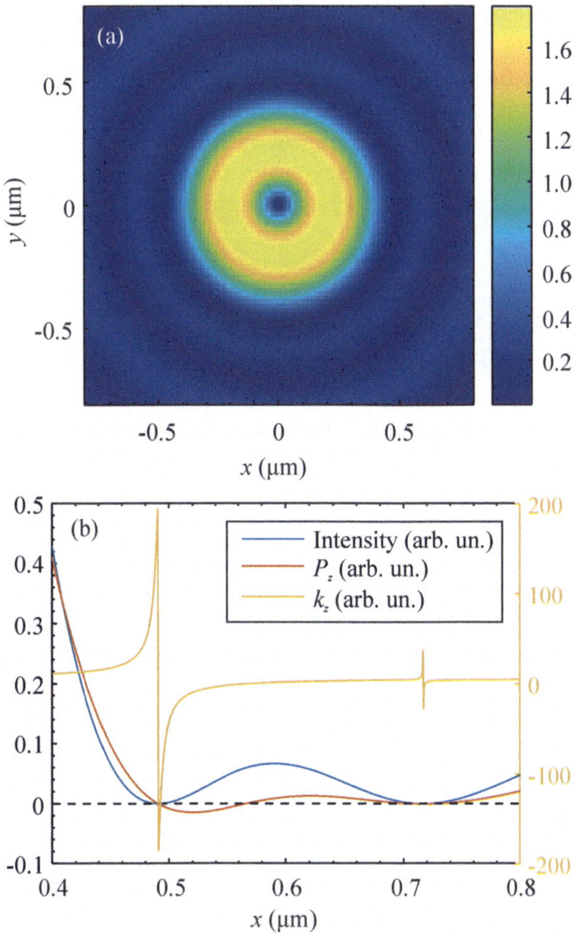

Fig. 3.12 Modulus of the azimuthal component of the electric field E_φ (3.64) (**a**) and the intensity cross-section (blue curve), the longitudinal component of the Poynting vector (brown curve) and the longitudinal component of the local wave vector (yellow curve) (**b**)

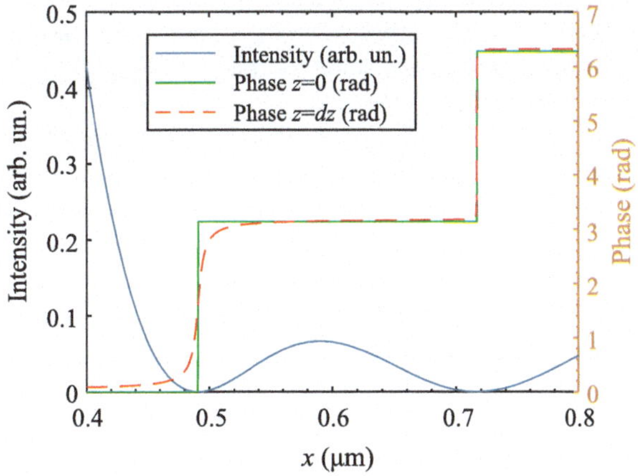

Fig. 3.13 Intensity (blue curve) and phase (green and red curves) cross-sections in a tightly focused azimuthally polarized beam at the focus (blue and green curves) and at a distance $dz = 8$ nm from the focus (red curve)

was also shown that, for azimuthally polarized light without an optical vortex, a reverse energy flow arises near the rings of zero intensity in the focal plane. Near these rings, the longitudinal component of the wave vector reaches gigantic values and changes sign when crossing zero intensity rings. Moreover, the canonical reverse flow near the optical axis before and after the focus was detected for the first time (Fig. 3.10). Previously, only negative values of the longitudinal component of the Poynting vector were detected on the optical axis at the focus [9].

3.4 Two-Dimensional Optical Vortices in Three-Dimensional, Nonvortex, Nonparaxial Beams

It is well-known that light fields can contain optical vortices in their cross-section. In such optical vortices, energy propagates along a spiral trajectory [45]. Ordinary scalar optical vortices carry orbital angular momentum [46] and have a topological charge [47]. Both quantities are conserved during free-space propagation. If a beam has isolated intensity zeros in its cross-section, these zeros do not disappear during propagation in free space. In this case, lines of zeros appear in the longitudinal section [37]. Structurally stable beams, such as Laguerre-Gaussian beams [48], exhibit not only an isolated intensity zero on the optical axis, but also a countable number of rings with zero intensity. These rings are conserved during free-space propagation and appear as lines of zero intensity in longitudinal radial planes [49]. It is also known that for non-paraxial light fields that are not modes (for example, Bessel-type beams) lines of intensity zeros in the cross-section yield isolated intensity zeros in longitudinal sections [50]. Similarly, vector beams at a tight focus can contain lines of intensity zeros (for example, rings), which result in isolated intensity zeros in the longitudinal section [42]. In contrast, 2D non-paraxial modal beams, such as Weber beams [51], possess no intensity zeros. Their complex amplitude is expressed through products of two Weber functions, which are in turn expressed through Hermite polynomials of a complex argument. It is known that Hermite polynomials have only real roots. Vector beams have lines of intensity zeros at the tight focus. However, some 2D paraxial light fields can exhibit isolated intensity zeros at the focus [24].

In this section, using the Richards-Wolf formalism [34] in a tight focus and the expansion of a non-paraxial light field into a plane-wave spectrum [21], it is shown that three-dimensional irrotational light fields, which contain lines of zero intensity in the cross section and isolated intensity zeros in the longitudinal section, form 2D optical vortices at these intensity zeros. An energy flow circulates around vortices. Moreover, along part of the closed circulation path, the energy flow is directed opposite to the optical axis.

3.4.1 Toroidal Canonical Energy Flow at the Focus of a Non-paraxial Beam

We consider a vector light field that, near the initial plane, is characterized by a single transverse component of the electric field along the Cartesian x-axis and a longitudinal component along the z-axis [21]:

$$\begin{aligned} E_x(r,\varphi,z) &= 2\pi \int_0^\infty A(\xi)\exp\left(ikz\sqrt{1-\xi^2}\right)J_0(kr\xi)\xi d\xi, \\ E_z(r,\varphi,z) &= -2\pi i\cos\varphi \int_0^\infty A(\xi)\exp\left(ikz\sqrt{1-\xi^2}\right)J_1(kr\xi)\frac{\xi^2}{\sqrt{1-\xi^2}}d\xi. \end{aligned} \tag{3.67}$$

In Eq. (3.67) $A(\xi)$ is the amplitude of the plane wave spectrum, which depends only on the radial coordinate $\xi = \sqrt{\xi_x^2 + \xi_y^2}$, (r, φ) are the polar coordinates in the beam cross-section, k is the wave number of monochromatic light, $J_0(x)$ and $J_1(x)$ are the Bessel functions of the zeroth and first orders. For simplicity, we choose in Eq. (3.67) the amplitude of the plane wave spectrum as *a circl*-function:

$$A(\xi) = circl\left(\frac{\xi}{a}\right) = \begin{cases} 1, & \xi < a, \\ 0, & \xi > a. \end{cases} \tag{3.68}$$

If $a > 1$, evanescent waves are present in the near field of the beam (3.67). Using Eq. (3.68), the integrals in Eq. (3.67) for the initial plane ($z = 0$) can be evaluated via known integral formulas [21]:

$$\begin{aligned} &\int_0^1 \sqrt{1-\xi^2}J_0(kr\xi)\xi d\xi = \frac{1}{(kr)^2}\left(\frac{\sin(kr)}{kr} - \cos(kr)\right), \\ &\int_0^a J_0(kr\xi)\xi d\xi = a^2\frac{J_1(kra)}{kra}. \end{aligned} \tag{3.69}$$

For the longitudinal component of the field (3.67) in the initial plane, two other known integral formulas can be used [21]:

$$\int_0^1 J_1(kr\xi)\frac{\xi^2}{\sqrt{1-\xi^2}}d\xi = \sqrt{\frac{\pi}{2kr}}J_{3/2}(kr) = \frac{1}{kr}\left(\frac{\sin(kr)}{kr} - \cos(kr)\right), \tag{3.70}$$
$$\int_0^a J_1(kr\xi)\xi^2 d\xi = a^2\frac{J_2(kra)}{kr},$$

For the vector field (3.67), the phase distribution must be analyzed by considering each of its components separately. We will first investigate the transverse component shifted along the optical axis over a distance z_0. Then the transverse component of such a shifted field

$$E_x(r,z;z_0) = 2\pi\int_0^a \exp\left(ik(z-z_0)\sqrt{1-\xi^2}\right)J_0(kr\xi)\xi d\xi,$$
$$E_x(r,z=z_0) = \pi a^2\left(\frac{2J_1(kra)}{kra}\right), \quad a \le 1. \tag{3.71}$$

From Eq. (3.71), it follows that the beam, which we will term a circular beam following Eq. (3.68), exhibits autofocus at a distance z_0 from the initial plane. The transverse intensity distribution in the focal plane takes the form of an Airy disk, described by the squared first-order Bessel function divided by its argument (the second expression in Eq. (3.71)). The light field is independent of the azimuthal angle φ and thus possesses circular symmetry, representing the simplest non-paraxial irrotational field. However, if we consider any longitudinal cross-section of this field (e.g., the horizontal xz-plane), the phase distribution within that plane reveals two-dimensional (2D) optical vortices of ± 1 orders. Indeed, in the focal plane, beam (3.71) possesses a countable number of null lines. These null lines constitute a set of concentric circles centered on the optical axis. Upon crossing each circle radially, the phase of the light field (3.71) undergoes a π-phase jump. Consequently, there are no optical vortices in the transverse cross-section due to the absence of isolated intensity nulls. In contrast, in longitudinal radial cross-sections of beam (3.71), isolated intensity nulls emerge, around which two-dimensional optical vortices are formed. Figure 3.14a shows the square root of the absolute value of the beam amplitude from Eq. (3.71) in the focal plane. The square root of the amplitude is plotted to clearly show the first few rings of the Airy pattern and the circular intensity nulls. The calculation parameters are as follows: wavelength $\lambda = 532$ nm, focal length $z_0 = 2\lambda$, upper integration limit $a = 1$, calculation domain $-2\lambda \le x \le 2\lambda$, $0 \le z \le 4\lambda$, number of points along each axis $N_x = N_z = 1025$, and number of points for the integration over variable ξ $N_\xi = 200$. The radius of the first zero-intensity circle is $r_0 = \gamma_{11}/(ka) \approx 0.76\lambda$, which coincides with the figure ($r_0 = 0.6\lambda$).

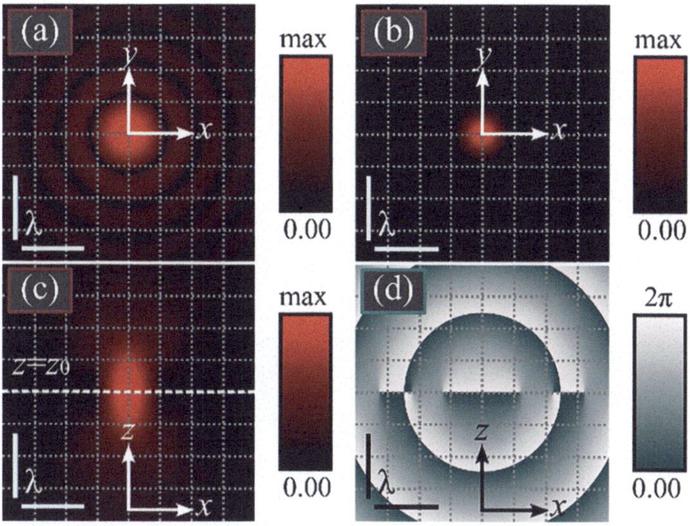

Fig. 3.14 The square root of the amplitude modulus in the focal plane (xy coordinates) at a distance $z_0 = 2\lambda$ from the initial plane for beam (3.71) with parameters $a = 1$, $\lambda = 532$ nm (**a**); intensity at the same location (**b**); amplitude (**c**) and phase (**d**) in the longitudinal horizontal section of beam (3.71) in the xz plane. Size of transverse frames is $4\lambda \times 4\lambda$ (**a**, **b**), size of longitudinal frames is $4\lambda \times 4\lambda$ (**c**, **d**)

Figure 3.14c demonstrates that isolated intensity nulls are located on the horizontal x-axis. Around these zeros, the phase distribution (Fig. 3.14d) changes from 0 to 2π upon a full circuit, indicating the formation of 2D optical vortices. Specifically, Fig. 3.14d reveals three optical vortices with a topological charge of -1 for $x > 0$ and three with a topological charge of +1 for $x < 0$. These two-dimensional vortices are unusual because they are located in the longitudinal plane. Energy flow circulates around the centers of these vortices. For intensity zeros associated with optical vortices of topological charge -1, the circulation is clockwise. In Fig. 3.14d, such vortices are located on the right-hand side of the horizontal axis. Conversely, for optical vortices with a topological charge of +1, the energy flow circulates counter-clockwise around the intensity zero. In Fig. 3.14d, these zeros are found on the left-hand side of the horizontal axis. However, when the energy flow in the longitudinal xz plane rotates, then along some part of its trajectory the energy flow is directed back. We demonstrate that this local reverse energy flow is linked to the intensity zeros at the focus of beam (3.71). For beam (3.67), considering Eq. (3.68), an expression for the longitudinal component of the canonical energy flow can be derived. The longitudinal z-component of the canonical energy flow is given by [8]:

$$P_{or,z} = \frac{1}{k}\,\mathrm{Im}\left(E_x^* \frac{\partial}{\partial z} E_x + E_y^* \frac{\partial}{\partial z} E_y + E_z^* \frac{\partial}{\partial z} E_z\right). \tag{3.72}$$

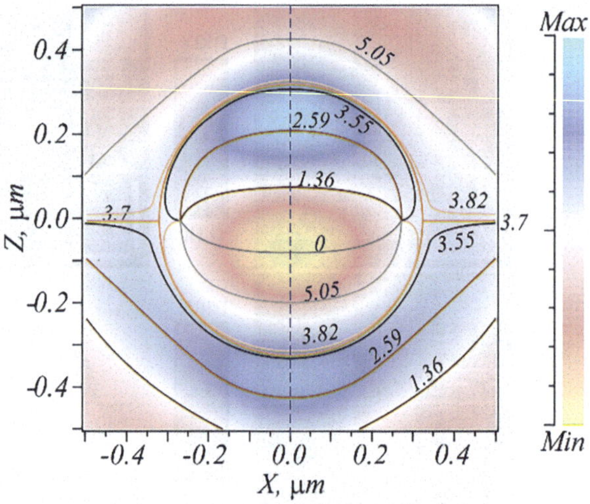

Fig. 3.15 Instantaneous amplitude of the light field (3.71) in the xz plane with lines of constant phase. Near the focus at $z_0 = 2\lambda$ in a central frame of size $\lambda \times \lambda$. The corresponding phase values are annotated adjacent to their contours

By substituting the projections of the electric field vector of the beam (5) into Eq. (3.6), and evaluating them using the integrals (3) and (4), we obtain in the focal plane ($z = z_0$):

$$P_{\mathrm{or},z}\bigg|_{z=0} = \frac{4\pi^2 a^2}{(kr)^2}\left(\frac{\sin(kr)}{kr} - \cos(kr)\right)\times \left[\frac{J_1(kra)}{kra} + \cos^2\varphi J_2(kra)\right]. \tag{3.73}$$

Eq. (3.73) indicates that the radius of the first intensity zero will be slightly larger than the first zero of the Bessel function $J_1(3.8) = 0$, yielding an approximate value of $r = 0.6\lambda$. The reverse flow region is shown in more detail in Fig. 3.15, which depicts lines of constant phase for the transverse component of the field (3.71) in the longitudinal plane xz within the central area $\lambda \times \lambda$. The center of the focal spot is set as the coordinate origin. From Fig. 3.15 it is evident that at $x = 0.2$ µm the energy propagates in the positive direction, since the phase values sequentially increase: 1.36, 2.59, 3.55, 3.82, 5.05, 2π. Then, at $z = 0$, the phase undergoes 2π jump and continuing upward along the same line $x = 0.2$ µm, the phases increase again and repeat: 1.36, 2.59, 3.55, 3.82, 5.05. In contrast along the line $x = 0.3$ µm, the phase initially increases (1.36, 2.59, 3.55, 3.82), then decreases upon crossing the horizontal axis (3.82, 3.55), and finally increases again (3.55, 3.82, 5.05). A local decrease in phase with increasing z distance signifies a negative projection of the wave vector onto the z-axis. From Fig. 3.15, the approximate subwavelength area of this reverse energy flow region: 0.26 µm $< x <$ 0.31 µm; -0.1 µm $< z <$ 0.1 µm. Thus, the diameter of the energy circulation region around zero intensity is approximately 0.2λ. A similar region with reverse energy flow exists for negative x values.

The existence of a reverse energy flow leads to the question of energy conservation in such light propagation. The answer can be gained from the analogy of a river with containing a whirlpool. In the whirlpool water circulates around the center of the vortex, with local motion both along and against the main flow. In a steady state, the overall amount of water in the river remains constant, since "new" water bypasses the vortex while the same "old" water recirculates within it. We derive the differential law of energy conservation from the Helmholtz equation. The Helmholtz equation for a scalar field is:

$$\left(\frac{\partial^2}{\partial^2 x}+\frac{\partial^2}{\partial^2 y}+\frac{\partial^2}{\partial^2 z}+k^2\right)E(x,y,z)=0,$$
$$E(x,y,z)=A(x,y,z)\exp[i\psi(x,y,z)]. \tag{3.74}$$

Instead of using Eq. (3.74) for the complex amplitude, we can express the result as two equations for the modulus and for the argument (phase) $\psi(x, y, z)$ of the complex amplitude $A(x, y, z)$:

$$\frac{\partial^2}{\partial x^2}A+\frac{\partial^2}{\partial z^2}A-A\left[\left(\frac{\partial\psi}{\partial x}\right)^2+\left(\frac{\partial\psi}{\partial z}\right)^2\right]+k^2A=0,$$
$$2\frac{\partial A}{\partial x}\frac{\partial\psi}{\partial x}+A\frac{\partial^2}{\partial x^2}\psi+2\frac{\partial A}{\partial y}\frac{\partial\psi}{\partial y}+A\frac{\partial^2}{\partial y^2}\psi+ \tag{3.75}$$
$$2\frac{\partial A}{\partial z}\frac{\partial\psi}{\partial z}+A\frac{\partial^2}{\partial z^2}\psi=0.$$

The second equation in (3.75) can be rewritten in a clearer form:

$$\frac{\partial}{\partial x}\left(A^2\frac{\partial}{\partial x}\psi\right)+\frac{\partial}{\partial y}\left(A^2\frac{\partial}{\partial y}\psi\right)+\frac{\partial}{\partial z}\left(A^2\frac{\partial}{\partial z}\psi\right)=$$
$$div(\mathbf{P}_{or,z})=0,$$
$$\mathbf{P}_{or,z}=\left(A^2\frac{\partial}{k\partial x}\psi\right)\mathbf{x}+\left(A^2\frac{\partial}{k\partial y}\psi\right)\mathbf{y}+ \tag{3.76}$$
$$\left(A^2\frac{\partial}{k\partial z}\psi\right)\mathbf{z}.$$

In Eq. (3.76), **x, y** and **z** denote the unit vectors along the Cartesian coordinate axes. From Eq. (3.76), it follows that the divergence of the canonical energy flow vector is zero. This expression represents the differential law of energy conservation in stationary form. Eq. (3.76) remains valid irrespective of the sign of the phase derivative $\partial\psi/\partial z$, that is, it remains valid both in the presence and in the absence of a reverse energy flow in the light field.

3.4.2 Toroidal Canonical Flow at Tight Focus

As shown in Ref. [42] for a tightly focused optical vortex with cylindrical polarization, the Jones vector in the initial plane has the form

$$\mathbf{E} = A(\theta)\exp(im\varphi)\begin{pmatrix} -\sin(n\varphi) \\ \cos(n\varphi) \end{pmatrix}, \tag{3.77}$$

where $A(\theta)$ is the amplitude of the initial field in an aplanatic system with focal length $f = r\cos\theta$, (r, φ) are the polar coordinates in the beam cross-section, m is the topological charge of the optical vortex, n is the order of the cylindrical vector beam, the longitudinal component of the Poynting vector without irrelevant constants $\mathbf{P} = \mathrm{Re}[\mathbf{E}^* \times \mathbf{H}]$, Re is the real part, has the form:

$$P_z = \frac{1}{2}\left(|I_{0,m+n}|^2 + |I_{0,m-n}|^2 - |I_{2,m+n-2}|^2 - |I_{2,m-n+2}|^2\right). \tag{3.78}$$

In (3.78) the following notation is used for the functions $I_{\nu,\mu}$, which depend only on the radial variable r near the focal plane ($z = 0$):

$$I_{\nu,\mu} = 2kf \int\limits_0^{\alpha} \sin^{\nu+1}\left(\frac{\theta}{2}\right)\cos^{3-\nu}\left(\frac{\theta}{2}\right)\cos^{1/2}(\theta)A(\theta)e^{ikz\cos\theta}J_\mu(kr\sin\theta)d\theta, \tag{3.79}$$

where α is the maximum angle of inclination of the rays to the optical axis θ, which determines the numerical aperture of the aplanatic lens NA = sin α, J_ν (ξ) is the Bessel function of the first kind of the νth order. In the special case of an optical vortex with linear polarization along the y-coordinate (at $n = 0$), instead of Eq. (3.78) we obtain:

$$P_z = |I_{0,m}|^2 - |I_{2,m-2}|^2. \tag{3.80}$$

From Eq. (3.80) it follows that the toroidal energy flow, determined using the Poynting vector, will be formed around circles with a center on the optical axis. The radii of these circles are defined from the equations for the equality of the longitudinal component of the Poynting vector to zero:

$$|I_{0,m}| = |I_{2,m-2}|. \tag{3.81}$$

Interestingly, the centers of toroidal energy flows will also located on these same radii in another special case—when the topological charge of the field (3.77) is zero ($m = 0$). Then, instead of Eq. (3.78), we obtain

$$P_z = |I_{0,n}|^2 - |I_{2,n-2}|^2 = (I_{0,n} - I_{2,n-2})(I_{0,n} + I_{2,n-2}). \tag{3.82}$$

And the radii of the circles containing the centers of toroidal energy flows are determined by an equation similar to Eq. (3.81):

$$|I_{0,n}| = |I_{2,n-2}|. \tag{3.83}$$

From a comparison of (3.81) and (3.83), it is evident that when $n = m$, the distributions of the longitudinal component of the Poynting vector in the focal plane are identical. Using (3.72), an expression for the longitudinal component of the canonical energy flow vector in the focal plane can be derived for the initial field (3.77) with $m = 0$:

$$\begin{gathered} P_{or,z}(r, \varphi) = \left[I_{0,n}\bar{I}_{0,n} + I_{2,n-2}\bar{I}_{2,n-2} + 2I_{1,n-1}\bar{I}_{1,n-1} - \right. \\ \left. \cos(2(n-1)\varphi)\left(I_{0,n}\bar{I}_{2,n-2} + I_{2,n-2}\bar{I}_{0,n} + 2I_{1,n-1}\bar{I}_{1,n-1}\right)\right]. \end{gathered} \tag{3.84}$$

In Eq. (3.84) the functions $I_{\nu,\ \mu}$ are defined in Eq. (3.79), and the functions $\bar{I}_{\nu,\mu}$ are slightly different from Eq. (3.79) and are equal to:

$$\bar{I}_{\nu,\mu} = 2kf \int_0^{\alpha} \sin^{\nu+1}\left(\frac{\theta}{2}\right) \cos^{3-\nu}\left(\frac{\theta}{2}\right) \cos^{3/2}(\theta) A(\theta) e^{ikz\cos\theta} J_\mu(kr\sin\theta) d\theta. \tag{3.85}$$

From a comparison of (3.82) and (3.84), it is evident that the longitudinal component of the Poynting vector at the focus of an azimuthal field of order n differs significantly from the longitudinal component of the canonical energy flow. The energy flow distribution in Eq. (3.84) has circular symmetry, and the energy flow in Eq. (3.84) depends on the azimuthal angle φ, and the maximum of the reverse flow is located on the horizontal axis at $\varphi = 0$ and $\varphi = \pi$. The energy flow (3.84) on the horizontal axis is equal to:

$$P_{or,z}(r, \varphi = 0) = (I_{0,n} - I_{2,n-2})\left(\bar{I}_{0,n} - \bar{I}_{2,n-2}\right). \tag{3.86}$$

From the comparison of (3.82) and (3.86) it follows that both energy flows vanish at the same radius, given by the expression $I_{0,n} = I_{2,n\text{-}2}$. It should be noted that that for $n = 1$ the dependence on the azimuthal angle φ in the canonical energy flow (3.84) vanishes. In this case both energy flows are identical, since the expressions are equal:

$$I_{0,1} - I_{2,1} = \bar{I}_{0,1} + \bar{I}_{2,1}. \tag{3.87}$$

That is, for azimuthal polarization in the focal plane, the longitudinal component of the Poynting vector coincides with the longitudinal component of the canonical energy flow vector. Figure 3.16 shows the intensity (a) and canonical energy flow

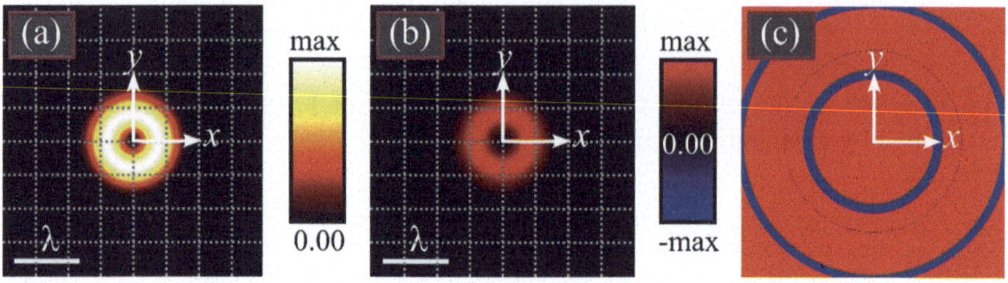

Fig. 3.16 Intensity distribution at the focus of the vector field (3.77) with azimuthal polarization ($m = 0$, $n = 1$) (**a**); distribution of the canonical energy flow (3.84) (**b**) and its sign (**c**): red color denotes positive sign, blue color denotes negative sign. Frames have dimensions of $2\lambda \times 2\lambda$, the wavelength is $\lambda = 532$ nm, the numerical aperture is 0.95

(b) distributions at the beam focus (3.77) for $m = 0$ and $n = 1$. Figure 3.16c shows that the reverse flow occurs on concentric circles centered on the optical axis.

Tight focusing of an azimuthal-polarized beam was simulated using an aplanatic objective with a numerical aperture of 0.95, assuming a flat wavefront and the wavelength of 532 nm. Figure 3.17 shows the distribution of the modulus (Fig. 3.17a) and phase (Fig. 3.17b) of the azimuthal component E_φ ($m = 0$, $n = 1$) of the electric field in the longitudinal plane yz. It also shows the direction of the canonical energy flow near the points with vortex behavior of E_φ. From Fig. 3.17b it is evident that at the points with zero intensity on the vertical axis there are four 2D optical vortices: two at the top with a topological charge of +1 and two at the bottom with a topological charge of −1. The energy flow circulates around these vortices in the vertical longitudinal plane yz. Figure 3.17c shows that the energy flow rotates counterclockwise around the two upper intensity zeros. Along part of this trajectory, the energy flow is directed opposite to the optical axis (reverse energy flow). The energy flow near the 3D focal plane circulates around 2D optical vortices along a circumference of radius approximately 0.1λ.

Figure 3.17 shows that the energy flow propagates in the forward direction between two 2D optical vortices with identical topological charges of +1, separated by a small subwavelength distance of approximately 200 nm. Interestingly, when two adjacent vortices with opposite topological charges appear in the path of the light, the forward light passes between them if the optical vortex to the left of the forward flow has a topological charge of +1, and the one to the right has a topological charge of −1, as in Fig. 3.17b. If the signs of the left and right 2D optical vortices are reversed (the left has a topological charge of −1, and the right has a topological charge of +1), the energy flow between the vortices will be directed in the opposite direction (the light wall is formed). Consequently, the forward energy flow cannot pass between the intensity zeros, but will be forced to circulate around these zeros. The latter case is illustrated in Fig. 3.17d, which shows the phase for the azimuthal field (3.77) E_φ ($m = 0$, $n = 1$) in the longitudinal yz plane behind the focus at a distance of approximately two wavelengths. It is evident from Fig. 3.17d that the direct energy flow does not pass between two vortices with topological charges −1 (upper) and + 1 (lower), separated by a distance of approximately 300 nm. This

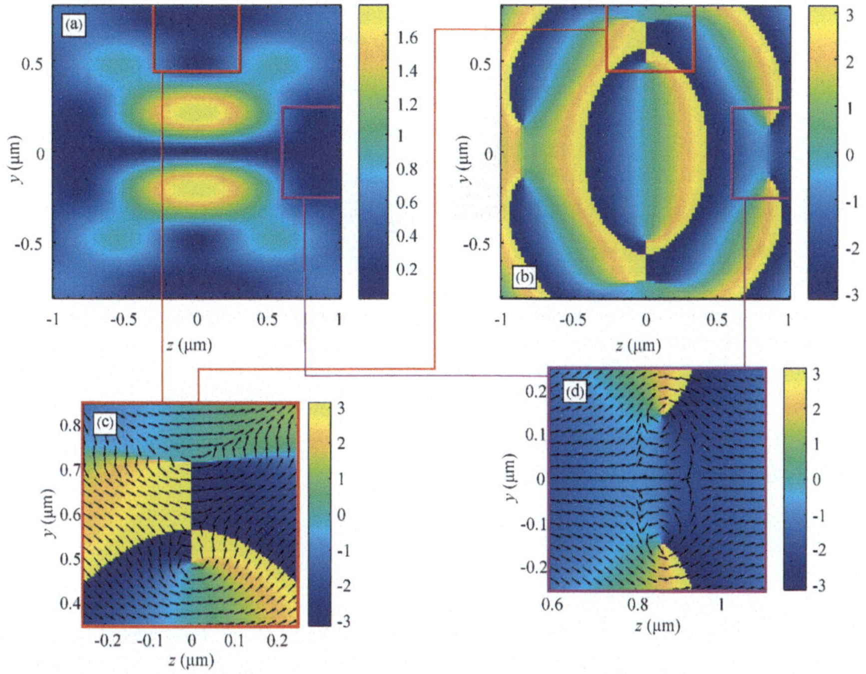

Fig. 3.17 Distributions of the modulus (**a**) and phase (**b**) of the azimuthal component of the electric field E_φ in the longitudinal plane yz. Panels (**c**) and (**d**) show magnified views near 2D optical vortices. The arrows indicate the direction of energy flow in the vicinity of two 2D optical vortices with a topological charge of $n = +1$. Panel (**c**) shows vortex with centers located on the vertical axis at points with zero intensity; panel (**d**) shows vortices located at a distance from 0.5 μm to 1 μm behind the focus (**d**)

occurs because in the gap between the two 2D vortices, the energy flow is directed in the opposite direction, blocking the direct light path between the intensity zeros.

3.4.3 Optical Vortices at the Focus of a Non-paraxial Field with Azimuthal Polarization

A similar pattern of 2D optical vortices occur near the focus of a nonparaxial beam with azimuthal polarization. Such a beam, at any distance from the optical axis, has only one component of the light field. Its amplitude is:

$$E_\varphi(r, z) = 2\pi i \int_0^\infty A(\xi) \exp\left(ik(z - z_0)\sqrt{1 - \xi^2} \right) J_1(kr\xi)\xi d\xi. \tag{3.88}$$

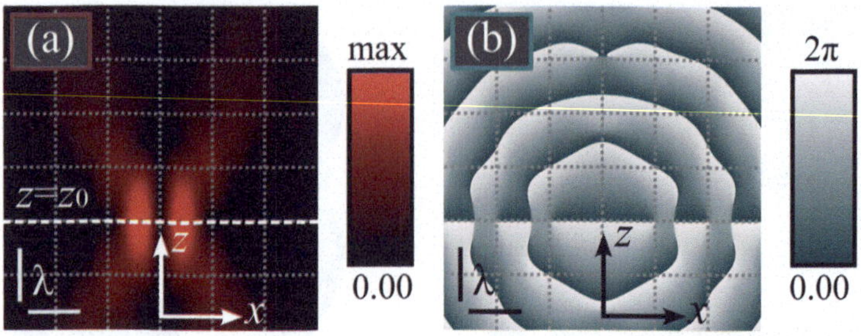

Fig. 3.18 Distribution of intensity (**a**) and phase (**b**) in the xz plane for a light field with azimuthal polarization (3.88) and (3.89)

If in Eq. (3.88) we choose the amplitude of the plane wave spectrum in the form $A\,(\xi) = \xi^{\,n}\,\mathrm{circl}(\xi)$, where $\mathrm{circl}(\xi) = 1,\ 0 < \xi < 1;\ 0,\ \xi > 1$, then the integral (3.88) can be calculated in the focal plane ($z = z_0$):

$$E_\varphi(r, z = z_0) = 2\pi i \int_0^1 \xi^2 J_1(kr\xi)\xi d\xi = 2\pi i \frac{J_2(kr)}{kr}. \tag{3.89}$$

It has a ring of zero intensity at radius $r = 0.82\lambda$. A longitudinal cross-section through this ring, for example, in the xz plane, reveals two intensity nulls on the horizontal x-axis. 2D optical vortices form around these nulls. Figure 3.18 shows the intensity (a) and phase (b) distribution of the section of beam (3.88) with amplitude (3.89) by the xz plane. Figure 3.18b shows that on the horizontal axis at $x > 0$ there are vortices with a topological charge of -1, and at $x < 0$—with a topological charge of +1. Figure 3.19a shows the circulation of the energy flow around one of the intensity zeros in the focal plane of the beam (3.88). It is also evident from Fig. 3.19b that near the optical axis, at a distance of approximately 3λ from the plane of focus, two optical vortices are formed, in which the topological charges relative to the optical axis have opposite signs: at $x > 0$—charge +1, and at $x < 0$—charge -1. This leads to the fact that at a distance of 3λ from the focus, near the optical axis, in a circle of radius approximately $0.2\ \lambda$, a reverse flow is formed, similar to Fig. 3.19b.

This section demonstrates that a non-paraxial light field with a circular plane-wave spectrum has isolated intensity zeros in the longitudinal plane [52, 53]. At these points longitudinal 2D optical vortices are formed (Fig. 3.14). The optical vortices located to the right of the optical axis, exhibit a topological charge of -1, while vortices located to the left exhibit a topological charge of +1. The phase distribution around these optical vortices shows that to the right of optical vortex with a charge of -1, the phase increases in the opposite direction (Fig. 3.15). This indicates that the light propagates in the opposite direction in this region. Similarly, using the Richards-Wolf formalism, it is shown that at the tight focus of azimuthal-polarized light, a ring of zero intensity with the center on the optical axis is formed (Fig. 3.16). In the longitudinal section of this ring, isolated intensity zeros are

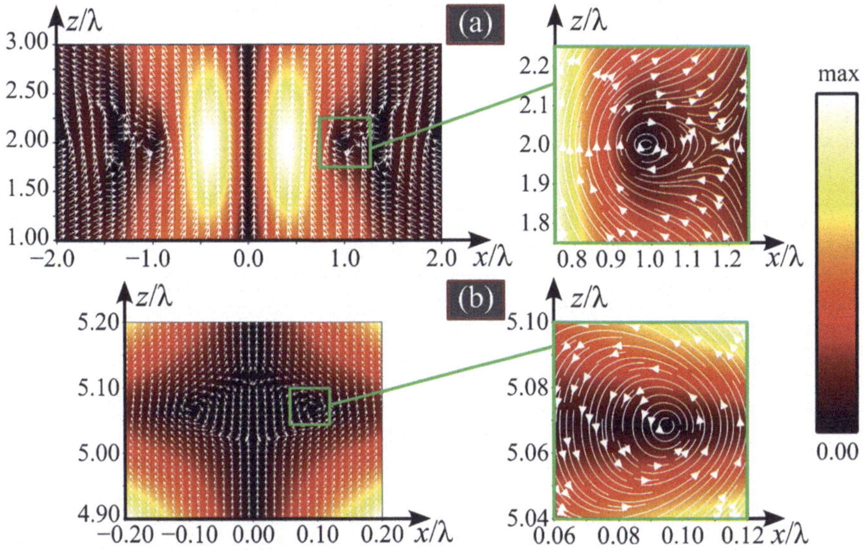

Fig. 3.19 Intensity distribution (identical to that in Fig. 3.18a) with arrows indicating the direction of wave vectors in the longitudinal plane xz. Panel (**a**) shows the region near two optical vortices at the focal plane ($z = 2\lambda$), panel (**b**) shows the region near the optical axis at a distance $z = 3.1\lambda$ behind the focal plane (**b**)

formed, around which 2D optical vortices appear (Fig. 3.17). An energy flow circulates around these longitudinal optical vortices (Fig. 3.17a). It is also shown that behind the focal plane and near the optical axis, two additional 2D optical vortices appear. The vortex to the right of the optical axis has a topological charge of +1, while the vortex to the left has a topological charge of −1. Consequently, the energy flow circulation around these adjacent vortices is such that forward-propagating light is blocked between them (Fig. 3.17c). Similar 2D optical vortices are formed when analyzing the focusing of azimuthal-polarized light via its decomposition into a plane-wave spectrum (Figs. 3.18 and 3.19).

References

1. A. Y. Bekshaev, M. S. Soskin, M. V. Vasnetsov. J. Opt. Soc. Am. A. **20**(8), 1635–1643 (2003)
2. M.V. Berry, J. Opt. A Pure Appl. Opt. **11**, 094001 (2009)
3. A. Bekshaev, K.Y. Bliokh, M. Soskin, J. Opt. **13**, 053001 (2011)
4. K.Y. Bliokh, A.Y. Bekshaev, F. Nori, Nat. Commun. **5**, 3300 (2014)
5. A.Y. Bekshaev, M.S. Soskin, Opt. Commun. **271**, 332 (2007)
6. A.Y. Bekshaev, J. Opt. **15**, 044004 (2013)
7. P. Vaveliuk, O. Martinez-Matos, Opt. Express **20**, 26913 (2012)
8. B. Ghosh, A. Daniel, B. Gorzkowski, A.Y. Bekshaev, R. Lapkiewicz, K.Y. Bliokh, J. Opt. Soc. Am. B **41**, 1276 (2024)

9. V.V. Kotlyar, S.S. Stafeev, A.G. Nalimov, Phys. Rev. A **99**, 033840 (2019)
10. A.V. Novitsky, D.V. Novitsky, J. Opt. Soc. Am. A **24**, 2844 (2007)
11. M.A. Salem, H. Bağcı, Opt. Express **19**, 8526 (2011)
12. V.V. Kotlyar, S.S. Stafeev, Optik (Stuttg). **245**, 167703 (2021)
13. Y.E. Geints, I.V. Minin, O.V. Minin, Opt. Commun. **524**, 128779 (2022)
14. V.V. Kotlyar, A.A. Kovalev, A.G. Nalimov, Comput. Opt. **33**, 369 (2009)
15. P. Saari, I.M. Besieris, Photonics **11**, 1129 (2024)
16. I. Besieris, P. Saari, Phys. Rev. A **107**, 033502 (2023)
17. G. Yuan, E.T.F. Rogers, N.I. Zheludev, Light Sci. Appl. **8**, 2 (2019)
18. M.V. Berry, S. Popescu, J. Phys. A Math. Gen. **39**, 6965 (2006)
19. E.T.F. Rogers, N.I. Zheludev, J. Opt. **15**, 094008 (2013)
20. N.I. Zheludev, G. Yuan, Nat. Rev. Phys. **4**, 16 (2021)
21. V.V. Kotlyar, S.S. Stafeev, Y. Liu, L. O'Faolain, A.A. Kovalev, Appl. Opt. **52**, 330 (2013)
22. A.G. Nalimov, V.V. Kotlyar, Y.V. Khanenko, Comput. Opt. **47**, 734 (2023)
23. P. Saari, I. Besieris, Eur. J. Phys. **42**, 55301 (2021)
24. V.V. Kotlyar, A.A. Kovalev, A.G. Nalimov, A.M. Telegin, J. Opt. Soc. Am. A **42**, 1396 (2025)
25. A.V. Ustinov, A.P. Porfirev, S.N. Khonina, Photonics **11**, 962 (2024)
26. V.V. Kotlyar, S.S. Stafeev, A.G. Nalimov, A.A. Kovalev, A.P. Porfirev, Phys. Rev. A **101**, 033811 (2020)
27. S.N. Khonina, A.P. Porfirev, A.V. Ustinov, M.S. Kirilenko, N.L. Kazanskiy, Photonics **9**, 121 (2022)
28. W. Lu, J. Zhou, Y. Zhang, P. Shi, X. Yuan, C. Min, Opt. Lett. **49**, 6009 (2024)
29. A. Daniel, B. Ghosh, B. Gorzkowski, R. Lapkiewicz, New J. Phys. **24**, 123011 (2022)
30. B. Ghosh, A. Daniel, B. Gorzkowski, R. Lapkiewicz, Optica **10**, 1217 (2023)
31. Y.E. Geints, I.V. Minin, O.V. Minin, Opt. Lett. **47**, 1786 (2022)
32. X. Xie, P. Shi, C. Min, X. Yuan, Photonics Res. **13**, 2033 (2025)
33. V.S. Ignatowsky, Trans. Opt. Inst. Petrogr. **1**, 4 (1920)
34. B. Richards, E. Wolf, Proc. R. Soc. A Math. Phys. Eng. Sci. **253**, 358 (1959)
35. H. Li, C. Wang, M. Tang, X. Li, Opt. Express **28**, 18607 (2020)
36. R. Chen, T. Song, Y. Luo, H. Li, X. Li, Photonics **10**, 743 (2023)
37. A.V. Volyar, V.G. Shvedov, T.A. Fadeeva, Opt. Spectrosc. **91**, 235 (2001)
38. M.V. Berry, J. Phys. A Math. Theor. **43**, 415302 (2010)
39. M.V. Bashevoy, V.A. Fedotov, N.I. Zheludev, Opt. Express **13**, 8372 (2005)
40. G. Chen, Z.Q. Wen, C.W. Qiu, Light Sci. Appl. **8**, 56 (2019)
41. V.V. Kotlyar, A.G. Nalimov, S.S. Stafeev, J. Opt. Soc. Am. B **36**, 2850 (2019)
42. V.V. Kotlyar, S.S. Stafeev, A.A. Kovalev, Opt. Express **27**, 16689 (2019)
43. V.V. Kotlyar, S.S. Stafeev, A.A. Kovalev, Comput. Opt. **43**, 337 (2019)
44. K.Y. Bliokh, E.A. Ostrovskaya, M.A. Alonso, O.G. Rodríguez-Herrera, D. Lara, C. Dainty, Opt. Express **19**, 26132 (2011)
45. M.J. Padgett, L. Allen, Opt. Commun. **121**, 36 (1995)
46. L. Allen, M.W. Beijersbergen, R.J.C. Spreeuw, J.P. Woerdman, Phys. Rev. A **45**, 8185 (1992)
47. J.F. Nye, M.V. Berry, Proc. R. Soc. A Math. Phys. Eng. Sci. **336**, 165 (1974)
48. L. Allen, M.J. Padget, M. Babiker, Prog. Opt. **39**, 291 (1999)
49. J. Hu, W. Wei, X. Li, Y. Shen, Q. Zhan, G. Li, Y. Qian, Laser Photon. Rev. **19**, 2401064 (2025)
50. Y. Zhang, Z. Mo, D. Xu, S. He, Y. Ding, Q. Huang, Z. Lu, D. Deng, Opt. Lett. **47**, 3059 (2022)
51. Z. Yang, W.-P. Zhong, M. Belić, Phys. Lett. A **560**, 130955 (2025)
52. V.V. Kotlyar, A.A. Kovalev, A.G. Nalimov, Appl. Phys. B Lasers Opt. **132**, 11 (2026)
53. S.S. Stafeev, V.D. Zaitcev, K. Guo, B. Liu, Z. Guo, V.V. Kotlyar, Appl. Opt. **65**, 1042 (2026)

Chapter 4
Local Wavevector Near the Region with Reverse Energy Flow

4.1 Local Wavevector of a Tightly Focused Optical Vortex with Linear Polarization

Light beams with non-uniform phase distribution—optical vortices—have attracted increasing interest from optics in recent decades. Optical vortices are characterized by a spiral wavefront. This shape leads to the fact that in the beam cross-section there are points with the phase uncertainty, i.e. phase singularity points. M. Berry in the work [1] introduces the concept of a local wavevector as a phase gradient for studying non-uniform scalar light fields, demonstrating that in the phase singularity points of optical vortices the absolute value of the local wavevector tends to infinity. Later, N.I. Zheludev [2] shows that this large local wavevector is often observed in focal spots obtained using the superoscillation effect [3–6]. Due to superoscillations it is possible to obtain compact focal spots with dimensions much smaller than the diffraction limit, but in this case, most of the light energy does not go into the focal spot itself but in the side lobes [7]. The superoscillation effect has recently attracted the attention of researchers. In particular there are created lenses that use this effect [8, 9] as well as microscopes based on such lenses [10, 11], although it should be noted that the sizes of the experimental spots are currently larger than the theory predicts. Another effect associated with the study of the behavior of the local wavevector is the reverse energy flow. In the paper [2] it was shown that zones of reverse energy flows are connected with regions with large values of the local wavevector. Usually these large values of wavevectors are obtained in plasmonics, due to this reason the aforementioned effect is called "plasmonics in free space" [12]. Previously, we investigated tight focusing of optical vortices and found the conditions for the formation of energy backflow on the axis under tight focusing condition [13–16]; however, the behavior of local wavevector was not investigated. Moreover, the behavior of singularity points in the case of focusing light into an extremely small region is of interest. For example, it was previously shown that there

V. Kotlyar et al., *Inverse Canonical Energy Flow in Laser Beams*,
https://doi.org/10.1007/978-3-032-27289-8_4

is an analog of the diffraction limit for the minimum distance between two points of phase singularity [17].

In this section, we investigated tight focusing of linearly polarized optical vortex. It was demonstrated that in contrast to focusing of scalar optical vortex, there are several centers of optical vortices under tight focusing conditions. As a consequence, several regions with a giant local wavevector can arise. If the topological charge of optical vortex is equal to two, then energy backflow region arises on the optical axis. This region is bounded by two optical vortices with unit topological charges shifted from the optical axis. If the topological charge exceeds two, then three optical vortices are formed in the focal plane, two of which have a unit topological charge and are shifted from the optical axis. It was analytically shown that with increasing of topological charge the centers of shifted optical vortices move away from the optical axis. Moreover, it was demonstrated that the centers of optical vortices limit the area of reverse energy flow, and the increasing of the distance between the vortices is connected with broadening of reverse energy flow zone.

4.1.1 Theoretical Background

The reverse energy flow has been previously recorded for various beams: cylindrical vector beams [13], optical vortices with circular [14] and linear [15] polarization, Poincaré beams [16], etc. In this paper, we investigated the simplest situation—optical vortices with linear polarization:

$$\mathbf{E} = e^{in\varphi}\begin{pmatrix} 1 \\ 0 \end{pmatrix}, \tag{4.1}$$

where n is the topological charge of the optical vortex. Previously it was shown that the linearly polarized optical vortices under tight focusing conditions demonstrate behavior different from the scalar approximation [18–20]. The behavior of light in the tight focal spot can be described using Richards-Wolf formula [21]:

$$\begin{aligned} \mathbf{U}(\rho, \psi, z) = & -\frac{if}{\lambda} \int_0^{\alpha} \int_0^{2\pi} B(\theta, \varphi) T(\theta) \mathbf{P}(\theta, \varphi) \times \\ & \times \exp\{ik[\rho \sin\theta \cos(\varphi - \psi) + z \cos\theta]\} \sin\theta \, d\theta \, d\varphi, \end{aligned} \tag{4.2}$$

where $\mathbf{U}$ (ρ,ψ,z) is the electric or magnetic field at the focus, B (θ, φ) is the electric or magnetic incident field, θ is the polar angle, φ is the azimuthal angle, T (θ) is apodization function, f is a focal length, $k = 2\pi/\lambda$ is a wavenumber, λ is a wavelength, α is a maximum polar angle determined by the numerical aperture of the lens (NA $= n\sin\alpha$, where $n = 1$ is refractive index of medium), $\mathbf{P}$ (θ, φ) is the polarization vector, for the electric field (4.1) it is equal to [22]:

$$\mathbf{P}(\theta,\varphi)=\begin{bmatrix} 1+\cos^2\varphi(\cos\theta-1) \\ \sin\varphi\cos\varphi(\cos\theta-1) \\ -\sin\theta\cos\varphi \end{bmatrix} \tag{4.3}$$

Using Richards-Wolf integral (4.2) in the Ref. [15] there were obtained electric and magnetic field components in tight focus of an optical vortex with topological charge n and circular polarization. Adapted to linear polarization the equations from [15] have the form:

$$\begin{aligned} E_x &= \frac{i^{n-1}}{2}e^{in\varphi}\left(2I_{0,n}+e^{i2\varphi}I_{2,n+2}+e^{-i2\varphi}I_{2,n-2}\right),\\ E_y &= \frac{i^n}{2}e^{in\varphi}\left(e^{-i2\varphi}I_{2,n-2}-e^{i2\varphi}I_{2,n+2}\right),\\ E_z &= -i^n e^{in\varphi}\left(e^{i\varphi}I_{1,n+1}-e^{-i\varphi}I_{1,n-1}\right), \end{aligned} \tag{4.4}$$

$$\begin{aligned} H_x &= \frac{-i^n}{2}e^{in\varphi}\left(e^{i2\varphi}I_{2,n+2}-e^{-i2\varphi}I_{2,n-2}\right),\\ H_y &= \frac{i^{n-1}}{2}e^{in\varphi}\left(2I_{0,n}-e^{i2\varphi}I_{2,n+2}-e^{-i2\varphi}I_{2,n-2}\right),\\ H_z &= i^{n+1}e^{in\varphi}\left(e^{i\varphi}I_{1,n+1}+e^{-i\varphi}I_{1,n-1}\right). \end{aligned} \tag{4.5}$$

Where $I_{0,\nu}$, $I_{1,\nu}$, $I_{2,\nu}$ are integrals depending on the radial coordinate only:

$$\begin{aligned} I_{0,\nu} &= \left(\frac{\pi f}{\lambda}\right)\int_0^\alpha \sin\theta(1+\cos\theta)T(\theta)A(\theta)e^{ikz\cos\theta}J_\nu(x)d\theta,\\ I_{1,\nu} &= \left(\frac{\pi f}{\lambda}\right)\int_0^\alpha \sin^2\theta T(\theta)A(\theta)e^{ikz\cos\theta}J_\nu(x)d\theta,\\ I_{2,\nu} &= \left(\frac{\pi f}{\lambda}\right)\int_0^\alpha \sin\theta(1-\cos\theta)T(\theta)A(\theta)e^{ikz\cos\theta}J_\nu(x)d\theta, \end{aligned} \tag{4.6}$$

where $J_{\mu,\nu}(x)$ is the Bessel function of the first kind, $x=kr\sin\theta$, (x, y, z) and (r, φ, z) are Cartesian and polar coordinates. Any radially symmetric function, for example, Bessel-Gauss function [23] can be used as the amplitude of the input function $A(\theta)$:

$$A(\theta)=J_1\left(2\beta\frac{\sin\theta}{\sin\alpha}\right)\exp\left[-\beta^2\left(\frac{\sin\theta}{\sin\alpha}\right)^2\right], \tag{4.7}$$

where β is the ratio of the radius of the aplanatic lens to the radius of the Gaussian beam.

Also in the paper [15] it was shown that the energy flow in the focal plane is equal to:

$$P_z = \frac{1}{2}\left(I_{0,n}^2 - I_{2,n+2}^2 - I_{2,n-2}^2\right) \tag{4.8}$$

and under tight focusing conditions, when the influence of the last two terms is large, it can take negative values (i.e. the effect of energy backflow is occurred). It was noted previously [2] that these areas of reverse energy flows are accompanied by areas with a large values of the local wavevector:

$$\mathbf{k} = \nabla\Phi, \tag{4.9}$$

where Φ is the phase of the scalar light field. Consequently, the region of the reverse energy flow (4.8) should also be accompanied by zones with large values of the local wavevector. It should be noted that in the Ref. [2] local wavevectors for vector fields were not investigated.

In the scalar approximation, regions with a large local wavevector (gradient of phase) correspond to the centers of optical vortices. Let us assume that the scalar optical vector has the form:

$$E = r^n e^{i\Phi} = (x + iy)^n, \tag{4.10}$$

where n is a topological charge of the vortex. The phase of the optical vortex has the form:

$$\Phi = \arg E = n\,\arctan\left(\frac{y}{x}\right). \tag{4.11}$$

It could be shown that the absolute value of the phase gradient is equal to:

$$|\nabla\Phi| = \frac{n}{r}, \tag{4.12}$$

i.e. it has a singularity point at $r = 0$, coinciding with the center of the vortex.

4.1.2 Numerical Simulation

In our research we investigated focusing of a beam (4.1) with a wavelength $\lambda = 633$ nm aplanatic objective ($T(\theta) = \cos^{1/2}\theta$) with a numerical aperture NA = 0.95. The wavefront of the focused light in the simulation is uniform $A(\theta) = 1$. We investigated the behavior of phase and local wavevector (4.9) under

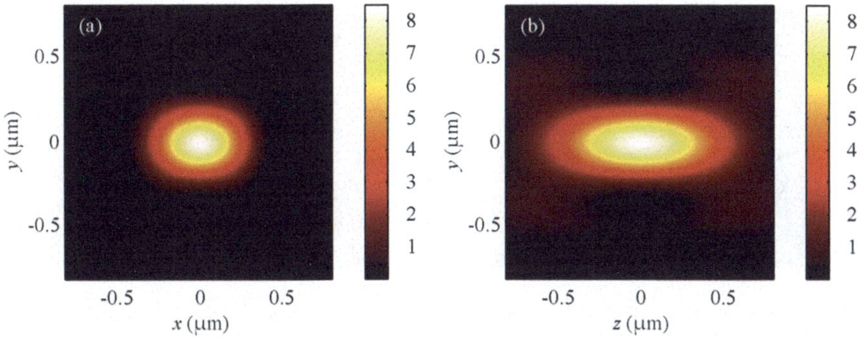

Fig. 4.1 Intensity in the transverse (focal) (**a**) and longitudinal plane (**b**) in a tightly focused linearly polarized beam ($n = 0$)

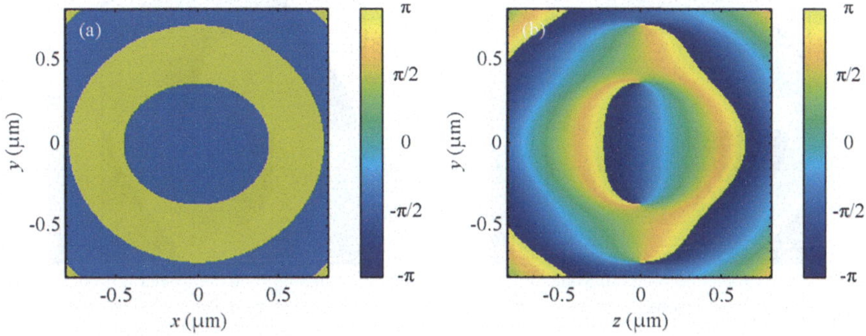

Fig. 4.2 Phase of E_x component of the focused linearly polarized light in the transverse (focal) (**a**) and longitudinal plane (**b**) ($n = 0$)

tight focusing of the optical vortex (4.1), when the vector effects (4.2) in the focal spot should be taken into account.

First we simulate the focusing of linearly polarized light without an optical vortex (i.e., the topological charge n in Eq. (4.1) was equal to zero). The results of calculation are shown in Figs. 4.1 and 4.2. Figure 4.1 shows the intensity distribution $I = I_x + I_y + I_z = |E_x|^2 + |E_{xy}|^2 + |E_{xz}|^2$ in the focal plane (Fig. 4.1a) and in the longitudinal plane yz (Fig. 4.1b). In this case, the intensity distribution has the shape of a peak (Fig. 4.1). The main contribution to the formation of the focal spot in this case is made by the component E_x, therefore the behavior of this component is investigated below. Figure 4.2 shows the distribution of the phase arg(E_x). As can be seen from Fig. 4.2, there are no singularity points.

Next we investigated focusing of an optical vortex with a topological charge equal to unity. From (4.8) it follows that in this case the longitudinal energy flow on the axis is positive. Similar to Figs. 4.1 and 4.2 in Figs. 4.3 and 4.4 there are shown intensity and phase distribution of E_x component, consequently.

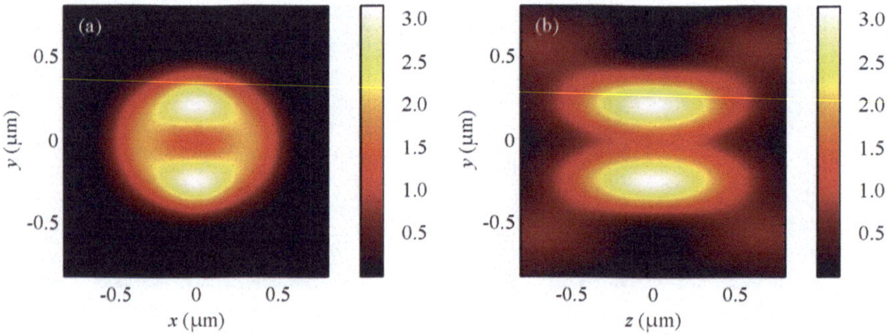

Fig. 4.3 Intensity in the transverse (focal) (**a**) and longitudinal plane (**b**) for an optical vortex with a topological charge of $n = 1$

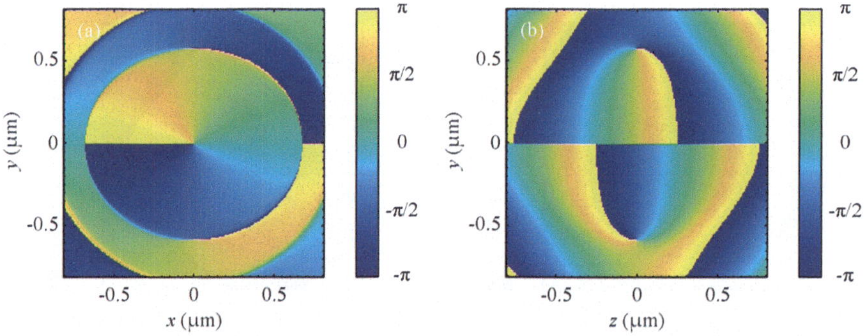

Fig. 4.4 Phase of E_x component of the focused optical vortex with a topological charge of $n = 1$ in the transverse (focal) (**a**) and longitudinal plane (**b**)

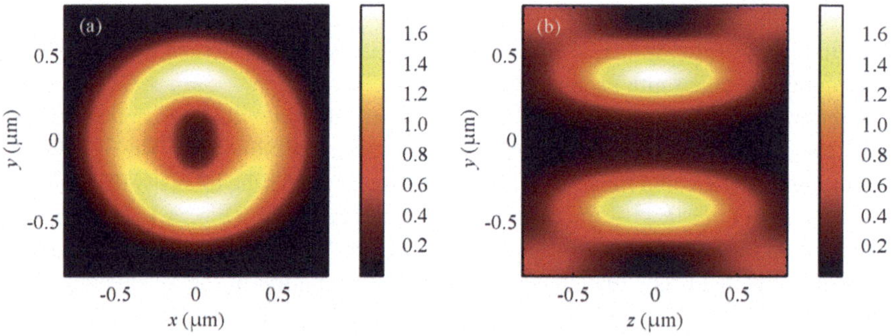

Fig. 4.5 Distribution of intensity in the transverse (focal) (**a**) and longitudinal (parallel to the direction of beam propagation) (**b**) planes for an optical vortex with a topological charge of $n = 2$

From Figs. 4.3 and 4.4 it follows that an optical vortex with a center on the optical axis is formed. Figures 4.5, 4.6, 4.7 and 4.8 show the results of focusing of a vortex with a topological charge equal to two ($n = 2$). It was shown previously that for this

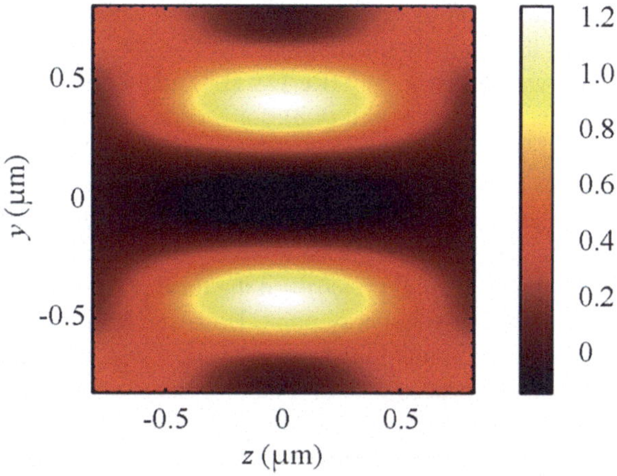

Fig. 4.6 Distribution of the longitudinal component of the Poynting vector P_z in the plane parallel to the direction of propagation for an optical vortex with a topological charge of $n = 2$

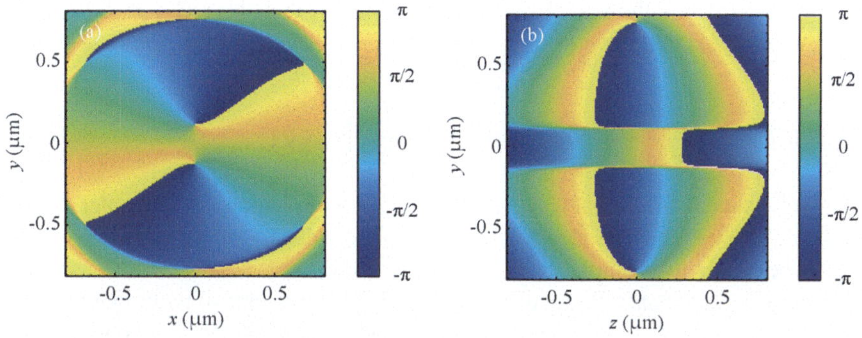

Fig. 4.7 Distribution of the phase of the x-component of the arg(E_x) field in a focused optical vortex with a topological charge of $n = 2$ in the transverse (focal) (**a**) and longitudinal (**b**) planes

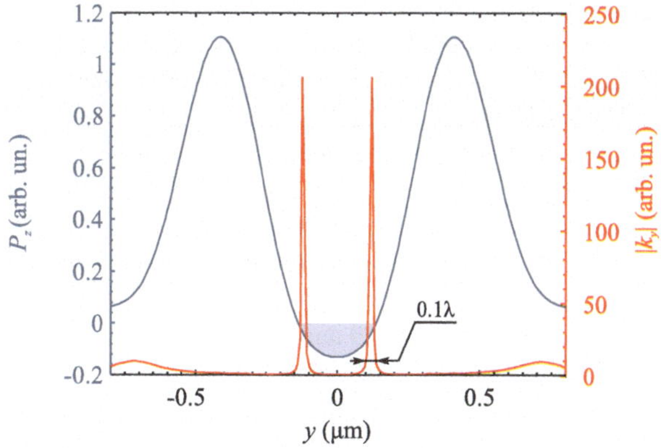

Fig. 4.8 Cross-section of the longitudinal component of the Poynting vector P_z (blue curve) and y-component of the local wavevector $|k_y|$ (red curve) at $z = 0.1$ μm

case the region of reverse energy flow is observed on the optical axis. Figure 4.5 shows the intensity distribution, and Fig. 4.6 shows the longitudinal component of the Poynting vector $P_z = (\mathrm{Re}[\mathbf{E} \times \mathbf{H}^*])_z$ in the plane parallel to the beam propagation axis. In accordance with (4.8), for a vortex with a topological charge equal to two, regions with negative values of the longitudinal component of the Poynting vector are observed on the optical axis (Fig. 4.6). Figure 4.7 shows the phase distribution of the electric field component E_x. It should be noted that in scalar approximation optical vortex (4.1) with an arbitrary topological charge n conserves its structure (helical phase $\arg(\exp(in\varphi))$—in the focal plane as well as in every transverse plane it has only one singularity point. However, from Fig. 4.7 it follows that under tight focusing conditions initial optical vortex with topological charge of two produces two optical vortices with unit topological charges.

From comparison of Figs. 4.6 and 4.7 it follows that the boundary of the region of reverse energy flow approximately coincides with the centers of optical vortices. To check this assumption in detail there are shown on Fig. 4.8 the cross-sections (at $z = 0.1$ μm) of the energy flow P_z (blue curve) and the y-component of the local wavevector (red curve):

$$\left|k_y\right| = \left|\frac{\partial(\arg E_x)}{\partial y}\right| \tag{4.10}$$

The large values of the local wavevector on Fig. 4.8 correspond to the centers of the vortices. The width of the area where the absolute value of the local wavevector in Fig. 4.8 exceeds wavenumber $k = 2\pi/\lambda$ is equal to 0.1λ.

Finally we calculate tight focusing of linearly polarized optical vortices with large topological charges. Below there are phase distribution of the component E_x of focused optical vortices with topological charge $n = 3$ (Fig. 4.9a), $n = 4$ (Fig. 4.9b), $n = 5$ (Fig. 4.9c) and $n = 6$ (Fig. 4.9d). To compare the behavior of the centers of optical vortices with energy flows we have added two cross-sections of these distributions in two single plots for optical vortices with two different topological charges: $n = 3$ (Fig. 4.10a) and $n = 5$ (Fig. 4.10c). Moreover, Fig. 4.11 shows phase of tightly focused vortices with topological charge equals to 10 (Fig. 4.11a) and 15 (Fig. 4.11b).

The width of the area where the absolute value of the local wavevector in Fig. 4.10 exceeds wavenumber $k = 2\pi/\lambda$ is equal to 0.1λ for shifted vortices. For central vortex this width is larger 0.15λ for $n = 3$ (Fig. 4.10a) and 0.21λ for $n = 5$ (Fig. 4.10b), i.e. the width increases with increasing topological charge.

In contrast to the scalar approximation, in which only one singularity point ($e^{in\varphi}$) is observed, from Figs. 4.9 and 4.10 it follows that for $n \geq 3$, three singularity points are observed in the focal plane (i.e. E_x component is the sum of three individual vortices). One vortex has the center on the optical axis, the centers of two other vortices are shifted along the axis perpendicular to the direction of polarization. This effect could be explained using Eq. (4.4), which shows that E_x component is a sum of three terms, and every term has helical phase. The behavior (Figs. 4.9 and 4.11) also

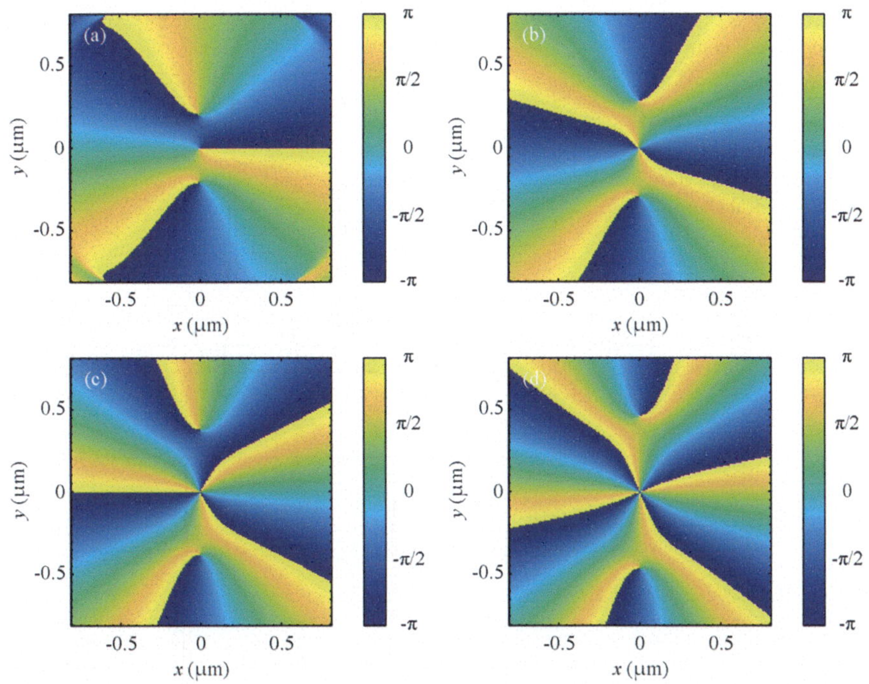

Fig. 4.9 Distribution of the phase of the E_x-component in a focused optical vortex with topological charge n: 3 (**a**), 4 (**b**), 5 (**c**) and 6 (**d**) in the focal plane

differs from vortex splitting due to astigmatism or propagation through crystals, when the number of vortices coincides with the topological charge of the incident vortex [24, 25]. It should be noted that, in fact, single tightly focused optical vortex generates optical vortex lattice [26–28].

From the phase distribution of the E_x field, it also follows that in the center there is an optical vortex with a topological charge of $n-2$, and the remaining two optical vortices with topological charges of 1 are located on the vertical axis and move away from the center with increasing n. This effect can also be explained using expression (4.4) for the projection of the E_x field:

$$E_x = \frac{i^{n-1}}{2} e^{in\varphi} \left(2I_{0,n} + e^{i2\varphi} I_{2,n+2} + e^{-i2\varphi} I_{2,n-2} \right). \tag{4.11}$$

From (4.11) it follows that since the first term is greater than the second term and third term, the topological charge of the entire superposition of optical vortices (4.11) is equal to n. In the center an optical vortex is formed, which has the smallest exponent of the radial variable. For small values of $kr << 1$ from (4.6) it follows that (4.11) can be approximately written:

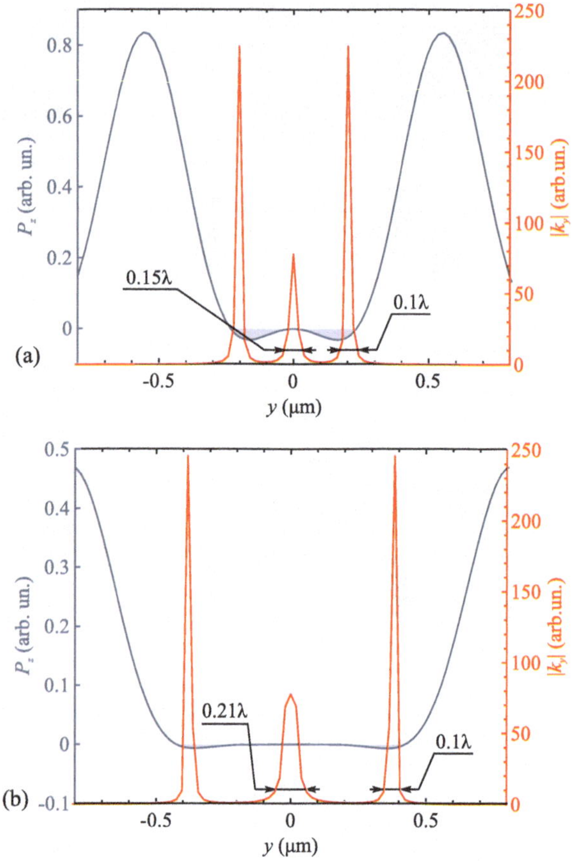

Fig. 4.10 Cross-section of the longitudinal component of the Poynting vector P_z (blue curve) and y-component of the local wavevector $|k_y|$ (red curve) in a focused optical vortex with topological charge $n = 3$ (**a**) and $n = 5$ (**b**) in the focal plane

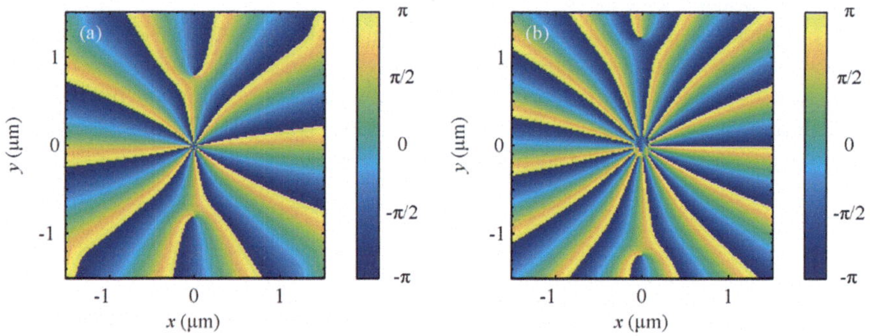

Fig. 4.11 Distribution of the phase of the E_x-component in a focused optical vortex with topological charge $n = 10$ (**a**), $n = 15$ (**b**) in the focal plane

$$E_x \sim 2a(kr)^n e^{in\varphi} + \\ b(kr)^{n+2} e^{i(n+2)\varphi} + c(kr)^{n-2} e^{i(n-2)\varphi}, \tag{4.12}$$

where *a*, *b*, *c* are constants proportional to the integrals (4.6), after the powers of *kr* have been taken out of them. When obtaining (4.12), the expression for the Bessel function with small arguments was used:

$$J_n(x << 1) \approx (x/2)^n (n!)^{-1}. \tag{4.13}$$

From (4.12) it follows that in the center there will be an optical vortex with a topological charge of $n-2$. The two remaining optical vortices will be located on the vertical Cartesian axis, because two zeros of intensity are formed where the sum of the first and third terms in (12) is equal to zero:

$$\left|2a(kr)^n e^{in\varphi} + c(kr)^{n-2} e^{i(n-2)\varphi}\right|^2 = \\ 4a^2(kr)^{2n} + c^2(kr)^{2(n-2)} + 4ac(kr)^{2(n-1)}\cos(2\varphi). \tag{4.14}$$

From (4.14) it follows that the zeros of intensity will be at azimuthal angles $\varphi = +\pi/2$ and $\varphi = -\pi/2$, because at these angles $\cos(2\varphi) = -1$. The distance between these two intensity zeros from (4.14), taking into account (4.13), will be approximately equal to:

$$2r \approx \frac{\lambda}{\pi}\sqrt{n(n-1)}. \tag{4.15}$$

From (4.15) it follows that as *n* increases, the two zero points on the *y*-axis move away from the center. Moreover, from Fig. 4.10 it also follows that the centers of optical vortices still limit the area of reverse energy flow, and the increasing of the distance between the vortices is connected with broadening of reverse energy flow zone. This broadening of the reverse energy flow with increasing topological charge was demonstrated in [14].

In this section, we investigated the behavior of the phase and local wave vector of a tightly focused optical vortex with linear polarization [29]. It is shown that the regions of phase uncertainty correspond to giant values of the local wave vector. Unlike a scalar optical vortex, several centers of optical vortices and, as a consequence, several regions with a giant wave vector can arise. If the topological charge of the focused vortex is equal to one, then the center of the focused vortex is located on the optical axis. If the topological charge is equal to two, then two optical vortices with unit topological charges shifted from the optical axis are observed in the focal plane. If the topological charge exceeds two, then three optical vortices are formed in the focal plane, two of which have a unit topological charge and are shifted from the optical axis. It was analytically shown that with increasing of topological charge the centers of shifted optical vortices move away from the optical axis. The centers of optical vortices limit the area of reverse energy flow, and the increasing of the distance between the vortices is connected with broadening of reverse energy flow zone.

4.2 Phase Singularities in Tightly Focused Optical Vortices Near Reverse Energy Flows

Research into optical vortices continues to expand, which is confirmed by the growth of publications in this area, including roadmaps [30–32]. The reason for the interest is in the numerous applications of these beams, for example, in microscopy [33], optical micromanipulation [34], metrology [35], optical communications in free space [36, 37] and waveguides [38, 39], etc. However, significant gaps persist in understanding their behavior. Studies of effects like spin-orbit conversion [40, 41] and interactions with matter or microparticles [42] remain active. Another unexplored area is the connection between optical vortices and reverse energy flows, i.e. regions where the Poynting vector is opposed to the beam propagation direction. While work in [2, 12, 43] demonstrated that reverse flow regions coincide with areas of giant wavevector [1, 44, 45] (which correspond to vortex center), these studies did not establish whether this spatial correlation constitutes a general principle.

It is important to note that the Poynting vector is the sum of two components [46]. One of them is the canonical energy flow, which is responsible for the transfer of energy in space, i.e., it corresponds to the classical definition of energy flow [47]. The second component is the spin energy flow, which is responsible for local energy circulations. This value does not transfer energy, but can interact with matter [48]. The canonical energy flow is always positive. The total energy flow is negative only if the spin component is negative [49]. Thus, in the reverse flow zone, the energy still propagates in the direction of beam propagation; however, the term "reverse flow" of energy has already become established in the literature [50, 51]. Our previous investigations [14, 52] into tightly focused phase and polarization vortices shows that reverse energy flow magnitude increases substantially for specific topological charges.

In this section we investigated tight focusing of optical vortices which demonstrates reverse energy flow regions in the focal plane. We analyzed both phase vortices and cylindrical vector beams (polarization vortices). For phase vortices, reverse energy flow regions are located between phase singularity centers (screw dislocations). Increasing the topological charge radially expands the reverse flow domain while simultaneously displacing vortex centers further from the optical axis. In contrast, focused cylindrical vector beams exhibit reverse flow regions without accompanying vortex dislocations (only edge dislocations are appeared in this case).

4.2.1 Focusing of Circularly Polarized Optical Phase Vortices

In this section, the numerical simulation was carried out using the Richards-Wolf formalism, which allows calculating the components of all projections of the electric and magnetic field at the focus of the lens [13]:

$$\mathbf{U}(\rho,\psi,z) = -\frac{if}{\lambda}\int_0^{\alpha}\int_0^{2\pi} B(\theta,\varphi)T(\theta)\mathbf{P}(\theta,\varphi)\times \\ \exp\{ik[\rho\sin\theta\cos(\varphi-\psi)+z\cos\theta]\}\sin\theta\,\mathrm{d}\theta\,\mathrm{d}\varphi, \tag{4.16}$$

where $\mathbf{U}(\rho,\psi,z)$ is the electric or magnetic field at the focus, $B(\theta,\varphi)$ is the amplitude of the electric or magnetic field in the entrance pupil of a lens with high numerical aperture (θ is the polar angle, φ is the azimuthal angle), $T(\theta)$ is the lens apodization function, f is the focal length, $k = 2\pi/\lambda$ is the wavenumber, λ is the wavelength (in the simulation it was assumed to be equal to 532 nm), α is the maximum polar angle determined by the numerical aperture of the lens (NA = sin α), $\mathbf{P}(\theta,\varphi)$ is the polarization vector for the electric field, which has the form:

$$\mathbf{P}(\theta,\varphi) = \begin{bmatrix} 1+\cos^2\varphi(\cos\theta-1) \\ \sin\varphi\cos\varphi(\cos\theta-1) \\ -\sin\theta\cos\varphi \end{bmatrix} a(\theta,\varphi) + \\ \begin{bmatrix} \sin\varphi\cos\varphi(\cos\theta-1) \\ 1+\sin^2\varphi(\cos\theta-1) \\ -\sin\theta\sin\varphi \end{bmatrix} b(\theta,\varphi), \tag{4.17}$$

where $a(\theta,\varphi)$ and $b(\theta,\varphi)$ are functions describing the state of polarization of the x- and y-components of the initial beams. The circular polarization investigated in this section is equal to:

$$\mathbf{E} = \begin{pmatrix} a(\theta,\varphi) \\ b(\theta,\varphi) \end{pmatrix} = \frac{e^{im\varphi}}{\sqrt{2}}\begin{pmatrix} 1 \\ \sigma i \end{pmatrix}, \tag{4.18}$$

where σ specifies the direction of the input polarization: at $\sigma = 1$ the polarization will be right circular, at $\sigma = -1$ it is left circular, and at $\sigma = 0$ it is linear. In the Ref. [21] it was shown that by substituting (4.18) into (4.16) it is possible to integrate over the azimuthal angle and obtain the values of the electric and magnetic field at the focus:

$$E_x = \frac{i^{m-1}}{\sqrt{2}} e^{im\varphi} \left(I_{0,m} + \gamma_+ e^{i2\varphi} I_{2,m+2} + \gamma_- e^{-i2\varphi} I_{2,m-2} \right),$$
$$E_y = \frac{i^{m}}{\sqrt{2}} e^{im\varphi} \left(\sigma I_{0,m} - \gamma_+ e^{i2\varphi} I_{2,m+2} + \gamma_- e^{-i2\varphi} I_{2,m-2} \right), \tag{4.19}$$
$$E_z = -\frac{2i^{m}}{\sqrt{2}} e^{im\varphi} \left(\gamma_+ e^{i\varphi} I_{1,m+1} - \gamma_- e^{-i\varphi} I_{1,m-1} \right),$$

$$H_x = \frac{-i^{m}}{\sqrt{2}} e^{im\varphi} \left(\sigma I_{0,m} + \gamma_+ e^{i2\varphi} I_{2,m+2} - \gamma_- e^{-i2\varphi} I_{2,m-2} \right),$$
$$H_y = \frac{i^{m-1}}{\sqrt{2}} e^{im\varphi} \left(I_{0,m} - \gamma_+ e^{i2\varphi} I_{2,m+2} - \gamma_- e^{-i2\varphi} I_{2,m-2} \right), \tag{4.20}$$
$$H_z = \frac{2i^{m+1}}{\sqrt{2}} e^{im\varphi} \left(\gamma_+ e^{i\varphi} I_{1,m+1} + \gamma_- e^{-i\varphi} I_{1,m-1} \right),$$

where $\gamma_\pm = (1 \pm \sigma)/2$ for circular polarization and $\gamma_+ = \gamma_- = 1/\sqrt{2}$ for linear polarization (i.e., when $\sigma = 0$). In (4.19) and (4.20) the following integrals are used:

$$I_{0,\nu} = \left(\frac{\pi f}{\lambda}\right) \int_0^\alpha \sin\theta \cos^{1/2}\theta (1 + \cos\theta) A(\theta) e^{ikz\cos\theta} J_\nu(x) d\theta,$$
$$I_{1,\nu} = \left(\frac{\pi f}{\lambda}\right) \int_0^\alpha \sin^2\theta \cos^{1/2}\theta A(\theta) e^{ikz\cos\theta} J_\nu(x) d\theta, \tag{4.21}$$
$$I_{2,\nu} = \left(\frac{\pi f}{\lambda}\right) \int_0^\alpha \sin\theta \cos^{1/2}\theta (1 - \cos\theta) A(\theta) e^{ikz\cos\theta} J_\nu(x) d\theta,$$

where $x = kr\sin\theta$, (x, y, z) and (r, φ, z) are Cartesian and polar coordinates. In (4.21) the integrals depend only on the radial variable. From (4.19) it follows that individual transverse components of the electric field E_x and E_y are the sum of three optical vortices with topological charges m, $m + 2$ and $m - 2$. From (4.19) and (4.20) it could be obtained equation for the longitudinal component of the Poynting vector (energy flow) [52]:

$$\mathbf{S} = [c/(8\pi)] \ \mathrm{Re}\,[\mathbf{E} \times \mathbf{H}^*], \tag{4.22}$$

where c is the speed of light in vacuum, Re() is the real part of the number, $\mathbf{E} \times \mathbf{H}$ is the vector product, * is the complex conjugation (we omit the constant $c/(8\pi)$ below). We obtain in the focal plane:

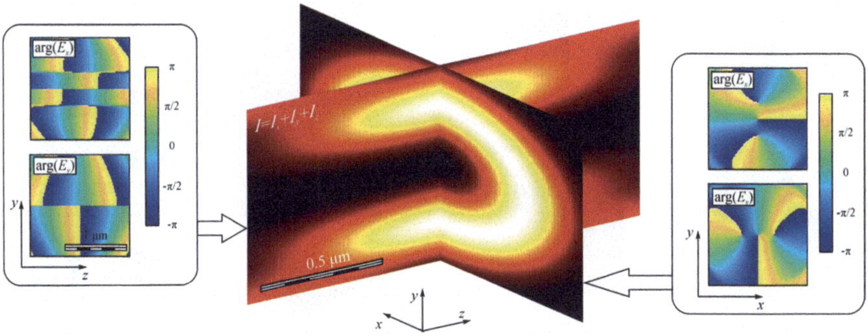

Fig. 4.12 Phase of the transverse components of the electric field E_x and E_y and intensity in the longitudinal and transverse planes when focusing an optical vortex with right-handed circular polarization and topological charge equals to minus three (the directions of rotation of the polarization and vortex phase are different)

$$S_z = \left(\frac{1+\sigma^2}{2}\right) I_{0,m}^2 - \gamma_+^2 I_{2,m+2}^2 - \gamma_-^2 I_{2,m-2}^2 \tag{4.23}$$

Our previous paper [14] demonstrated that in tight focusing of circularly polarized optical vortices, regions with a Poynting vector opposing the beam propagation direction emerge only when the handedness of the circular polarization contrasts with the vortex phase rotation direction. Conversely, no reverse energy flow is observed at the focus when these rotation directions coincide. We now investigate the behavior of the phase and local wave vector under tight focusing of circularly polarized optical vortices. Using Eq. (4.16) we simulate the focusing of an optical vortex (4.18) with $\lambda = 633$ nm by an aplanatic lens (NA = 0.95). Figure 4.12 presents the intensity and phase distributions for an optical vortex with opposing rotation directions: topological charge $m = -3$ and right-handed circular polarization (RCP). Figure 4.13 shows the corresponding distributions for the same rotation directions: RCP with $m = +3$.

It is evident from Figs. 4.12 and 4.13 that in the first case, three centers of optical vortices are observed for the transverse components of the electric field. In the second case, one center of the optical vortex is observed on the optical axis. In the first case, regions of reverse energy flow are observed at the focus, but not in the second case. The question arises: why, in the case when the orbital and spin moments of the initial field are subtracted, as $m + \sigma = 3–1 = 2$ in Fig. 4.12, does the optical vortex at the focus disintegrate? And in the case when the orbital and spin moments of the initial field are added, as $m + \sigma = 3 + 1 = 4$ in Fig. 4.13, does the optical vortex at the focus remain? Let us write down the transverse projections of the electric field for $m = 3$ and $\sigma = -1$ based on 3:

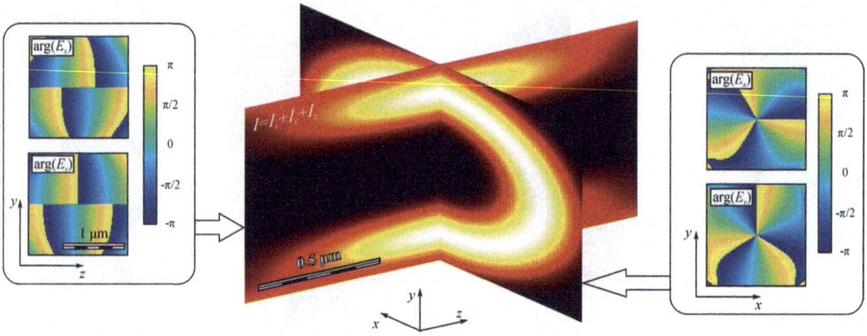

Fig. 4.13 Phase of the transverse components of the electric field and intensity in the longitudinal and transverse planes when focusing an optical vortex with right-handed circular polarization and topological charge equals to three (the directions of rotation of the polarization and vortex phase coincide)

$$\begin{aligned} E_x &= \frac{-1}{\sqrt{2}}\left(e^{i3\varphi} I_{0,3} + e^{i\varphi} I_{2,1}\right), \\ E_y &= \frac{-i}{\sqrt{2}}\left(-e^{i3\varphi} I_{0,3} + e^{i\varphi} I_{2,1}\right), \end{aligned} \tag{4.24}$$

and transverse projections of the electric field for $m = 3$ and $\sigma = 1$:

$$\begin{aligned} E_x &= \frac{-1}{\sqrt{2}}\left(e^{i3\varphi} I_{0,3} + e^{i5\varphi} I_{2,5}\right), \\ E_y &= \frac{-i}{\sqrt{2}}\left(e^{i3\varphi} I_{0,3} - e^{i5\varphi} I_{2,5}\right). \end{aligned} \tag{4.25}$$

From (4.24) it is evident that in the plane of focus near the optical axis the optical vortex with topological charge 1 will dominate, since the first-order Bessel function in the integral $I_{2,1}$ of the first order and at $kr << 1$ takes on larger values than the third-order Bessel function in the integral $I_{0,3}$. That is, the initial optical vortex with left circular polarization ($m + \sigma = 3–1 = 2$) in the focus near the optical axis formed a first-order optical vortex with right circular polarization ($m + \sigma = 1 + 1 = 2$). In this case, the angular momentum of the beam is preserved. And the other two zeros of the intensity of the transverse coordinate I_x (after the decay of the optical vortex of order $m = 3$ in the focus) are located on the vertical axis, since the expression for the intensity has the form:

$$I_x = \frac{1}{2}\left(I_{0,3}^2 + I_{2,1}^2 + 2 I_{0,3} I_{2,1} \cos 2\varphi\right). \tag{4.26}$$

From (4.26) it is evident that there will be two zeros of intensity at $\varphi = \pi/2$ and $\varphi = -\pi/2$. From (4.25) it is evident that the optical vortex with topological charge $m = 3$ will dominate in the focal plane, since the third-order Bessel function in the

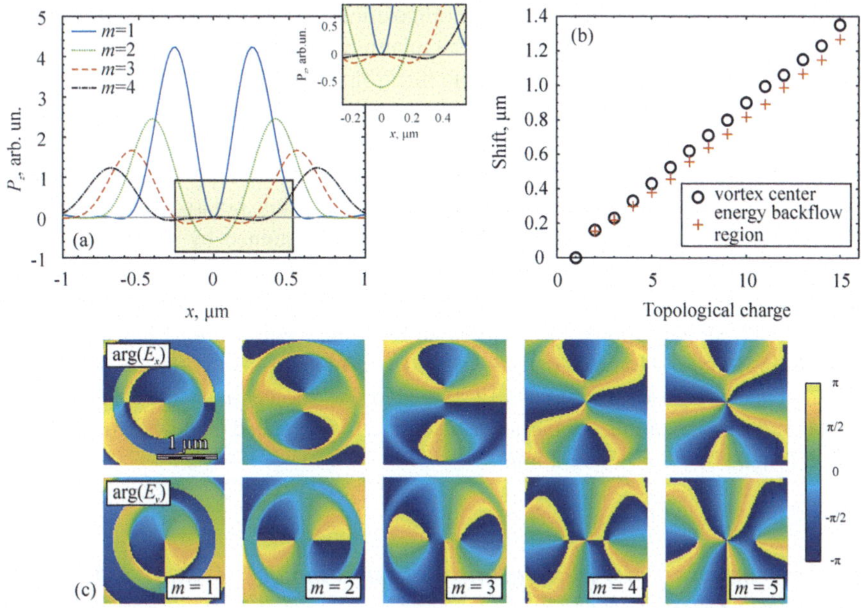

Fig. 4.14 Section of the longitudinal component of the Poynting vector (**a**), shift of the centers of optical vortices (**b**) (crosses indicate the centers of optical vortices, circles indicate the region of reverse energy flow) and distribution of the phases of the electric field (**c**) for different topological charges of the optical vortex

integral $I_{0.3}$ near the optical axis has larger values than the fifth-order Bessel function in the integral $I_{2.5}$. That is, the initial optical vortex with angular momentum ($m + \sigma = 3 + 1 = 4$) in the focus near the optical axis will be preserved.

Figure 4.14 shows the cross-sections of the longitudinal component of the Poynting vector (Fig. 4.14a) for different topological charges of the vortices, as well as the dependence of the displacement of the vortex centers (Fig. 4.14b) on the topological charge (the centers of the optical vortices are shown by crosses, the beginning of the region of the reverse energy flow is shown by circles). The centers of the vortices (or vortex dislocations) in Figs. 4.12 and 4.13 correspond to the region with a giant local wave vector [2, 29, 44].

It is evident from Fig. 4.14 that the regions of reverse flows are located between the centers of optical vortices. When the topological charge is unity, no reverse flow is observed. When the initial focused vortex has a topological charge equal to two, a splitting of the initial vortex into two vortices with unit topological charges is observed at the focus, between which the region of reverse energy flow is located (it is located on the axis). If the topological charge exceeds two, then three vortices are observed; the region of reverse energy flow is also located between them, and now has the form of a tube. With a further increase in the topological charge of the vortex, the region of reverse energy flow widens, while the centers of the optical vortices shift further from the optical axis.

It should also be noted that vortex decomposition leads to a decrease in the orbital angular momentum, because the orbital angular momentum of a single vortex with a topological charge n is always larger than the orbital angular momentum of the sum of vortices with topological charges less than n. Thus, different handedness of polarization and phase rotation lead to the decomposition of the initial vortex (Figs. 4.12 and 4.14) and, as a consequence, to a spin-orbit interaction in which the orbital angular momentum is less than for coinciding polarization and phase directions. This fact has previously been repeatedly obtained by various research groups, including experimental investigations. For example, in [41, 42], scanning near-field optical microscopy was used to show that focal spots formed by optical vortices with single topological charges and different polarizations have different sizes. In [53], the motion of a particle in the focus of an optical vortex with a single topological charge was experimentally studied; it was shown that in the case of right-hand circular polarization, the particle rotated at a higher speed than when using light with left-hand circular polarization.

4.2.2 Focusing of Polarization Optical Vortices (Cylindrical Vector Beams)

It was shown above that, depending on the direction of rotation of the circular polarization, the reverse energy flow at the focus will either be observed or absent. Linear polarization corresponds to an intermediate state—the reverse energy flow at the focus is observed, but not so great [21]. Previously, the behavior of the phase in tightly focused optical vortices with linear polarization was already investigated by us in [15]. Unlike circular polarization, in which two transverse components contribute to the formation of the focal spot, in a linearly polarized optical vortex only one transverse component makes a significant contribution (the influence of longitudinal components is not considered in this work). Just as when focusing an optical vortex with a circular polarization direction opposite to the direction of rotation of the vortex phase, in a linearly polarized optical vortex, a splitting of one vortex into three is observed.

Cylindrical vector beams are a superposition of two circularly polarized phase optical vortices. The polarization in the beam cross-section is linear, but the polarization direction changes depending on the azimuthal angle. In [50], we have demostrated that a polarization optical vortex also has a zone with a reverse energy flow at the focus, but unlike a classical phase vortex, the flow at the focus is not rotated in the transverse plane. The cylindrical vector beam is described by equation:

$$\mathbf{E}=\begin{pmatrix} a(\theta,\varphi) \\ b(\theta,\varphi) \end{pmatrix}=\begin{pmatrix} \cos m\varphi \\ \sin m\varphi \end{pmatrix}=\frac{1}{2}e^{im\varphi}\begin{pmatrix} 1 \\ -i \end{pmatrix}+\frac{1}{2}e^{-im\varphi}\begin{pmatrix} 1 \\ i \end{pmatrix}, \tag{4.27}$$

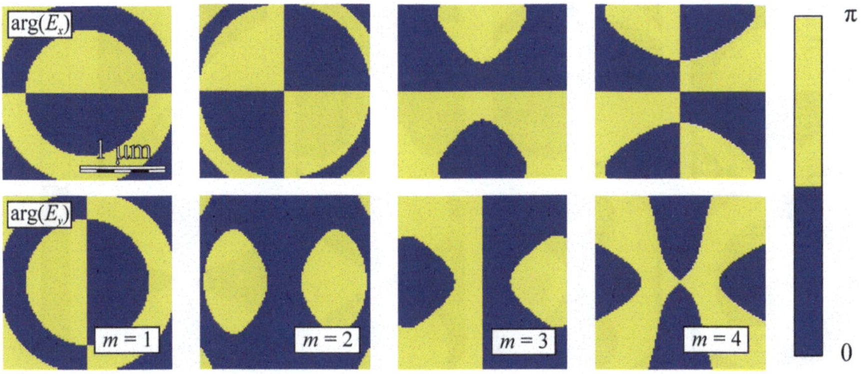

Fig. 4.15 Distribution of phase E_x (**a**, **b**, **c**, **d**) and E_y (**e**, **f**, **g**, **h**) at the focus during focusing of cylindrical vector beams with orders m = 1 (**a**, **e**), 2 (**b**, **f**), 3 (**c**, **g**), 4 (**d**, **h**)

where m is the beam order. Figure 4.15 shows the phase distribution at the focus when focusing cylindrical vector beams of different orders $m = 1, 2, 3$, and 4. As for the simulations above, it was assumed that light with a wavelength of 633 nm is focused by an aplanatic objective with a numerical aperture of NA = 0.95.

It is evident from Fig. 4.15 that in this case there are no vortex dislocations—only edge dislocations. As a consequence, gigantic values of local wave vectors will not be observed. It follows from Fig. 4.15 that the regions of reverse flows do not necessarily have to be accompanied by phase optical vortices. In this case, the displacement of the reverse flow region is accompanied by the displacement of edge dislocations.

Finally, we investigated the behavior of anti-CVB in the focal plane. Anti-CVB are CVB of negative orders, i.e. for $m \to -m$:

$$\mathbf{E} = \begin{pmatrix} a(\theta, \varphi) \\ b(\theta, \varphi) \end{pmatrix} = \begin{pmatrix} \cos m\varphi \\ -\sin m\varphi \end{pmatrix} = \frac{1}{2} e^{im\varphi} \begin{pmatrix} 1 \\ i \end{pmatrix} + \frac{1}{2} e^{-im\varphi} \begin{pmatrix} 1 \\ -i \end{pmatrix}, \tag{4.28}$$

From (4.28) it follows that the direction of rotation of polarization and phase in each of the vortices are the same. Therefore, it can be expected that no vortex splitting will be observed. Figure 4.16 shows the phase distributions for the beam (4.28) of different orders.

There is no vortex splitting observed in Fig. 4.16, as well as there is no reverse energy flow in this case. From the first expression in Eq. (4.19) it follows that at the focus the transverse projection of the electric vector can be expressed as a superposition of two optical vortices with topological charges $m > 0$ and $m - 2$ for light with left circular polarization:

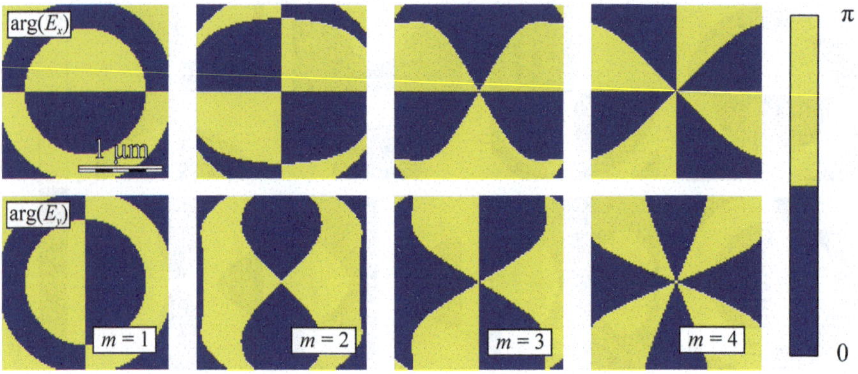

Fig. 4.16 Distribution of phase E_x (**a**, **b**, **c**, **d**) and E_y (**e**, **f**, **g**, **h**) at the focus during focusing of anti-CVB with orders $m = 1$ (**a**, **e**), 2 (**b**, **f**), 3 (**c**, **g**), 4 (**d**, **h**)

$$E_x(r,\varphi) = \frac{i^{m-1}}{\sqrt{2}}\left(e^{im\varphi}I_{0,m}(r) + e^{i(m-2)\varphi}I_{2,m-2}(r)\right). \tag{4.29}$$

At a given radius r, both functions in Eq. (4.21) are constant, and the topological charge of the entire superposition equals to the topological charge of the optical vortex with the larger coefficient [54]. Near the optical axis, the magnitude of the function $I_{2,m-2}$ exceeds the magnitude of the function $I_{0,m}$, since the Bessel functions in the integrals in Eq. (4.21) grow at small r as r^{m-2} and r^m. Therefore, at the center of the focal plane there is an optical vortex with topological charge of $m - 2$. At larger radii r, on the contrary, the magnitude of $I_{0,m}$ is larger than the magnitude of $I_{2,m-2}$, and the topological charge of the entire superposition (4.29) is equal to m. This is possible if at some radius r there are two optical vortices with topological charges of 1. These two vortices are formed at a distance r, at which $|I_{0,m}(r)| = |I_{0,m-2}(r)|$. Moreover, this radius r, at which these two optical vortices appear, increases with the topological charge m, as illustrated in Fig. 4.15. For an optical vortex with right-hand circular polarization, the expression derived from Eq. (4.19) differs from Eq. (4.29):

$$E_x(r,\varphi) = \frac{i^{m-1}}{\sqrt{2}}\left(e^{im\varphi}I_{0,m}(r) + e^{i(m+2)\varphi}I_{2,m+2}(r)\right). \tag{4.30}$$

From Eq. (4.30), it follows that this projection of the electric field vector is a superposition of two optical vortices with topological charges m and $m + 2$. Similarly to the previous case, it can be stated that near the optical axis, the topological charge of the entire superposition is equal to m, since at small r, the m-th order Bessel function grows faster than the $(m + 2)$-th order Bessel function in Eq. (4.21). Consequently, splitting of the initial optical vortex with topological charge m at the focus does not occur. At large r, the magnitude of $I_{0,m}$ exceeds the magnitude of $I_{2,m+2}$, and the topological charge of the entire superposition in Eq. (4.30) remains.

In this section we investigated tight focusing of optical vortices, for which areas with the Poynting vector directed in the direction opposite to the beam propagation direction are observed in the focal plane, i.e. areas of reverse energy flow [55]. Focusing of phase optical vortices, as well as polarization optical vortices (cylindrical vector beams) was theoretically and numerically investigated using Richards-Wolf formalism. It is shown that when focusing optical vortices in which the direction of rotation of polarization and phase are different, splitting of the initial focused vortex into several optical vortices and the appearance of a zone with a reverse energy flow are observed. Moreover, the regions with a reverse energy flow are located between the centers of displaced phase singularities (vortex dislocations). With an increase in the topological charge of the vortex, the region of the reverse flow along the radial coordinate widens, and with it the distance from the centers of the vortices to the optical axis increases. If the directions of rotation of the phase and polarization in the optical vortex coincide, then neither splitting of the vortex nor a reverse flow at the focus is observed. When focusing cylindrical vector beams, the area of reverse flow is observed, but vortex dislocations are absent. In this case, the displacement of the reverse flow region is accompanied by the displacement of edge dislocations.

4.3 Canonical Energy Flow of the Electric Hertzian Dipole

Distribution of the electric field of a point or linear electric dipole in the far field is well-known for a long time [56, 57]. The dipole, as a model of a point emitter, is widely employed in nowadays works [58–64] for designing nanostructures, like emitters [65], switches [66] and detectors [67]. However, in all these works, the dipole irradiance is considered only in the far-field. This is because the Umov-Poynting vector of a point dipole in the spherical coordinates has only radial component both in the near and in the far field. Therefore, there was no need in additional studies of the near field of the electric dipole. It is however known that the Umov-Poynting vector is represented as a sum of the orbital and spin constituents [1, 47, 68, 69]. These works justify that the true energy flow should be related with the vector of the canonical (or orbital) energy flow, rather than with the Umov-Poynting vector. In a recent paper [70], H.-I. Lee was probably the first who investigated the orbital energy flow and the spin flow in the near field of an electric dipole. In [70], explicit analytical expressions were obtained for the components of the spin flow vector of a linear dipole. It was also shown in [70] that in the near field of a dipole, backward canonical energy flow arises. However, [70] does not contain analytical expressions for the components of the canonical energy flow vector in the near field of a linear dipole.

In this section, we derive explicit analytical expressions for the radial and meridional component of the canonical energy flow vector in the near field of a linear electric dipole. We demonstrate theoretically that at a subwavelength distance from the dipole, backward canonical energy flow arises. We also define the shape of a curve that bounds the backflow area. In any point on this curve, the energy flow is zero. In one half of this backflow area, energy circulates clockwise, while in the other half—counterclockwise. In addition, we obtain in this work explicit expressions for the azimuthal components of the spin angular momentum and orbital angular momentum vectors of the dipole.

4.3.1 Canonical Energy Flow in the Near Field of a Dipole

We consider here an electric dipole with the dipole moment Il, where I is the dipole current and l is the dipole length. The dipole is oriented along the axis z and the angular frequency of the dipole oscillations is ω. Then, in the spherical coordinates (r, θ, φ), the dipole radiation has well-known complex amplitudes [57]:

$$\begin{aligned} E_r(r,\theta) &= \frac{-iIlk^3 e^{-ikr}\cos\theta}{2\pi\omega\varepsilon\varepsilon_0}\left(\frac{1}{(kr)^3}+\frac{i}{(kr)^2}\right), \\ E_\theta(r,\theta) &= \frac{-iIlk^3 e^{-ikr}\sin\theta}{4\pi\omega\varepsilon\varepsilon_0}\left(\frac{1}{(kr)^3}+\frac{i}{(kr)^2}-\frac{1}{(kr)}\right), \\ H_\varphi(r,\theta) &= \frac{Ile^{-ikr}k^2\sin\theta}{4\pi(kr)}\left(\frac{1}{(kr)}+i\right). \end{aligned} \tag{4.31}$$

In Eq. (4.31), ε and ε_0 are respectively the dielectric permittivity of medium and vacuum permittivity, k is the wavenumber, $k=(\omega/c)\sqrt{\varepsilon\mu}$, c is the speed of light in vacuum, μ is the magnetic permeability of medium. Figure 4.17 illustrates the dipole location and orientation in the Cartesian coordinates, as well as the spherical coordinates and directions of the electromagnetic field components. The Umov-Poynting vector $\mathbf{S} = (1/2)\,\mathrm{Re}\,(\mathbf{E} \times \mathbf{H}^*)$ for the dipole (1) is well-known [57] and has only the radial nonzero coordinate:

$$S_r = \frac{I^2 l^2 k^5 \sin^2\theta}{32\pi^2\omega\varepsilon\varepsilon_0 (kr)^2}. \tag{4.32}$$

As seen from Eq. (4.32), maximal energy flow is directed perpendicularly to the dipole (i.e. at $\theta = \pi/2$), equally for all polar angles φ.

Below we consider distribution of the canonical energy flow of the dipole (4.31) and demonstrate that, in the near field, backward energy flow arises. Up to insignificant constants, the Umov-Poynting vector $\mathbf{S}$ is the sum of two vectors [1, 68–70]: orbital energy flow vector (canonical energy flow) $\mathbf{S}_{or}$ and spin flow $\mathbf{S}_{sp}$:

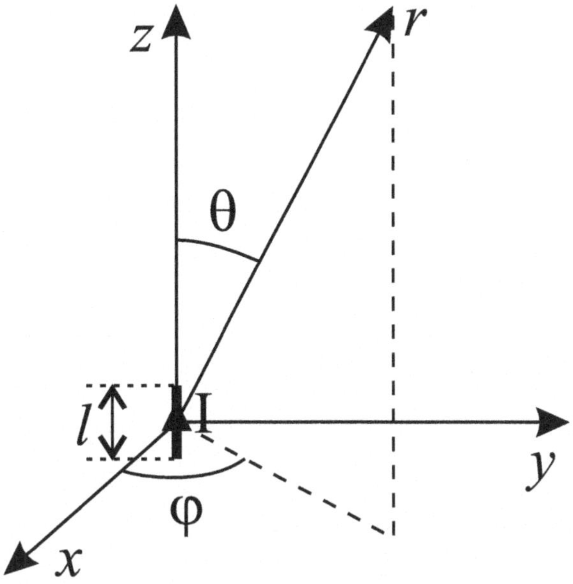

Fig. 4.17 Dipole configuration under study: the electric dipole with the dipole current I and with the dipole length l is located in the origin and oriented along the vertical axis z. In addition to the Cartesian coordinates (x, y, z), spherical coordinates (r, θ, φ) are shown

$$\begin{aligned}&\mathbf{S}=\mathbf{S}_{or}+\mathbf{S}_{sp}=\frac{1}{k}\mathrm{Im}(\mathbf{E}^{*}\cdot(\nabla)\mathbf{E})+\frac{1}{2k}(\nabla\times\overline{\mathbf{S}}),\\&\overline{\mathbf{S}}=\mathrm{Im}(\mathbf{E}^{*}\times\mathbf{E}).\end{aligned}\tag{4.33}$$

In Eq. (4.33), $\overline{\mathbf{S}}$ stands for the spin angular momentum vector (up to a constant multiplier). In the spherical coordinates, the canonical energy flow vector $\boldsymbol{S}_{or}$ for the dipole (4.31) has only two nonzero components, radial and meridional:

$$\begin{aligned}S_{or,r}&=\frac{\omega\varepsilon_0\varepsilon}{2k^2}\mathrm{Im}\left(E_r\frac{\partial}{\partial r}E_r^*+E_\theta\frac{\partial}{\partial r}E_\theta^*\right),\\S_{or,\theta}&=\frac{\omega\varepsilon_0\varepsilon}{2k^2 r}\mathrm{Im}\left(E_r\frac{\partial}{\partial\theta}E_r^*+E_\theta\frac{\partial}{\partial\theta}E_\theta^*-E_rE_\theta^*+E_r^*E_\theta\right).\end{aligned}\tag{4.34}$$

Substituting the components of the strength vector of the electric field from Eq. (4.31) into expressions (4.34), we obtain that only the radial component of the canonical energy flow is nonzero:

$$\begin{aligned}S_{or,r}&=\left(\frac{I^2l^2}{8\pi^2\omega\varepsilon\varepsilon_0}\right)\frac{k^5}{(kr)^2}\left(\frac{1}{(kr)^2}-\frac{3\sin^2\theta}{2(kr)^2}+\frac{\sin^2\theta}{4}\right),\\S_{or,\theta}&=\left(\frac{I^2l^2}{16\pi^2\omega\varepsilon\varepsilon_0}\right)\frac{k^5\sin(2\theta)}{(kr)^4}.\end{aligned}\tag{4.35}$$

In the far field, when $kr >> 1$, only the last term remains in the expression (4.35):

$$S_{or,r} \approx \left(\frac{Il}{2\pi\omega\varepsilon\varepsilon_0}\right)^2 \frac{k^6 \sin^2\theta}{4(kr)^2}, \quad kr >> 1. \tag{4.36}$$

In the near field, at $kr << 1$, on the contrary, the first two terms in Eq. (4.35) are overwhelming:

$$S_{or,r} \approx \left(\frac{Il}{2\pi\omega\varepsilon\varepsilon_0}\right)^2 \frac{k^6}{(kr)^4}\left(1 - \frac{3\sin^2\theta}{2}\right). \tag{4.37}$$

It is seen from Eq. (4.37) that if $\sqrt{2/3} < |\sin\theta| < 1$ then the canonical energy flow is negative, i.e. $S_{or,r} < 0$. At angles $0 < \theta < 55°$, the energy flow goes away from the dipole, while at angles $55° < \theta < 125°$, the energy flow returns back to the dipole. At angles $125° < \theta < 180°$, the energy flow also goes away from the dipole. Equation (4.35) indicates that on the horizontal axis ($\theta = 90°$), at a distance $kr = \sqrt{2}$ from the dipole, a critical point exist. Before this pint, the energy flow is negative, but beyond this point it becomes positive. Making the right part of Eq. (4.35) equal to zero, we derive an expression for coordinates (r, θ) of points with zero radial canonical energy flow:

$$|\sin\theta| = \frac{2}{\sqrt{6-(kr)^2}} < 1, \quad kr < \sqrt{2}, \quad r < \frac{\lambda}{4.4}. \tag{4.38}$$

According to Eq. (4.38), the points of zero radial energy flow can exist only at $kr \leq \sqrt{2}$. For instance, at $kr = 1$, two such points exist with the coordinates $\theta = \arcsin(2/\sqrt{5}) = 63°$ and $\theta = 180°$—$\arcsin(2/\sqrt{5}) = 117°$. The backflow area in the coordinates (r, θ) is bounded by the curve (4.38). This area of energy backflow is located within the angles $55° < \theta < 125°$ and bounded by the maximal radius $r = \sqrt{2}/k$. As seen from the last inequality (4.38), this is a subwavelength area with the size being a little smaller than the quarter of the wavelength λ. We note that the energy backflow of a linear dipole was found numerically in work [70]. However, no explicit analytical expressions were obtained in this work for the canonical flow (4.35) and (4.37). It is also interesting that the radial component of the Umov-Poynting vector has the same expression (4.32) both in the near and in the far field. At the same time, in the near field, the radial component of the Umov-Poynting vector is significantly different from the radial component of the canonical energy flow (4.35). This is because the linear dipole in the near field has the spin $\overline{\mathbf{S}}$ (3). Indeed, using Eq. (4.33), we can obtain the spin angular momentum components for the linear dipole (4.31). In the spherical coordinates, vector $\overline{\mathbf{S}}$ has only a single polar (or azimuthal) component:

$$\overline{S}_\varphi = \mathrm{Im}(\mathbf{E}\times\mathbf{E}^*)|_\varphi = 2\mathrm{Im}\left(E_r E_\varphi^*\right) = \\ -\left(\frac{Il}{2\pi\omega\varepsilon\varepsilon_0}\right)^2 \frac{\sin(2\theta)k^6}{2(kr)^3}. \tag{4.39}$$

Now we find the components of the spin flow vector $\mathbf{S}_{sp} = (2k)^{-1}\left(\nabla\times\overline{\mathbf{S}}\right)$ for the dipole (4.31). Using Eqs. (4.33) in the spherical coordinates, we get:

$$\begin{aligned} S_{sp,r} &= \frac{1}{2kr\sin}\frac{\partial}{\partial\theta}\left(\sin\theta\overline{S}_\varphi\right) \\ &= -\left(\frac{Il}{2\pi\omega\varepsilon\varepsilon_0}\right)^2 \frac{k^6}{(kr)^4}\left(1-\frac{3}{2}\sin^2\theta\right), \\ S_{sp,\theta} &= -\frac{1}{2kr}\frac{\partial}{\partial r}\left(r\overline{S}_\varphi\right) \\ &= -\left(\frac{Il}{2\pi\omega\varepsilon\varepsilon_0}\right)^2 \frac{k^6\sin 2\theta}{(kr)^4}. \end{aligned} \tag{4.40}$$

Equations (4.40) allow drawing two interesting conclusions. First, it is clear why the Umov-Poynting vector both in the near and in the far field has the same form (4.32) and does not indicate the presence of energy backflow in the near field. This is because in the near field of the linear dipole (4.31), distributions of the radial components of the canonical energy flow vector (4.36) and of the spin flow vector (4.40) compensate each other:

$$S_{sp,r} + S_{or,r} = \left(\frac{Il}{2\pi\omega\varepsilon\varepsilon_0}\right)^2 \frac{k^6\sin^2\theta}{4(kr)^2} = S_r. \tag{4.41}$$

Second, as seen from the second Eq. (4.40), the spin flow vector in the near field has a nonzero meridional component $S_{sp,\theta}$, which is for some reason absent in the Umov-Poynting vector (4.32). This contradiction can be resolved the following way. Since the sum of the orbital energy flow and of the spin flow should be equal to the Umov-Poynting vector (4.33) and the latter has zero meridional component (4.32), then the sum of the meridional components of the spin flow and of the orbital energy flow should be zero:

$$S_{or,\theta} = -S_{sp,\theta} = \left(\frac{I^2l^2}{16\pi^2\omega\varepsilon\varepsilon_0}\right)\frac{k^5\sin 2\theta}{(kr)^4} \tag{4.42}$$

although the meridional component of the canonical energy flow (4.42) cannot be obtained directly from the field components (4.31), since this yields zero (4.35). Thus, the canonical energy flow in the near field of the linear dipole has both the radial and the meridional components:

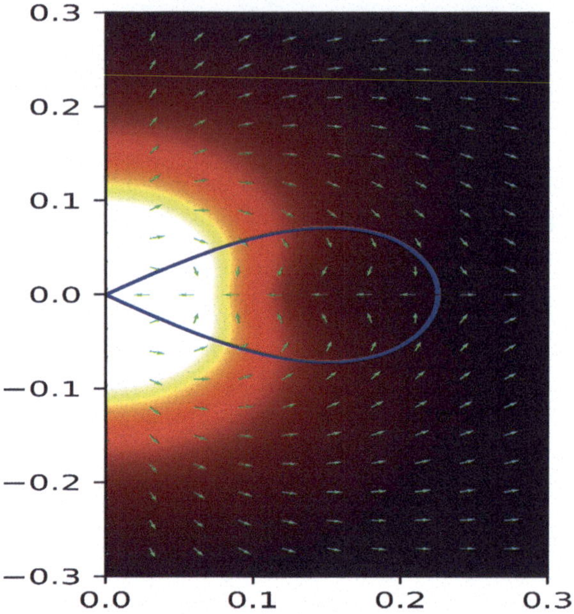

Fig. 4.18 Distribution of the amplitude (colors) and of the canonical energy flow vector (4.35) (4.43) (arrows) in the near field of the linear dipole (4.31) in the zx-plane ($\varphi = \pi/2$). The coordinates along the axes are given in wavelength units (wavelength is $\lambda = 532$ nm). Computation area is $0 \le x \le 0.3\lambda$, −0/ $3\lambda \le z \le 0.3\lambda$. The blue curve is the curve (4.38) of zero radial energy flow. The same pattern would be for any other azimuthal angle φ

$$\begin{aligned} S_{or,r} &= \left(\frac{I^2 l^2}{8\pi^2 \omega \varepsilon \varepsilon_0}\right) \frac{k^5}{(kr)^4} \left(1 - \frac{3\sin^2\theta}{2}\right), \\ S_{or,\theta} &= \left(\frac{I^2 l^2}{16\pi^2 \omega \varepsilon \varepsilon_0}\right) \frac{k^5 \sin 2\theta}{(kr)^4}. \end{aligned} \tag{4.43}$$

Since the sine of the double angle in Eq. (4.43) is negative at $\theta > 90°$, the energy flow at $55° < \theta < 90°$ circulates clockwise, while at angles $90° < \theta < 125°$—counterclockwise. However, this energy flow does not circulate around the singular points, where the flow is zero, as stated in [70]. Rather, it behaves a different way. Since the points with zero radial energy flow reside on the curve (4.38), lines of the energy flow outgo from the origin (forward flow from the dipole), then they bend and reach the points with zero radial energy flow. After this point, energy flow lines go back to the dipole, tangent to the horizontal straight line. This is seen in Fig. 4.18. It is interesting that the dipole has nonzero orbital angular momentum density $\mathbf{L} = \mathbf{r} \times \mathbf{P}_{or}$. The only nonzero component of the orbital angular momentum of the dipole in the near field is the azimuthal component:

$$L_\varphi = r S_{or,\theta} = \left(\frac{I^2 l^2}{16\pi^2 \omega \varepsilon \varepsilon_0}\right) \frac{k^4 \sin 2\theta}{(kr)^3}. \tag{4.44}$$

Comparison of Eqs. (4.39) and (4.44) reveals that the azimuthal components of the spin angular momentum and orbital angular momentum vectors have mutually opposite directions. When passing from $z > 0$ to $z < 0$, both vectors change their direction to opposite one.

4.3.2 Numerical Simulation

Figure 4.18 illustrates distribution of amplitude (in pseudocolors), while arrows indicate directions of the canonical energy flow in the near field, computed by the formulae (4.35) and (4.43) for the linear dipole (4.31) in the zx-plane ($\varphi = \pi/2$). Additionally Fig. 4.19 shows by lines and arrows the direction of canonical energy flow of dipole in near field (4.43). As seen in Figs. 4.18 and 4.19, the energy flow bends clockwise at $z > 0$ and counterclockwise at $z < 0$. On the blue curve (4.38), the radial component of the energy flow is zero, and inside this curve the backward energy flow arises. The blue curve in Fig. 4.18 reminds a lasso. Some part of light is "trapped" into this lasso and returns back to the dipole. Other part of light from the dipole, "not trapped", bypasses this lasso and goes away to infinity, and, the farther from the origin, the energy flow becomes more and more radial (as seen in Fig. 4.20). The energy flow lines no longer bend and the light no longer carry the orbital angular momentum. The energy backflow arises in the near field of the dipole for the following reason. The electric and magnetic strength vectors of a plane wave oscillate in phase and thus, in any moment of time, the Umov-Poynting vector has a real value and the energy propagates in space. The magnetic vector of the dipole (4.31) is retarded from the electric vector by a quarter-period. Thus, the Umov-Poynting vector is pure imaginary (reactive). This means that during a one quarter of the period, energy emits from the dipole, while during another quarter of the period, the energy returns back to the dipole [56, 57]. Therefore, the Umov-Poynting vector in the near field, averaged over the period, has the backward flow (4.35).

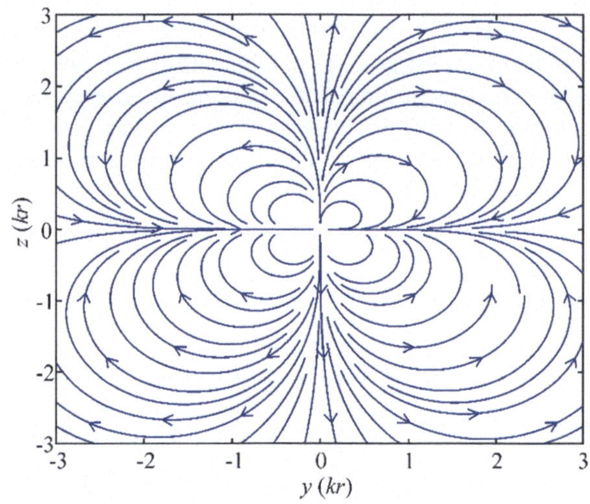

Fig. 4.19 The direction of canonical energy flow of dipole (4.31) in near field, calculated using Eq. (4.40)

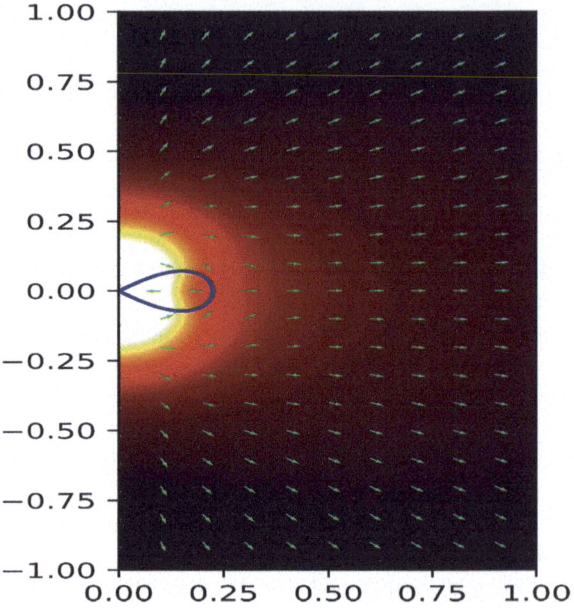

Fig. 4.20 The same as Fig. 4.18, but for a larger area: $0 \le x \le \lambda$, $-\lambda \le z \le \lambda$. The blue curve is the curve (4.38) of zero radial energy flow. The same pattern would be for any other azimuthal angle φ

In this section, we have obtained the following results. It was demonstrated theoretically for a linear dipole, oriented along the vertical axis z (Fig. 4.17), that the lines of the canonical energy flow (4.35), outgoing from the origin along the radial coordinate at an angle exceeding 55°, bend due to the meridional component of the canonical flow (4.43), intersect the curve (4.38) of zero radial energy flow, and then return back to the dipole into the origin (backward energy flow). Such a curvilinear energy motion generates the azimuthal component of the orbital angular momentum (4.44) and of the spin angular momentum (4.40), having a different direction. Lines of the radial energy flow, outgoing from the origin by the angles, smaller than 55° and greater than 125° in the near field, also bend, but, instead of turning backward, go away to infinity (Fig. 4.20), in consistence with Eqs. (4.32) and (4.35). It is interesting that analyzing the energy flow by employing the Umov-Poynting vector (4.32) does not allow discovering that the dipole in the near field has a backward energy flow (4.35), spin angular momentum (4.39) and orbital angular momentum (4.44).

References

1. M.V. Berry, J. Opt. A Pure Appl. Opt. **11**, 094001 (2009)
2. N.I. Zheludev, G. Yuan, Nat. Rev. Phys. **4**, 16 (2021)
3. Y. Aharonov, D.Z. Albert, L. Vaidman, Phys. Rev. Lett. **60**, 1351 (1988)
4. M.V. Berry, J. Opt. **15**, 044006 (2013)
5. M.V. Berry, M.R. Dennis, J. Phys. A Math. Theor. **42**, 022003 (2009)

6. G. Chen, Z.Q. Wen, C.W. Qiu, Light Sci. Appl. **8**, 56 (2019)
7. E.T.F. Rogers, N.I. Zheludev, J. Opt. **15**, 094008 (2013)
8. A. Yu, G. Chen, Z. Zhang, Z. Wen, L. Dai, K. Zhang, S. Jiang, Z. Wu, Y. Li, C. Wang, X. Luo, Sci. Rep. **6**, 38859 (2016)
9. M. Li, W. Li, H. Li, Y. Zhu, Y. Yu, Sci. Rep. **7**, 1 (2017)
10. T. Liu, G. Li, J. Hu, K. Liu, T. He, C. Wan, J. Wu, S. Yang, Opt. Commun. **548**, 129829 (2023)
11. E.T.F. Rogers, S. Quraishe, K.S. Rogers, T.A. Newman, P.J.S. Smith, N.I. Zheludev, APL Photonics **5**, 066107 (2020)
12. G. Yuan, E.T.F. Rogers, N.I. Zheludev, Light Sci. Appl. **8**, 2 (2019)
13. S.S. Stafeev, V.V. Kotlyar, A.G. Nalimov, E.S. Kozlova, IEEE Photonics J. **11**, 4500810 (2019)
14. V.V. Kotlyar, S.S. Stafeev, A.G. Nalimov, Phys. Rev. A **99**, 033840 (2019)
15. V.V. Kotlyar, S.S. Stafeev, A.G. Nalimov, Comput. Opt. **44**, 5 (2020)
16. V.V. Kotlyar, S.S. Stafeev, V.D. Zaitsev, A.M. Telegin, Photonics **9**, 969 (2022)
17. F.S. Roux, Opt. Commun. **223**, 31 (2003)
18. Z. Lu, Y. Meng, D. Deng, Opt. Commun. **545**, 129738 (2023)
19. F. Ye, J. Zou, D. Deng, Ann. Phys. **532**, 1 (2020)
20. J. Zhuang, L. Zhang, D. Deng, Opt. Lett. **45**, 296 (2020)
21. B. Richards, E. Wolf, Proc. R. Soc. A Math. Phys. Eng. Sci. **253**, 358 (1959)
22. S.F. Pereira, A.S. van de Nes, Opt. Commun. **234**, 119 (2004)
23. K.S. Youngworth, T.G. Brown, Opt. Express **7**, 77 (2000)
24. T. Fadeyeva, Y. Egorov, A. Rubass, G.A. Swartzlander, A. Volyar, Opt. Lett. **32**, 3116 (2007)
25. M.V. Bretsko, Y.E. Akimova, A.V. Volyar, S.I. Khalilov, Y.A. Egorov, A.O. Ivakhnenko, Opt. Mem. Neural Netw. **32**, S75 (2023)
26. L. Zhu, M. Tang, H. Li, Y. Tai, X. Li, Nanophotonics **10**, 2487 (2021)
27. X. Qin, H. Zhang, M. Tang, Y. Zhou, Y. Tai, X. Li, Opt. Lett. **49**, 2213 (2024)
28. X.B. Yang, H. Zhang, M.M. Tang, H.X. Ma, Y.P. Tai, X.Z. Li, Appl. Phys. Lett. **125** (2024)
29. S.S. Stafeev, V.V. Kotlyar, Opt. Commun. **591**, 132111 (2025)
30. Y. Shen, Q. Zhan, L.G. Wright, D.N. Christodoulides, F.W. Wise, A.E. Willner, K. Zou, Z. Zhao, M.A. Porras, A. Chong, C. Wan, K.Y. Bliokh, C.-T. Liao, C. Hernández-García, M. Murnane, M. Yessenov, A.F. Abouraddy, L.J. Wong, M. Go, S. Kumar, C. Guo, S. Fan, N. Papasimakis, N.I. Zheludev, L. Chen, W. Zhu, A. Agrawal, M. Mounaix, N.K. Fontaine, J. Carpenter, S.W. Jolly, C. Dorrer, B. Alonso, I. Lopez-Quintas, M. López-Ripa, Í.J. Sola, J. Huang, H. Zhang, Z. Ruan, A.H. Dorrah, F. Capasso, A. Forbes, J. Opt. **25**, 093001 (2023)
31. H. Rubinsztein-Dunlop, A. Forbes, M.V. Berry, M.R. Dennis, D.L. Andrews, M. Mansuripur, C. Denz, C. Alpmann, P. Banzer, T. Bauer, E. Karimi, L. Marrucci, M. Padgett, M. Ritsch-Marte, N.M. Litchinitser, N.P. Bigelow, C. Rosales-Guzmán, A. Belmonte, J.P. Torres, T.W. Neely, M. Baker, R. Gordon, A.B. Stilgoe, J. Romero, A.G. White, R. Fickler, A.E. Willner, G. Xie, B. McMorran, A.M. Weiner, J. Opt. **19**, 013001 (2017)
32. K.Y. Bliokh, E. Karimi, M.J. Padgett, M.A. Alonso, M.R. Dennis, A. Dudley, A. Forbes, S. Zahedpour, S.W. Hancock, H.M. Milchberg, S. Rotter, F. Nori, Ş.K. Özdemir, N. Bender, H. Cao, P.B. Corkum, C. Hernández-García, H. Ren, Y. Kivshar, M.G. Silveirinha, N. Engheta, A. Rauschenbeutel, P. Schneeweiss, J. Volz, D. Leykam, D.A. Smirnova, K. Rong, B. Wang, E. Hasman, M.F. Picardi, A.V. Zayats, F.J. Rodríguez-Fortuño, C. Yang, J. Ren, A.B. Khanikaev, A. Alù, E. Brasselet, M. Shats, J. Verbeeck, P. Schattschneider, D. Sarenac, D.G. Cory, D.A. Pushin, M. Birk, A. Gorlach, I. Kaminer, F. Cardano, L. Marrucci, M. Krenn, F. Marquardt, J. Opt. **25**, 103001 (2023)
33. S. Fürhapter, A. Jesacher, S. Bernet, M. Ritsch-Marte, Opt. Express **13**, 689 (2005)
34. D.G. Grier, Nature **424**, 810 (2003)
35. V. D'Ambrosio, N. Spagnolo, L. Del Re, S. Slussarenko, Y. Li, L.C. Kwek, L. Marrucci, S.P. Walborn, L. Aolita, F. Sciarrino, Nat. Commun. **4**, 2432 (2013)
36. B. Wang, X. Zhang, J. Tian, B. Merabet, Z. Li, S.A.A. Shah, Y. Lei, B. Liu, K. Guo, Z. Guo, Sensors **24**, 944 (2024)

37. A.D. Bulygin, Y.E. Geints, I.Y. Geints, Photonics **10** (2023)
38. M. Zhang, L. Zhu, T. Hao, G. Zhou, C. Qiu, Z. Zhao, N. Hartmann, B. Xiao, Y. Zou, W. Feng, H. Zhu, M. Zhang, Y. Zhang, Y. Li, T.P. Russell, F. Liu, Adv. Mater. **33**, 1 (2021)
39. Y. Yu, Y. Lian, Q. Hu, L. Xie, J. Ding, Y. Wang, Z. Lu, Photonics **9** (2022)
40. Z. Bomzon, M. Gu, J. Shamir, Appl. Phys. Lett. **89**, 241104 (2006)
41. Y. Iketaki, T. Watanabe, N. Bokor, M. Fujii, Opt. Lett. **32**, 2357 (2007)
42. N. Bokor, Y. Iketaki, T. Watanabe, M. Fujii, Opt. Express **13**, 10440 (2005)
43. G.H. Yuan, N. Zheludev, *2017 Conference on Lasers and Electro-Optics Europe & European Quantum Electronics Conference* (IEEE, 2017), p. 1
44. M. Berry, N. Zheludev, Y. Aharonov, F. Colombo, I. Sabadini, D.C. Struppa, J. Tollaksen, E.T.F. Rogers, F. Qin, M. Hong, X. Luo, R. Remez, A. Arie, J.B. Götte, M.R. Dennis, A.M.H. Wong, G.V. Eleftheriades, Y. Eliezer, A. Bahabad, G. Chen, Z. Wen, G. Liang, C. Hao, C.-W. Qiu, A. Kempf, E. Katzav, M. Schwartz, J. Opt. **21**, 053002 (2019)
45. M.V. Berry, P. Shukla, New J. Phys. **23**, 113014 (2021)
46. A. Bekshaev, K.Y. Bliokh, M. Soskin, J. Opt. **13**, 053001 (2011)
47. B. Ghosh, A. Daniel, B. Gorzkowski, A.Y. Bekshaev, R. Lapkiewicz, K.Y. Bliokh, J. Opt. Soc. Am. B **41**, 1276 (2024)
48. A.Y. Bekshaev, J. Opt. **15**, 044004 (2013)
49. V.V. Kotlyar, S.S. Stafeev, Optik (Stuttg) **245**, 167703 (2021)
50. I. Besieris, P. Saari, Phys. Rev. A **107**, 033502 (2023)
51. R. Chen, T. Song, Y. Luo, H. Li, X. Li, Photonics **10**, 743 (2023)
52. L. Han, C. Gao, J. Qi, F. Li, Phys. Rev. A **111**, 032206 (2025)
53. Y. Zhao, J.S. Edgar, G.D.M. Jeffries, D. McGloin, D.T. Chiu, Phys. Rev. Lett. **99**, 073901 (2007)
54. V.V. Kotlyar, A.A. Kovalev, A.V. Volyar, Opt. Express **28**, 8266 (2020)
55. S.S. Stafeev, V.D. Zaitcev, K. Guo, B. Liu, Z. Guo, V.V. Kotlyar, Appl. Optics **65**, 1042 (2026)
56. J.D. Jackson, *Classical Electrodynamics*, 3rd edn. (Wiley, New York, 1999)
57. V.A. Neganov, S.B. Raevskii, G.P. Yarovoi, *Linear Macroscopic Electrodynamics* (Radio i Svyaz, Moscow, 2000)
58. J. Shu, X. Li, H.F. Arnoldus, J. Mod. Opt. **55**, 2457 (2008)
59. Y. Chen, Y. Zhang, A. Femius Koenderink, Opt. Express **25**, 21358 (2017)
60. M. Neugebauer, P. Banzer, S. Nechayev, Sci. Adv. **5**, eaav7588 (2026)
61. S. Nechayev, J.S. Eismann, M. Neugebauer, P. Woźniak, A. Bag, G. Leuchs, P. Banzer, Phys. Rev. A **99**, 41801 (2019)
62. M.F. Picardi, A.V. Zayats, F.J. Rodríguez-Fortuño, Laser Photon. Rev. **13**, 1900250 (2019)
63. X. Zambrana-Puyalto, N. Bonod, Nanoscale **8**, 10441 (2016)
64. X. Liu, W. Yan, Y. Liang, Z. Nie, Y. Wang, Z. Jiang, Y. Song, X. Zhang, Adv. Photonics Res. **3**, 2100117 (2022)
65. C. Zhang, X. Wang, L. Qiu, Front. Chem. **9** (2021)
66. P. Shi, L. Du, X. Yuan, Nanophotonics **10**, 3927 (2021)
67. Z. Shao, J. Zhu, Y. Zhang, Y. Chen, S. Yu, Opt. Lett. **43**, 1263 (2018)
68. K.Y. Bliokh, A.Y. Bekshaev, F. Nori, Phys. Rev. Lett. **119**, 73901 (2017)
69. K.Y. Bliokh, F. Nori, Phys. Rep. **592**, 1 (2015)
70. H.-I. Lee, Optics **3**, 313 (2022)

Chapter 5
Orbital Angular Momentum at the Sharp Focus of Laser Light

5.1 Spin-Orbital Conversion at Sharp Focus

It was Poynting himself who envisaged long ago that a circularly polarized light wave might be expected to carry spin angular momentum (SAM) [1]. Later on, SAM was experimentally observed by measuring the rotation angle of a quarter-wave plate, with the conversion of an incident right-handed circularly polarized wave into outgoing left-handed circularly polarized light [2]. The study of the free-space propagation of Laguerre-Gauss (LG) beams has shown that light also carries angular orbital momentum (OAM) [3]. Light waves that carry only SAM have been known to be capable of generating beams carrying OAM, leading to the familiar spin-to-orbit conversion. This effect was dealt with in numerous of publications, with most significant topic-related works outlined in more detail in this section. A link between spin and orbit was revealed as early as 2005. It has been experimentally demonstrated [4] that at the tight focus of a right-handed circularly polarized Laguerre-Gauss (LG) beam with topological charge TC $= 1$, a deep on-axis intensity minimum is observed, rather than a theoretically predicted intensity null. Meanwhile, an analogous left-handed circularly polarized LG beam with TC $= 1$ was found to generate just a small on-axis intensity dip in the focal intensity pattern. The conclusion was that given the opposite-sign longitudinal SAM and OAM components, spin partially compensated for the vortical component of the beam. Relationships for three E-vector projections at the tight focus have been derived [5] using Richards-Wolf theory [6], with a right-handed circularly polarized original optical beam shown to be transformed at the focus into a left-handed circularly polarized optical vortex with TC $= 2$. In other words, a beam with spin +1 was shown to form at the focus an extra optical vortex with spin -1 and TC $= +2$. It has also been shown [7] that when focusing a right-handed circularly polarized optical vortex with TC $= m$, extra optical vortices are generated: a left-handed circularly polarized vortex with TC $= m + 2$ and an on-axis optical vortex with TC $= m + 1$. It was revealed that when focusing a linearly polarized optical wave, alongside the above-mentioned

V. Kotlyar et al., *Inverse Canonical Energy Flow in Laser Beams*,
https://doi.org/10.1007/978-3-032-27289-8_5

extra optical vortices, right-handed circularly polarized optical vortices with T = $m - 2$ and an on-axis optical vortex with TC = m-1 were also generated. An experiment on an optical trapping and rotation of 3-um golden microparticles at the sharp focus of a circularly polarized LG beam with TC = 1 has been described [8]. The microparticles were found to move along a circular path about 25% faster in a right-handed circularly polarized beam than in a left-handed circularly polarized beam. The authors theorized that the effect was linked with spin-to-orbital conversion, however, failing to offer its detailed explanation. Not long after, it was revealed that when focusing even a conventional circularly polarized Gaussian beam, a transverse energy flow was formed at the focus [9]. It was also shown that with increasing numerical aperture, the longitudinal SAM component of the Gaussian beam was decreasing. Considering that the total angular momentum (SAM plus OAM) should conserve, the longitudinal OAM component might be expected to increase. The authors mistakenly presumed the longitudinal SAM to convert into the longitudinal OAM. However, this is not the case. In this work, we show that the longitudinal SAM is partially converted into the transverse SAM, and not longitudinal OAM. In Ref. [10], a gold microparticle trapped at the focus of a circularly polarized Gaussian beam was experimentally shown to move on a circular path centered at the optical axis clockwise or anticlockwise, depending on whether the Gaussian beam had left- or right-handed circular polarization. In Ref. [11], it was experimentally demonstrated that when illuminating a spiral structure synthesized on a metal substrate by a circularly polarized beam, a surface plasmon-polariton was generated with the topological charge of either TC = 2 or TC = 0, depending on whether the illuminating beam was right- or left-handed circularly polarized. In Ref. [12], it was theoretically shown that when sharply focusing a left-handed circularly polarized LG beam with TC = 2, a reverse energy flow was generated on the optical axis. However, the illumination by a right-handed circularly polarized beam did not result in the generation of the reverse on-axis energy flow. In Ref. [13] a spin optical Hall effect was experimentally demonstrated by scattering light from a microparticle placed at the tight focus. If a microparticle whose center was displaced from the focal spot center by a fraction of wavelength was illuminated by a linearly polarized beam, the field of the microparticle image was found to be partly left- and partly right-handed circularly polarized. A theory similar to Richards-Wolf theory has been developed for a dipole source [14]. It was shown that in a non-paraxial imaging system, with increasing numerical aperture, the ratio of the longitudinal components of the SAM vector to the OAM vector was decreasing, meaning that the longitudinal SAM component decayed while the longitudinal OAM component grew. A general theory of spin-to-orbital conversion upon focusing / scattering of light and in imaging systems has been proposed [15]. Relationships for longitudinal SAM and OAM projections were derived and their sum was shown to conserve. However, the transverse SAM and OAM projections were beyond the scope of study in [15]. An analysis of the tight focusing of a cylindrically (or high-order radially) polarized optical vortex reported in Ref. [16] revealed that regions with the maximum values of the longitudinal SAM and longitudinal OAM components were spatially separated in the focal spot. Such a separation was given the name a spin-orbital optical

Hall effect. Here, we notice that it is not quite clear why the location of the longitudinal SAM and OAM should be expected to coincide, as this is known to be the case when focusing only a right-handed circularly polarized optical vortex. For a left-handed circularly polarized incident vortex, the longitudinal SAM and OAM components are already directed oppositely. It has been experimentally demonstrated [17] that a 1-um polystyrene bead trapped at the tight focus of a circularly polarized optical vortex with TC = 10 starts moving on a circular path, with the circle radius being larger for the right-handed circularly polarized incident beam and smaller for the left-handed circularly polarized beam. This can be due to the right-handed circularly polarized vortex with TC = m generating an extra optical vortex with TC = $m + 2$ at the focus, with the left-handed circularly polarized vortex creating an extra vortex with TC = $m - 2$ at the focus. With the vortex radius being proportional to the square root of its TC, the circle radius for the vortex with TC = $m + 2$ appears to be larger than that for the vortex with TC = $m - 2$. A review of works concerned with the interaction of spin-orbit angular momentum can be found in Ref. [18]. Tight focusing of two time-varying optical beams was analyzed in Ref. [19] and an optical spin Hall effect was observed at the focus, with the positive and negative longitudinal SAM components found to be spatially separated. The authors related the spin Hall effect with a geometrical (or Berry) phase [20]. In the review dealing with spin-orbit interactions of light [20], the SAM was presumed to partly convert into the OAM. However, this is not the case. In this work, we show that upon tight focusing of light, the total SAM conserves, only experiencing a partial transformation: with the longitudinal SAM being partly transformed into the transverse SAM. Alongside works concerned with spin-to-orbital conversion, for completeness sake, it is worth mentioning other remarkable optical phenomena that occur at the focus, which include photonic wheels [21–24], polarization Möbius strips [24–30], spin and orbital Hall effects [31–34], optical magnetization [35–37], reverse energy flow [38–41], and optical skyrmions [42–44].

In this section, employing the Richard-Wolf formalism and numerical simulation, we show that when tightly focusing a right-handed circularly polarized optical vortex, the cross-section-averaged longitudinal SAM vector projection is somewhat decreased whereas the cross-section-averaged longitudinal OAM vector projection is increased by the same magnitude. However, we argue that it would be wrong to assume that the longitudinal SAM is partly transformed into the longitudinal OAM. On the contrary, we demonstrate that the cross-section averaged SAM vector conserved upon focusing, with the longitudinal SAM component partly converting into the transverse SAM component. Hence, relative to its original value, the cross-section-averaged longitudinal OAM at the focus is exactly increased by the magnitude of the increase of the total azimuthal orbital (or canonical) energy flow in the focal plane.

5.1.1 Sharply Focusing a Right-Handed Circularly Polarized Optical Vortex

For the convenience of readers, below we give formulae from Refs. [45, 46] for basic relationships that describe the light field characteristics in the focal plane. In the original plane, the Jones vector of a right-handed circularly polarized vortex beam is given by

$$\mathbf{F}(\theta,\varphi)=\frac{B(\theta)}{\sqrt{2}}\exp(im\varphi)\begin{pmatrix}1\\ i\end{pmatrix} \tag{5.1}$$

where φ is the azimuthal angle in the output pupil plane of an aplanatic optical system, θ is the polar angle defining the tilt of the light rays to the optical axis, $B(\theta)$ is the amplitude of the original light field (real function) which is radially symmetric relative to the optical axis, and m is the (integer) topological charge of the optical vortex. Using the Richards-Wolf formalism [6], we can derive for all projections of the energy flow vector in the focal plane. In the general case, the Poynting vector **P** is defined by the well-known expression, $\mathbf{P}=c\,\mathrm{Re}\,(\mathbf{F}^{*}\times\mathbf{H})/(2\pi)$, where **F** and **H** are the electric and magnetic field vectors, "*" and "×" denote the complex and vector products, Re is the real part of the complex number, and c is the speed of light in vacuum (the constant $c/(2\pi)$ will be dropped). Projections of the energy flow vector at the focus are given by [45]:

$$\begin{aligned}&P_r=0,\\&P_\varphi=\Omega(r),\\&P_z=J_{0,m}^2-J_{2,m+2}^2,\\&\Omega(r)=J_{1,m+1}\left(J_{0,m}+J_{2,m+2}\right).\end{aligned} \tag{5.2}$$

The functions $J_{\alpha,\beta}$ in (5.2) present the following integrals, with the first and second indices, respectively, defining the integral type ($\alpha=0,1,2$) and the order of the Bessel function $J_\beta(\xi)$ under the integral ($\beta=0,\pm1,\pm2,\ldots$):

$$J_{\alpha,\beta}=\left(\frac{4\pi f}{\lambda}\right)\int_0^{\theta_0}\sin^{\alpha+1}\left(\frac{\theta}{2}\right)\cos^{3-\alpha}\left(\frac{\theta}{2}\right)\cos^{1/2}(\theta)B(\theta)e^{ikz\cos\theta}J_\beta(\xi)d\theta, \tag{5.3}$$

where $k=2\pi/\lambda$ is the wave number of light of wavelength λ, f is the focal length of an aplanatic system, (r,φ,z) are the cylindrical coordinates, and θ_0 is the maximum angle of the tilt of rays to the optical axis, which defines the numerical aperture of the aplanatic lens, $NA=\sin(\theta_0)$, $\xi=kr\sin(\theta)$. The Poynting vector is composed of two terms [47], the orbital energy flow, P_{or} and spin flow, P_{sp}:

$$\begin{aligned}&\mathbf{P} = \mathbf{P}_{or} + \mathbf{P}_{sp} = \mathrm{Im}(\mathbf{F}^* \cdot (\nabla)\mathbf{F}) + \frac{1}{2}(\nabla \times \mathbf{S}),\\&\mathbf{S} = \mathrm{Im}(\mathbf{F}^* \times \mathbf{F}).\end{aligned} \tag{5.4}$$

From (5.4), the azimuthal projection of the orbital energy flow (canonical momentum [47]) at the focus can be expressed as follows:

$$\begin{aligned}&P_{or,\varphi} = \mathrm{Im}(\mathbf{F}^* \cdot (\nabla)\mathbf{F})_\varphi =\\&\frac{1}{r}\mathrm{Im}\left(F_x^* \frac{\partial F_x}{\partial \varphi} + F_y^* \frac{\partial F_y}{\partial \varphi} + F_z^* \frac{\partial F_z}{\partial \varphi}\right) =\\&\frac{1}{r}\left(mJ_{0,m}^2 + (m+2)J_{2,m+2}^2 + 2(m+1)J_{1,m+1}^2\right).\end{aligned} \tag{5.5}$$

In (5.5), F_x, F_y, F_z denote projections of the electric field **F** of Eq. (5.1) in the focal plane. In view of Eqs. (5.4) and (5.5), the longitudinal component of the angular momentum component is expressed as

$$\begin{aligned}&J_z = rP_\varphi = rP_{or,\varphi} + rP_{sp,\varphi} =\\&r\mathrm{Im}(\mathbf{F}^* \cdot (\nabla)\mathbf{F})_\varphi + \frac{r}{2}(\nabla \times \mathbf{S})_\varphi = L_z + S_z.\end{aligned} \tag{5.6}$$

From (5.6), the azimuthal component of the orbital energy flow is seen to contribute to the longitudinal OAM component at the focus. Knowing the azimuthal projections of the energy flow at the focus (Eq. (5.2)) and orbital energy flow (Eq. (5.5)), we can derive the azimuthal projection of the spin flow at the focus of field (5.1):

$$\begin{aligned}&P_{sp,\varphi} = P_\varphi - P_{or,\varphi} =\\&J_{1,m+1}(J_{0,m} + J_{2,m+2}) -\\&\frac{1}{r}\left(mJ_{0,m}^2 + (m+2)J_{2,m+2}^2 + 2(m+1)J_{1,m+1}^2\right).\end{aligned} \tag{5.7}$$

In (5.7), no singularity occurs at $r = 0$, because at $m > 0$ the integrals (5.3) entering (5.7), tend to zero as r^2. At $r = 0$, the expression in (5.7) equals zero. By analogy with Eq. (5.2), we can also derive relationships for the SAM projections in the focal plane for the original field in (5.1). The SAM vector is defined by the familiar relationship, $\mathbf{S} =$ Im $(\mathbf{F}^* \times \mathbf{F})/(8\pi\omega)$, where Im denotes the imaginary part of the complex number and ω is the angular frequency of light. Below, the constant 1/(8πω) is dropped and relationships for the SAM vector components at the focus of original field (5.1) analogous to the Poynting vector components in (5.2) can be expressed as

$$\begin{aligned} &S_r = 0, \\ &S_\varphi = \Omega(r), \\ &S_z = J_{0,m}^2 - J_{2,m+2}^2, \\ &\Omega(r) = J_{1,m+1}\left(J_{0,m} + J_{2,m+2}\right). \end{aligned} \tag{5.8}$$

From a comparison of (5.2) and (5.8), the longitudinal and transverse projections of the Poynting vector and SAM vector are seen to be identical, which are codirectional and parallel.

The OAM vector in the focal plane of field (5.1) for the longitudinal projection can be expressed as [45].

$$\begin{aligned} L_z &= \mathrm{Im}\left(F_x^* \frac{\partial}{\partial\varphi} F_x + F_y^* \frac{\partial}{\partial\varphi} F_y + F_z^* \frac{\partial}{\partial\varphi} F_z\right) = \\ &= mJ_{0,m}^2 + (m+2)J_{2,m+2}^2 + 2(m+1)J_{1,m+1}^2. \end{aligned} \tag{5.9}$$

Note that Eq. (5.9) describes the term Lz in (5.6). An expression for the intensity distribution in the focal plane of field (5.1) reads as [45].

$$\begin{aligned} &I(r,\varphi,z=0) = |F_x|^2 + |F_y|^2 + |F_z|^2 = \\ &J_{0,m}^2 + J_{2,m+2}^2 + 2J_{1,m+1}^2. \end{aligned} \tag{5.10}$$

Averaged characteristics of the light field in the focal plane, including the beam total energy (power), $\overline{I}=W$, total longitudinal energy flow $\overline{P}_z$, total longitudinal SAM, $\overline{S}_z$, and total longitudinal OAM, $\overline{L}_z$can be expressed as

$$\begin{aligned} \widehat{I} &= \int_0^\infty \int_0^{2\pi} I r dr d\varphi = \int_0^\infty \int_0^{2\pi} \left(J_{0,m}^2 + J_{2,m+2}^2 + 2J_{1,m+1}^2\right) r dr d\varphi \\ &= W_0 + W_2 + 2W_1. \end{aligned} \tag{5.11}$$

$$\begin{aligned} \widehat{P}_z &= \int_0^\infty \int_0^{2\pi} P_z r dr d\varphi = \frac{1}{2} \int_0^\infty \int_0^{2\pi} r dr d\varphi \left(J_{0,m}^2 - J_{2,m+2}^2\right) = \\ &= W_0 - W_2 = W - 2W_2 - 2W_1. \end{aligned} \tag{5.12}$$

$$\begin{aligned} \widehat{S}_z &= \int_0^\infty \int_0^{2\pi} S_z r dr d\varphi = \frac{1}{2} \int_0^\infty \int_0^{2\pi} r dr d\varphi \left(J_{0,m}^2 - J_{2,m+2}^2\right) = \\ &= W_0 - W_2 = W - 2W_2 - 2W_1. \end{aligned} \tag{5.13}$$

$$
\begin{aligned}
\widehat{L}_z &= \int_0^\infty rdr \int_0^{2\pi} d\varphi L_z(r,\varphi) = \\
&= \int_0^\infty rdr \int_0^{2\pi} d\varphi \left(mJ_{0,m}^2 + (m+2)J_{2,m+2}^2 + 2(m+1)J_{1,m+1}^2\right) = \\
&= mW_0 + (m+2)W_2 + 2(m+1)W_1.
\end{aligned}
\tag{5.14}
$$

Equations (5.11)–(5.14) have been derived using the following integral [48]:

$$
\begin{aligned}
W_\alpha &= 2\pi \int_0^\infty \left|J_{\alpha,\beta}(r)\right|^2 rdr = \\
&= 4\pi f^2 \int_0^{\theta_0} \sin^{2\alpha+1}\left(\frac{\theta}{2}\right)\cos^{5-2\alpha}\left(\frac{\theta}{2}\right)|B(\theta)|^2 d\theta.
\end{aligned}
\tag{5.15}
$$

From (5.13) and (5.14), it follows that the sum of SAM and OAM (the angular momentum) equals a product of the beam total energy by the topological charge m plus one:

$$
\widehat{S}_z + \widehat{L}_z = (m+1)W. \tag{5.16}
$$

To look into how the average characteristics of original field (5.1) will change at the focus, let us write down these characteristics in the original plane:

$$
\widehat{S}_{z,0} = W, \quad \widehat{P}_{z,0} = W, \quad \widehat{L}_{z,0} = mW. \tag{5.17}
$$

From a comparison of (5.16) and (5.17), the sum of the longitudinal SAM and OAM for field (5.1) is seen to conserve upon focusing. At the focus, the longitudinal SAM is smaller in magnitude than the original value by $\widehat{S}_{z,0} - \widehat{S}_z = 2(W_2 + W_1)$. A comparison of (5.14) and (5.17) suggests that at the focus, the longitudinal OAM increases exactly by the same magnitude: $\widehat{L}_z - \widehat{L}_{z,0} = 2(W_2 + W_1)$. From a comparison of Eqs. (5.12) and (5.17), the longitudinal energy flow in the focal plane is seen to reduce by the same magnitude: $\widehat{P}_{z,0} - \widehat{P}_z = 2(W_2 + W_1)$. The question is where the said longitudinal components of the SAM and Poynting vectors have been redistributed to. Using (5.15) it can be shown that the cross-section-averaged longitudinal component of energy flow of (5.12) is formed immediately behind the spherical lens, conserves upon focusing and is defined by

$$\widehat{P}_z = \frac{W}{2}(1 + \cos\theta_0), \tag{5.18}$$

where the angle θ_0 defines the spherical lens numerical aperture NA = sin θ_0. In a similar way, the transverse energy flow component immediately behind the spherical lens can be derived as

$$\widehat{P}_\perp = \frac{W}{4}\left(\frac{2\theta_0 - \sin 2\theta_0}{1 - \cos\theta_0}\right). \tag{5.19}$$

While immediately behind the spherical lens the transverse energy flow is purely radial, an azimuthal component is forming as the beam propagates. In the focal plane, the transverse flow is seen to be purely azimuthal, Eqs. (5.5) and (5.7), with only a portion of this azimuthal flow, in the form of the azimuthal orbital energy flow, being converted into the generation of the longitudinal OAM component, Eq. (5.6). The azimuthal spin flow, $P_{sp,\varphi}$ is redistributed to the longitudinal spin component, S_z. Based on Eq. (5.2), it is not possible to derive an explicit expression to define an explicit expression for an average azimuthal energy flow which rotates on a circular path in the focal plane:

$$\widehat{P}_\varphi = \int_0^\infty \int_0^{2\pi} P_\varphi r dr d\varphi = \int_0^\infty \int_0^{2\pi} \Omega(r) r dr d\varphi. \tag{5.20}$$

Interestingly, at $m = 0$ (no optical vortex), the azimuthal energy flow conserves and equals $P_{\varphi,\, m=0} = \Omega(r) = J_{1,\,1}(J_{0,\,0} + J_{2,\,2})$, with the total azimuthal energy flow in (5.20) remaining unchanged, as it is defined not by the topological charge but by the numerical aperture. The azimuthal projection of the spin vector density, Eq. (5.4) is at a tangent to any optical-axis-centered circumference in the focal plane. An average azimuthal spin projection in the focal plane is described by a relationship similar to (5.20):

$$\widehat{S}_\varphi = \int_0^\infty \int_0^{2\pi} S_\varphi r dr d\varphi = \int_0^\infty \int_0^{2\pi} \Omega(r) r dr d\varphi. \tag{5.21}$$

Thus, we infer that for a light wave generating a transverse energy flow in the focal plane, the polarization vector also rotates in a plane perpendicular to the vector $\mathbf{S}_\varphi$. If $\Omega(r) > 0$ in (5.4), then the polarization vector rotates anticlockwise.

Hence, we can think of the following mechanism of spin-to-orbital conversion that occurs upon tightly focusing a light beam. In the initial plane, prior to reaching the focusing lens, the beam has longitudinal projections of spin, S_z, and energy flow, P_z, which are equal to the beam total energy, W, (Eq. (5.17)), and has no OAM at $m = 0$. Just after passing the lens, the longitudinal spin and energy flow projections decrease due to transverse projections occurring, Eq. (5.19). Thanks to the presence of the transverse energy flow component, the longitudinal component of averaged OAM is generated immediately behind the lens:

$$\widehat{L}_z = \frac{W}{2}(1 - \cos\theta_0) = 2(W_1 + W_2). \tag{5.22}$$

From (5.22) it is seen that the maximum total OAM can only reach half the beam energy, $L_z = W/2$ at $\theta_0 = \pi/2$. In the course of focusing, an azimuthal component is started forming and growing in the transverse spin and energy flow projections, with only the azimuthal components of spin vector, Eq.(8), and energy flow, Eq. (5.2) and no radial component remaining in the focal plane. In the course of focusing, the longitudinal OAM component, Eq. (5.22), conserves. So, in the focal plane, OAM (5.22) is contributed to only by the azimuthal component of the orbital energy flow in (5.5) and (5.6).

The effect of reducing longitudinal energy flow in the focal plane can also be substantiated in a different way by proving that the longitudinal SAM has decreased due to it partly converting to the transverse SAM. This approach uses explicit expressions for the electric field projections at the tight focus of original field (5.1):

$$\begin{aligned} F_{x,m} &= \frac{i^{m-1}}{\sqrt{2}} e^{im\varphi}\left(J_{0,m} + e^{2i\varphi} J_{2,m+2}\right), \\ F_{y,m} &= \frac{i^{m}}{\sqrt{2}} e^{im\varphi}\left(J_{0,m} - e^{2i\varphi} J_{2,m+2}\right), \\ F_{z,m} &= -\sqrt{2} i^{m} e^{i(m+1)\varphi} J_{1,m+1}. \end{aligned} \tag{5.23}$$

From Eq. (5.23) for the components of field (5.1) it is seen that after the original optical vortex, exp($im\varphi$), passed through the spherical lens, two extra vortices are generated, exp(i(m + 2)φ) and exp(i(m + 1)φ), with beam's energy partly going to their generation. It is these extra vortices that contribute to the transverse energy flow, because the terms $J_{1,\,m+1}$, $J_{2,\,m+2}$ enter both the azimuthal flow in Eq. (5.2) and the longitudinal OAM projection in Eq. (5.22). Even if the original field has no optical vortex ($m = 0$), optical vortices with topological charges 2 and 1 are nonetheless generated at the focus, creating the azimuthal energy flow of Eq. (5.2) and the longitudinal OAM component of Eq. (5.22). In a similar way, as a result of focusing, the longitudinal SAM projection appears to be partly transformed into the azimuthal transverse SAM projection, Eq. (5.21), being equal in magnitude to the transverse Poynting vector projection (5.20).

Interestingly, the phenomenon of increasing radius of circular movement of a microparticle at the tight focus of an optical vortex experimentally observed in Ref. [17] can be interpreted as a presence of an optical vortex with topological charge (m + 2) in (5.23). The authors showed that the radius if the microparticle's circular movement was larger for the right-handed circularly polarized beam than for the left-handed circularly polarized beam. However, no such a simple interpretation of the experimental results based on Eq. (5.23) was offered in Ref. [17].

5.1.2 Numerical Modeling

The numerical modeling was conducted using Richards-Wolf integrals [6] for original field (5.1), assuming a 633-nm wavelength and NA = 0.95.

It is seen from Fig. 5.1 that even with no optical vortex (m = 0), the transverse energy flow is directed towards the optical axis prior to the focus (a), away from the optical axis behind the focus (b), and rotates about the optical axis on circular paths at the focus due to spin-to-orbit conversion (c). Notably, at different distances from the optical axis, the transverse energy flow has a different handedness. The multitude of the blue arrows in Fig. 5.1 mark the presence of an inverse energy flow in the focus neighborhood. Figure 5.2 depicts a zoomed-up central fragment of Fig. 5.1b, with the energy flow seen to rotate anticlockwise within the focal spot (Fig. 5.2a). The distribution pattern of the azimuthal energy flow, marked red in Fig. 5.2b, is in the form of rings and equals zero at the center.

Figure 5.3 depicts distribution patterns for the (a) horizontal, S_x and (b) vertical, S_y components of the SAM vector at the focus for field (5.1) at $m = 0$. The transverse SAM flow is seen to be azimuthal, rotating anticlockwise, and coincident with the azimuthal energy flow in Fig. 5.2.

We have analyzed optical phenomena and underlying mechanisms that occur upon tightly focusing a right-handed circularly polarized optical vortex [49–51]. Relationships for the focal-spot-averaged longitudinal components of the SAM, OAM, and Poynting vector as well as averaged longitudinal transverse

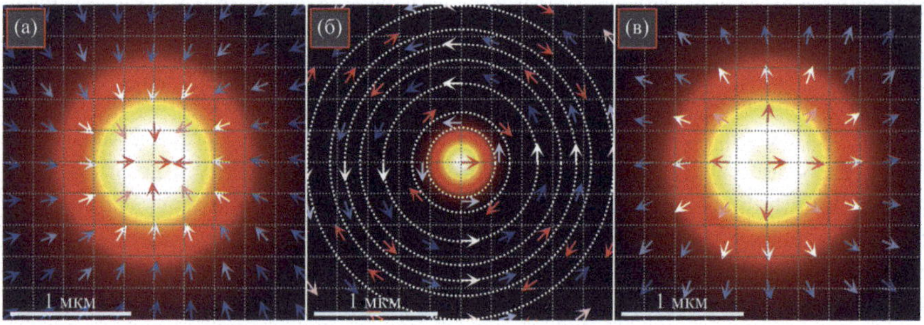

Fig. 5.1 Patterns of intensity (color) and energy flows (arrows) near the tight focus of a right-handed circularly polarized optical vortex ($m = 0$) at wavelength $\lambda = 633$ nm, focal length $f = 10$ um, vortex order $m = 0$, $NA = 0.95$, a uniform radial distribution of the focused field amplitude ($B(\theta) = 1$), in the transverse planes at: (a) $z = -\lambda$ (prior to the focus), (b) $z = 0$ (at the focus), and (c) $z = +\lambda$ (behind the focus). Red, blue, and white-colored arrows, respectively, denote a positive, negative, and near-zero longitudinal energy flow. Dotted circles mark boundaries between the positive and negative longitudinal energy flows

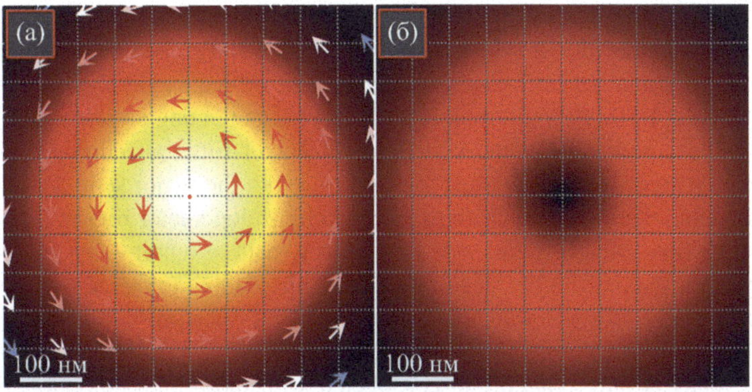

Fig. 5.2 Zoomed-up patterns of (**a**) intensity I (color) and azimuthal energy flow P_φ (arrows) at the tight focus of a right-handed circularly polarized optical vortex for the same parameters as in Fig. 5.1a and (**b**) the azimuthal Umov-Poynting vector component, P_φ

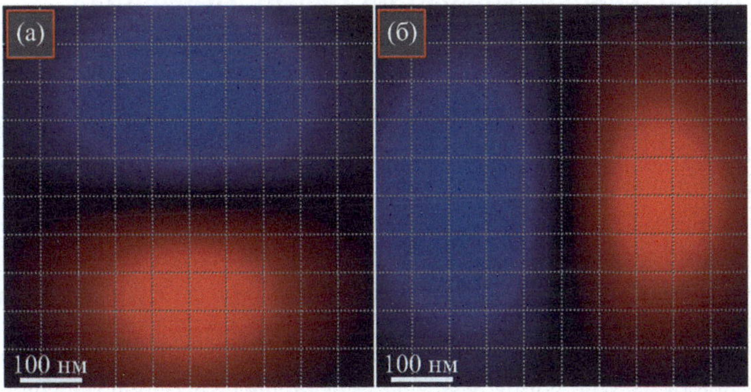

Fig. 5.3 Distribution patterns for the (**a**) horizontal S_x and (**b**) vertical S_y components of the SAM vector at the tight focus of a right-handed circularly polarized optical vortex for the same parameters as in Fig. 5.1 (red– positive value, blue– negative values)

components of the SAM and Poynting vector have been proposed. Upon tightly focusing, the original averaged longitudinal SAM component has been found to conserve, being only redistributed between the longitudinal and transverse components. Besides, in the course of focusing, the original longitudinal energy flow has also been found to conserve and be redistributed between the longitudinal and transverse components. It has also been revealed that at the focus the orbital component of the azimuthal energy flow partly goes to the generation of the beam longitudinal OAM. Because of this, it would be incorrect to suggest that in the course of focusing the longitudinal SAM component (spin) is partly converted into the OAM component (orbit) as a decrease in the longitudinal SAM component is due to it being partly converted to the transverse (azimuthal) SAM component. Interestingly, in Ref. [47] the authors counterposed the Poynting vector (energy flow) with the canonical (orbital) energy flow although the latter represents a part of the Poynting vector, bar the spin component. Here, we have demonstrated that only the part of the azimuthal component related with the orbital energy flow is converted to the OAM. In Ref. [9], a different mechanism for generating an azimuthal energy flow at the focus was offered. At every moment of time, an elliptical intensity distribution pattern is being formed at the focus, with the major semi-axis directed along the polarization vector. With time, the polarization vector rotates in the focal plane with a frequency of light, with the intensity pattern (energy density) rotating about the optical axis with the same frequency, thus leading to the generation of the azimuthal energy flow at the focus. We note that such an explanation of the mechanism of spin-to-orbital conversion serves to prove that the energy flow at the tight focus is only related and co-directed with the orbital component of the Poynting vector.

5.2 Orbital Angular Momentum Density Distribution at the Sharp Focus

Much attention in modern optics is paid to structured light fields with orbital angular momentum (OAM) and non-uniform polarization [52–54]. Interest in structured light appears due to its application in optical telecommunications [55], manipulation of micro-objects [56], quantum information [57], optical microscopy [58], and surface microstructuring [59]. An important task for structured light is the theoretical determination of its OAM. The first work on determining the OAM of Laguerre-Gaussian beams was performed back in 1992 [3]. In subsequent studies it was shown that the azimuthal energy flux (azimuthal projection of the Poynting vector) in the cross section of paraxial vortex beams depends on the polarization state and rotates in different directions (clockwise or counterclockwise) for light with left- and right-handed circular polarization [60–62]. It was experimentally shown in [63] that the trapped microparticles move in different directions at different distances from the optical axis in a Bessel beam with circular polarization. But it turns out that the Poynting vector is the sum of two vectors, one of which describes the canonical energy flux, and the second vector describes the spin flux [47]. It is shown in [47] that for a paraxial optical vortex with elliptical polarization, the azimuthal orbital energy flux (canonical flux) does not depend on the state of light polarization (left- or right-handed elliptical polarization), but depends only on the topological charge of the optical vortex. Therefore, the OAM density does also not depend on the rotation direction of the polarization vector. But in the case of paraxial light fields with non-uniform polarization this is not true. For example, it is shown in [64] that for a paraxial vector field (without an optical vortex) with non-uniform linear polarization (superposition of light with linear and cylindrical polarizations) the OAM density at some distances from the waist plane depends on $\sin(n\varphi)$, where n is the order of cylindrical polarization, φ is the azimuthal angle in the beam cross-section. That is, the OAM density depends on the beam polarization state and changes sign when the azimuthal angle changes. For non-paraxial light fields, due to the spin-orbit conversion effect, the longitudinal projection of the OAM depends on the polarization state [20, 65, 66].

In this section, we obtain analytical expressions for the longitudinal projection of the OAM vector in the plane of sharp focus for four different light fields. The expressions are obtained using the Richards-Wolf theory [6]. The OAM density for all four fields follows from the derived expressions to depend on the polarization state. We show that for two of the four fields that do not have an optical vortex, the OAM density at the focus changes sign when both coordinates change: radial and azimuthal.

Physically, the dependence of the OAM in the focus on the polarization state is explained by the spin-orbit conversion [20, 65, 66] and the spin Hall effect in the focus [64]. The spin-orbit conversion in the focus consists in the fact that light with an initial elliptical or circular polarization during focusing forms a rotation of the light energy in a circle in the focus and thereby creates a longitudinal projection of the OAM. Moreover, the rotation of the energy in a circle in the focus occur either clockwise or counterclockwise, depending on whether the initial light field has a left- or right-handed circular polarization. Therefore, for the first light field considered in this work, which is an optical vortex with elliptical polarization, the OAM in the focus due to the spin-orbit conversion increases if the light has a right-handed elliptical polarization or decreases if the polarization is a left-handed. The second light field considered in the work and which is a superposition of fields with cylindrical and elliptical polarizations has no optical vortex in the initial plane. Local optical vortices are formed in the focus of it. They are formed due to the spin and orbital Hall effects [64]. This effect consists in the fact that, although the light has no OAM in the initial plane, an even number of local regions are formed in the focus, in which the OAM signs are different. Moreover, the total OAM is zero except for the case when the cylindrical beam becomes a beam with linear polarization, and the polarization of entire superposition is elliptical. In this case, the second beam coincides with the first one at $n = 0$. The third beam considered in the work is an optical vortex with cylindrical polarization. Such a beam has no spin and therefore its total OAM in the focus and at the beginning is the same and does not depend on the polarization state. But the OAM density due to the orbital Hall effect has a different distribution of local OAM maxima and minima depending on the order of cylindrical polarization n. Finally, the fourth beam considered in the work is a beam with hybrid circular-cylindrical polarization. It does not have an OAM in the initial plane, but due to the spin and orbital Hall effects local regions are formed in the focus with different signs of the OAM. The results obtained in this work were not included in our book [67]. In [67], the orbital Hall effect in the focus is considered by calculating projections of the Poynting vector. In this work we calculated the OAM density, and not the projections of the Poynting vector. An evident application of the obtained results is the use of regions in the focus with different signs of OAM in micromechanics problems. If two dielectric microparticles in the form of gears are captured in two adjacent regions with OAM of different signs, they will be rotated by light in different directions and can bring in motion a micromechanical device.

5.2.1 OAM Density at the Focus of an Optical Vortex with Elliptical Polarization

The orbital angular momentum vector is given by the following formula with an accuracy of up to a constant 1/(8πω), where ω is the angular frequency of monochromatic light [67]:

$$\mathbf{J} = \mathrm{Im} \int\limits_0^{\infty} \int\limits_0^{2\pi} \sum_{p=x,y,z} E_p^*(\mathbf{r} \times \nabla) E_p r dr d\varphi,$$
$$\mathbf{j} = \mathrm{Im} \sum_{p=x,y,z} E_p^*(\mathbf{r} \times \nabla) E_p. \tag{5.24}$$

where the vector product of the radius vector **r** and the vector ∇ is written in brackets, Im is the imaginary part of the number, (r, φ) are the polar coordinates in the beam cross-section. The second formula describes the OAM density vector, and the first one describes the total OAM integrated over the entire beam cross-section. Next, we obtain analytical expressions for both the density and the total OAM, but only for one longitudinal projection of the vectors (5.24):

$$J_z = \mathrm{Im} \int\limits_0^{\infty} \int\limits_0^{2\pi} \left(E_x^* \frac{\partial}{\partial \varphi} E_x + E_y^* \frac{\partial}{\partial \varphi} E_y + E_z^* \frac{\partial}{\partial \varphi} E_z \right) r dr d\varphi,$$
$$j_z = \mathrm{Im} \left(E_x^* \frac{\partial}{\partial \varphi} E_x + E_y^* \frac{\partial}{\partial \varphi} E_y + E_z^* \frac{\partial}{\partial \varphi} E_z \right). \tag{5.25}$$

Let us consider a sharp focusing of an optical vortex with elliptical polarization, it's the initial field Jones vector can be written as:

$$\mathbf{E}_1(\varphi) = \frac{e^{im\varphi}}{\sqrt{1+\alpha^2}} \begin{pmatrix} 1 \\ i\alpha \end{pmatrix}, \tag{5.26}$$

where m is the topological charge of the optical vortex, α is a real parameter specifying the vector state of the initial beam (5.26): if $\alpha = 1$ there is a right-handed circular polarization, if $\alpha = 0$ there is a linear polarization, when $0 < \alpha < 1$ an elliptical polarization takes place. Three projections of the electric field strength vector at a sharp focus for the initial field (5.26) are known [68]:

$$\begin{aligned}
E_x &= \frac{i^{m-1}}{\sqrt{1+\alpha^2}} e^{im\varphi} \times \\
&\times \left(I_{0,m} + \frac{(1+\alpha)}{2} e^{2i\varphi} I_{2,m+2} + \frac{(1-\alpha)}{2} e^{-2i\varphi} I_{2,m-2} \right), \\
E_y &= \frac{i^{m}}{\sqrt{1+\alpha^2}} e^{im\varphi} \times \\
&\times \left(\alpha I_{0,m} - \frac{(1+\alpha)}{2} e^{2i\varphi} I_{2,m+2} + \frac{(1-\alpha)}{2} e^{-2i\varphi} I_{2,m-2} \right), \\
E_z &= \frac{-2i^{m}}{\sqrt{1+\alpha^2}} e^{im\varphi} \times \\
&\times \left((1+\alpha) e^{i\varphi} I_{1,m+1} - (1-\alpha) e^{-i\varphi} I_{1,m-1} \right).
\end{aligned} \tag{5.27}$$

where

$$\begin{aligned}
I_{\alpha,\beta} = \left(\frac{4\pi f}{\lambda}\right) \int_0^{\theta_0} \sin^{\alpha+1}\left(\frac{\theta}{2}\right) \cos^{3-\alpha}\left(\frac{\theta}{2}\right) \times \\
\times \cos^{1/2}(\theta) B(\theta) e^{ikz\cos\theta} J_\beta(\xi) d\theta,
\end{aligned} \tag{5.28}$$

where $k = 2\pi/\lambda$ is the wave number of light with the wavelengths λ, f is the focal distance of an aplanatic system, (r, φ, z) are cylindrical coordinates, θ_0 is the maximum tilt angle of light rays to the optical axis, which defines a numerical aperture of the aplanatic optical system $NA = \sin(\theta_0)$, $\xi = kr\sin(\theta)$, $B(\theta)$ is the real function, which depends on the polar angle only and describes an amplitude distribution of the focused light beam. For the field (5.26) $B(\theta) = 1$. The projections of the field at the focus (5.27) are obtained using the Richards-Wolf formalism [6]. Note that if we calculate the OAM density (5.25) for the initial field (5.26), but neglect the longitudinal coordinate (as is done in the paraxial approximation case), we obtain the well-known result: $j_z = m$. That is, in the paraxial case, the OAM density does not depend on the polarization parameter α. In the case of non-paraxial light fields, as in the case of sharp focusing, the OAM density depends on the polarization parameters of the initial field. Indeed, substituting the expressions for the field projections at the focus (5.27) into (5.25), we obtain an expression for the OAM density:

$$
\begin{aligned}
j_{1z}(r,\varphi) &= \frac{1}{1+\alpha^2}\times \\
&\times\Bigg\{\bigg[\left(1+\alpha^2\right)mI_{0,m}^2+\frac{(1+\alpha)^2}{2}(m+2)I_{2,m+2}^2+ \\
&+\frac{(1-\alpha)^2}{2}(m-2)I_{2,m-2}^2+ \\
&+(1+\alpha)^2(m+1)I_{1,m+1}^2+(1-\alpha)^2(m-1)I_{1,m-1}^2\bigg]+ \\
&+\left(1-\alpha^2\right)\cos(2\varphi)\left[(m+1)I_{0,m}I_{2,m+2}+\right. \\
&\left.+(m-1)I_{0,m}I_{2,m-2}-2mI_{1,m+1}I_{1,m-1}\right]\Bigg\}.
\end{aligned}
\tag{5.29}
$$

It is seen from (5.29) that the OAM density depends on both polar coordinates (r, φ). Therefore, at different azimuthal angles it can have different signs. From (5.29) it is possible to obtain simpler expressions for the OAM density of an optical vortex with linear and right-handed circular polarizations:

$$
\begin{aligned}
j_{1z,L}(r,\varphi) &= \bigg\{\Big[mI_{0,m}^2+\frac{1}{2}(m+2)I_{2,m+2}^2+ \\
&+\frac{1}{2}(m-2)I_{2,m-2}^2+(m+1)I_{1,m+1}^2+(m-1)I_{1,m-1}^2\Big]+ \\
&+\cos(2\varphi)\left[(m+1)I_{0,m}I_{2,m+2}+\right. \\
&\left.+(m-1)I_{0,m}I_{2,m-2}-2mI_{1,m+1}I_{1,m-1}\right]\bigg\},
\end{aligned}
\tag{5.30}
$$

$$
j_{1z,C}(r) = mI_{0,m}^2+(m+2)I_{2,m+2}^2+2(m+1)I_{1,m+1}^2.
\tag{5.31}
$$

It is seen that the OAM density at the focus for the optical vortex with circular polarization has radial symmetry (5.31), and the OAM density for linear polarization can change sign 4 times when going around the optical axis (5.30), since there is a term with $\cos(2\varphi)$. The squares of the integrals (5.28) in the first square brackets are always positive, and the products of these integrals in the second square brackets can change sign when the radial coordinate changes. That is, at some distance from the optical axis, at the same azimuthal angle, the expression for the OAM density (5.30) can change sign. There is no such effect in the paraxial case [47], since in this case, the OAM cannot change sign when the radial coordinate changes. The total OAMs of both beams (5.30) and (5.31) are equal to:

$$
J_{1z,L} = mW_0+mW_2+2mW_1 = mW,
\tag{5.32}
$$

$$
\begin{aligned}
J_{1z,C} &= mW_0+(m+2)W_2+2(m+1)W_1 = \\
&= mW+2(W_2+W_1).
\end{aligned}
\tag{5.33}
$$

Due to the spin-orbit conversion the OAM at the focus is clear from (5.33) to be different from zero even if there is no initial optical vortex ($m = 0$). That is, light only with right-handed elliptical polarization forms an azimuthal energy flow at the focus rotating counterclockwise, and therefore forms a longitudinal projection of the OAM. In (5.32) and (5.33) the following definitions are used [69]:

$$
\begin{aligned}
W_\alpha &= 2\pi \int\limits_0^\infty \left|I_{\alpha,\beta}(r)\right|^2 r dr = \\
&= 4\pi f^2 \int\limits_0^{\theta_0} \sin^{2\alpha+1}\left(\frac{\theta}{2}\right)\cos^{5-2\alpha}\left(\frac{\theta}{2}\right)|B(\theta)|^2 d\theta.
\end{aligned}
\tag{5.34}
$$

It is clear from a comparison of (5.32) and (5.33) that at $m = 0$ (no vortex) the total OAM of a linearly polarized field is zero, and for a field with circular polarization the total OAM is nonzero due to the spin-orbit effect at the sharp focus.

5.2.2 OAM Density of a Hybrid Light Field at the Focus

Let us consider other examples of vector light fields, in which the OAM density at the focus can change sign with changing of the radial variable. Let us consider the Jones vector of the initial vector field, which has regions with linear and circular polarization in its cross-section, which is why it is called hybrid [70]:

$$
\mathbf{E}_2(\varphi) = \frac{1}{\sqrt{1+\delta^2}}(\cos n\varphi i\delta + \sin n\varphi). \tag{5.35}
$$

At $\delta = 0$, field (5.35) is a cylindrical vector field of the n-th order, and at $n = 0$, field (5.35) is a plane wave with right-handed elliptical polarization. The light field (5.35) does not have an optical vortex, but we will show that at the focus the density of the longitudinal OAM is nonzero, and there are regions with positive and negative OAM. The projections of the electric field strength vector (5.35) at the focus have the form [70]:

$$
\begin{aligned}
E_x &= \frac{1}{\sqrt{1+\delta^2}} \times \left[i^{n-1}(I_{0,n}\cos n\varphi + \right. \\
&\quad \left. + I_{2,n-2}\cos(n-2)\varphi) + \delta I_{2,2}\sin 2\varphi\right], \\
E_y &= \frac{1}{\sqrt{1+\delta^2}} \times \left[i^{n-1}(I_{0,n}\sin n\varphi - \right. \\
&\quad \left. - I_{2,n-2}\sin(n-2)\varphi) + \delta(I_{0,0} - I_{2,2}\cos 2\varphi)\right], \\
E_z &= \frac{2}{\sqrt{1+\delta^2}}(i^n I_{1,n-1}\cos(n-1)\varphi - i\delta I_{1,1}\sin\varphi).
\end{aligned}
\tag{5.36}
$$

Substituting projections of the field at the focus (5.36) into the expression (5.25) for the density of the longitudinal OAM, we obtain for an even number $n = 2p$, $p = 0$, 1, 2, ... (if $n = 2p + 1$ the OAM density is zero):

$$
\begin{aligned}
j_{2z} &= \frac{\delta(-1)^p}{1+\delta^2}\left[I_{2,2}I_{0,n}(n+2)\cos(n-2)\varphi -\right. \\
&I_{2,2}I_{2,n-2}(n-4)\cos n\varphi - I_{0,0}I_{0,n}n\cos n\varphi - \\
&I_{0,0}I_{2,n-2}(n-2)\cos(n-2)\varphi - 4I_{1,1}I_{1,n-1}\times \\
&\left.((n-1)\sin\varphi\sin(n-1)\varphi + \cos\varphi\cos(n-1)\varphi)\right].
\end{aligned}
\tag{5.37}
$$

The OAM density is seen from (5.37) to depend on the azimuthal angle φ and the radial variable r. For some coordinate values (r, φ) the OAM is positive, and for some it is negative. This is explained by the orbital Hall effect. The orbital Hall effect is as follows. If the circular or cylindrical symmetry is broken in the initial light field, then light with different spins (with left- and right-handed circular polarization) deviates from the optical axis differently and forms local regions in the focus in which the spin has different signs. And because of the spin-orbit conversion, light fields with different signs of spins form regions with different signs of OAM. The OAM density is seen from (5.37) to be zero at $\delta = 0$. It is also seen from (5.37) that all the terms of the density are multiplied by trigonometric functions of the azimuthal angle. This means that when integrating over the angle all these terms give zero, and therefore the total longitudinal OAM of the beam (5.36) is zero: $J_{2z} = 0$. The exception is the case when $n = 0$. In this case the polarization of the field (5.35) is elliptical and due to the spin-orbit effect at the focus the total longitudinal OAM is nonzero and equal to: $J_{2z,\ n\ =\ 0} = 4\delta(1 + \delta^2)^{-1}(W_2 + W_1)$, the density of the longitudinal OAM at $n = 0$ is equal to:

$$
j_{2z,n=0} = \frac{4\delta}{1+\delta^2}\left(I_{2,2}^2 + I_{1,1}^2 + I_{2,2}I_{0,0}\cos 2\varphi\right).
\tag{5.38}
$$

All terms consisting of products of different integrals (5.28) is also seen from (5.37) and (5.38) can change their sign when the radial variable r changes. Therefore, the density of the longitudinal OAM can also change sign when the radial coordinate changes. This means that microparticles captured in the focus of the beam (5.35) at different distances will rotate in different directions along circles of different radii (clockwise and counterclockwise). We obtain a simple expression for the density of the OAM (5.37) at $n = 2$:

$$
\begin{aligned}
j_{2z,n=2} &= \frac{\delta}{1+\delta^2}\times \\
&\times\left[4I_{2,2}I_{0,2} - 4I_{1,1}^2 + 2\cos 2\varphi(I_{2,2}I_{2,0} - I_{0,0}I_{0,2})\right].
\end{aligned}
\tag{5.39}
$$

The total longitudinal OAM of the field (5.39) is zero, since it can be shown that it is equal to the expression:

$$J_{2z,n=2} = \frac{4\delta}{\left(1+\delta^2\right)}\left(W_{20} - W_1\right) = 0, \tag{5.40}$$

and both terms in (5.40) are equal to each other:

$$\begin{aligned} W_{20} &= 2\pi \int\limits_0^\infty I_{2,2}(r) I_{0,2}(r) r dr = \\ &= 4\pi f^2 \int\limits_0^{\theta_0} \sin^3\left(\frac{\theta}{2}\right)\cos^3\left(\frac{\theta}{2}\right)|B(\theta)|^2 d\theta = W_1. \end{aligned} \tag{5.41}$$

5.2.3 OAM Density at the Focus of an Optical Vortex with Cylindrical Polarization

Let us obtain an expression for the OAM density longitudinal projection in the sharp focus of an optical vortex with a topological charge of m and with a cylindrical polarization of order n. The Jones vector for the initial field can be expressed as:

$$\mathbf{E}_3(\varphi) = e^{im\varphi}\begin{pmatrix} \cos n\varphi \\ \sin n\varphi \end{pmatrix}. \tag{5.42}$$

The electric field strength vector projections in the focus plane can be written as [71]:

$$\begin{aligned} E_x &= \frac{i^{m+n-1}}{2} e^{i(m+n)\varphi}\left(I_{0,m+n} + e^{-2i\varphi} I_{2,m+n-2}\right) + \\ &\frac{i^{m-n-1}}{2} e^{i(m-n)\varphi}\left(I_{0,m+n} + e^{2i\varphi} I_{2,m+n+2}\right), \\ E_y &= \frac{i^{m+n}}{2} e^{i(m+n)\varphi}\left(-I_{0,m+n} + e^{-2i\varphi} I_{2,m+n-2}\right) + \\ &\frac{i^{m-n}}{2} e^{i(m-n)\varphi}\left(I_{0,m+n} - e^{2i\varphi} I_{2,m+n+2}\right), \\ E_z &= i^{m+n} e^{i(m+n-1)\varphi} I_{1,m+n-1} - i^{m-n} e^{i(m-n+1)\varphi} I_{1,m-n+1}. \end{aligned} \tag{5.43}$$

Substituting the field projections at the focus (5.43) into expression (5.25) for the longitudinal OAM density, we obtain:

$$\begin{aligned}
j_{3z}(r,\varphi) = & \left[\frac{(m+n)}{2}I_{0,m+n}^2 + \frac{(m-n)}{2}I_{0,m-n}^2 + \right.\\
& \frac{(m+n-2)}{2}I_{2,m+n-2}^2 + \frac{(m-n+2)}{2}I_{2,m-n+2}^2 + \\
& \left.(m+n-1)I_{1,m+n-1}^2 + (m-n+1)I_{1,m-n+1}^2\right] + \\
& (-1)^n\cos((2n-2)\varphi)[(m+1)I_{0,m+n}I_{2,m-n+2} + \\
& (m-1)I_{0,m-n}I_{2,m+n-2} - 2mI_{1,m-n+1}I_{1,m+n-1}].
\end{aligned} \tag{5.44}$$

There are regions in the focal plane where the OAM is proportional to sum of orders $m + n$, and there are regions, where OAM is proportional to the difference of orders m-n since the initial beam (5.42) contains two types of singularities: phase singularity of order m and polarization singularity of order n. It is seen from (5.44) that the OAM density at the focus of the field (5.42) depends on both numbers m and n, as well as on both cylindrical coordinates (r, φ). When going around a closed trajectory of a certain radius and with the center on the optical axis, the OAM density can have $4(n - 1)$ regions of different signs, since there is a term with $\cos(2n\text{-}2)\varphi$. Since Eq. (5.44) includes terms with the product of integrals (5.28) with different numbers, such terms change sign when the radial variable changes. Therefore, the entire OAM density (5.44) can change sign when the variable r changes with a constant variable φ. At $m = 0$ (no vortex), the OAM density (5.44) is zero. At $n = 1$, the OAM density (5.44) is radially symmetric and greater than zero ($m > 0$). If we integrate the OAM density (5.44) over the entire plane of the focus, we obtain the longitudinal projection of the entire OAM:

$$J_{3z} = m(W_0 + W_2 + 2W_1) = mW. \tag{5.45}$$

Expression (5.45) coincides with Eq. (5.32) and shows that the total OAM of the beam (5.42) is determined only by the topological charge of the optical vortex m and does not depend on the order of cylindrical polarization n.

5.2.4 OAM Density at the Focus of the Beam with Angular Dependence of Elliptical Polarization

Let us consider an initial field with hybrid polarization, which changes the polarization ellipse depending on the azimuthal angle and is described by a Jones vector of the form:

$$\mathbf{E}_4(\varphi) = \begin{pmatrix} -i\sin n\varphi \\ \cos n\varphi \end{pmatrix}. \tag{5.46}$$

Depending on the angle, the polarization state of the field (5.46) changes from linear to elliptical and right-handed circular polarization. Moreover, when going along the circle with the center on the optical axis linear and circular polarizations repeat n times. Therefore, we also call this hybrid polarization circular-cylindrical polarization. The projections of the electric field strength vector in the sharp focal plane of the field (5.46) can be written as [72]:

$$\begin{aligned}
E_x &= \frac{-i^n}{2}\left[I_{0,n}\sin n\varphi + \frac{1+i}{2}I_{2,m+2}\sin(m+2)\varphi + \right.\\
&\left. + \frac{1-i}{2}I_{2,m-2}\sin(m-2)\varphi\right],\\
E_y &= \frac{-i^n}{2}\left[iI_{0,n}\cos n\varphi - \frac{1+i}{2}I_{2,m+2}\times\right.\\
&\left.\times\cos(m+2)\varphi + \frac{1-i}{2}I_{2,m-2}\cos(m-2)\varphi\right],\\
E_z &= \frac{i^n}{2}\left[(1+i)I_{1,n+1}\sin(n+1)\varphi + \right.\\
&\left. +(1-i)I_{1,m-1}\sin(m-1)\varphi\right].
\end{aligned} \tag{5.47}$$

Substituting the projections of the field at the focus (5.47) into expression (5.25) we obtain the following expression for the density of the longitudinal OAM:

$$\begin{aligned}
j_{4z} &= \frac{1}{4}\left[-\sin 2\varphi\{(n+1)I_{0,n}I_{2,n+2} + \right.\\
&+(n-1)I_{0,n}I_{2,n-2} + 2nI_{1,n+1}I_{1,n-1}\} + \\
&\left. +2\sin 2n\varphi\{I_{2,n+2}I_{2,n-2} + 2I_{1,n+1}I_{1,n-1}\}\right].
\end{aligned} \tag{5.48}$$

The OAM density (5.48) depends on the doubled azimuthal angle 2φ. Therefore, the density distribution at the focus is a centrally symmetric (like a snowflake), i.e., the rays φ and $\varphi + \pi$ have identical OAM distributions. Because all terms in (5.48) are multiplied by trigonometric periodic functions, the total OAM will be zero, $J_{4z} = 0$. At $n = 0$, the initial field (5.46) have vertical linear polarization, and it follows from (5.48) that the OAM density of such a field at the focus is zero. At $n > 0$, in expression (5.48), all terms have a product of integrals (5.28) with different numbers, and therefore all terms can change sign when the radial variable r changes. Therefore, the expression for the OAM density (5.48) can also change sign when r changes and at a constant φ. These conclusions are confirmed below using numerical simulation. In this case, as in the case of field (5.35), the cylindrical symmetry of the initial field amplitude is broken. Therefore, due to the spin and orbital Hall effects, local regions with spin of different signs and with longitudinal OAM of different signs are formed in the focus.

5.2.5 Numerical Simulation

Shown in Fig. 5.4 are the intensity distributions (Fig. 5.4a,c,e,g) and the orbital angular momentum density (Fig. 5.4b,d,f,h) in the sharp focus of beams with hybrid polarization (the Jones vector is $[-i\sin n\varphi, \cos n\varphi]^T$ (5.46), Fig. 5.4a–d) and the superposition of beams with radial and linear polarization (the Jones vector is $[\cos n\varphi, i\alpha + \sin n\varphi]^T$ (5.35), Fig. 1e–h). The light in the focus is calculated using the Debye integrals within the Richards–Wolf formalism [6], and the OAM density is calculated using formula (5.25). The following parameters are used for the simulation: wavelength $\lambda = 532$ nm, the focal length $f = 20\lambda$, the polarization order $n = 2$ (Fig. 5.4a,b,e,f) and $n = 4$ (Fig. 5.4c,d,g,h), relative amplitude of the linearly polarized beam $\alpha = 1$, the numerical aperture $NA = 0.95$, the size of the calculated region is 2×2 μm.

It is seen from Fig. 5.4 that the OAM density changes sign both with a change in the radial variable r and with a change in the azimuthal angle φ. The distribution in Fig. 5.4b,d confirms the correctness of the expression (5.48), the type of OAM distributions in Fig. 5.4f,h confirms the correctness of expressions (5.37) and (5.39). The different sign of the OAM in different areas of the focal plane demonstrates the orbital Hall effect, i.e. the separation of photons with different OAM signs in space. At the same time, in areas with different OAM signs, the canonical energy flux, as well as the azimuthal projection of the Poynting vector, rotate in different directions (clockwise and counterclockwise).

Figure 5.5 illustrates transverse Poynting vector in the light field depicted in Fig. 5.4a,b. In Fig. 5.5, 8 subwave regions are visible, in which the Poynting vector rotates along closed trajectories either clockwise or counterclockwise. According to Fig. 5.5, the density of longitudinal angular momentum for the field (5.46) at $n = 2$ is positive in 4 regions and negative in the other 4 regions. Figure 5.5 differs from Fig. 5.4b because the Poynting vector (Fig. 5.5) includes not only the energy flux, but also the spin flux [47], and the OAM takes into account only the energy flow without taking into account the spin.

Figure 5.6 depicts the intensity distributions (Fig. 5.6a,c,e,g) and the orbital angular momentum density (Fig. 5.6b,d,f,h) in the sharp focus of vortex beams with circular polarization (the Jones vector is $e^{im\varphi}\exp[1, i]^T$ (3), Fig. 5.6a–d) and of cylindrically polarized optical vortices (the Jones vector is $e^{im\varphi}[\cos n\varphi, \sin n\varphi]^T$ (5.42), Fig. 5.6e–h). The light in the focus is also calculated using the Debye integrals within the Richards–Wolf formalism [6], and the OAM density is calculated using formula (5.25). The following parameters are used for the simulation: wavelength $\lambda = 532$ nm, the focal length $f = 20\lambda$, vortex topological charge $m = 2$ (Fig. 5.6a,b,e,f) and $m = 4$ (Fig. 5.6c,d,g,h), the polarization order $n = 4$ (Fig. 5.6e,f) and $n = 2$ (Fig. 5.6g,h), the numerical aperture $NA = 0.95$, the size of the calculated region is 2×2 μm.

The OAM density distributions in Fig. 5.6b,d are expectable, since the homogeneously polarized optical vortices with positive topological charges $m = 2$ and $m = 4$ should have positive OAM density. Circularly symmetric OAM distributions

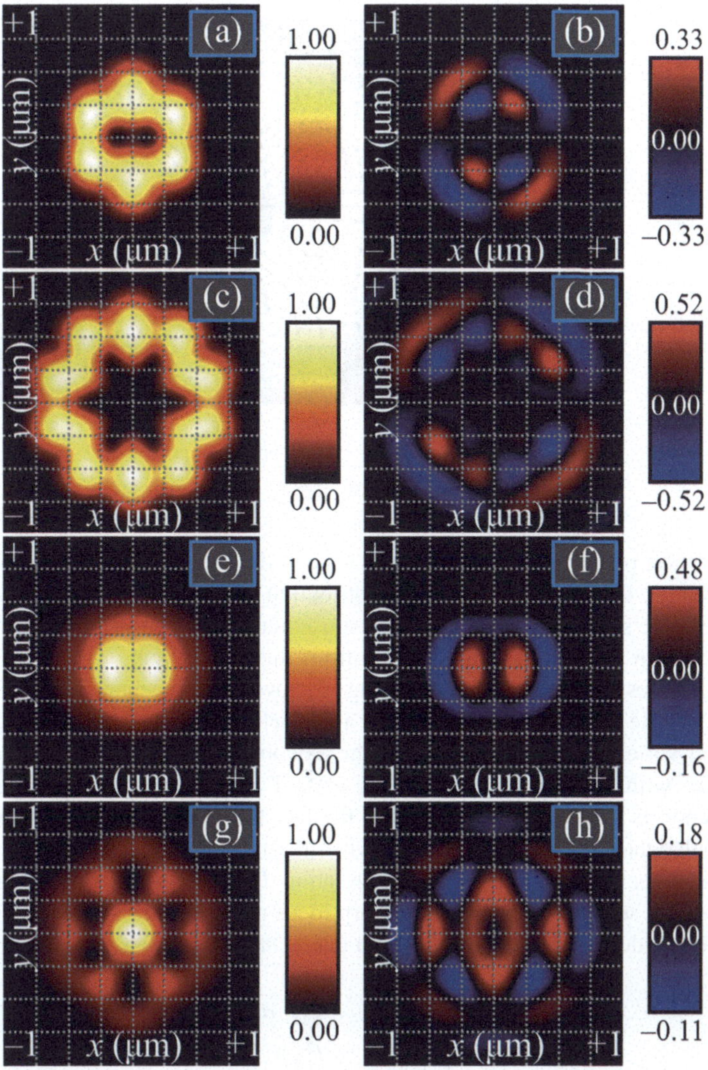

Fig. 5.4 Distributions of intensity (**a**, **c**, **e**, **g**) and density of longitudinal projection of OAM (**b**, **d**, **f**, **h**) in the sharp focus of beams with hybrid polarization (5.46) (**a–d**) and in the superposition of beams with radial and linear polarization (5.35) (**e–h**) at different polarization orders: $n = 2$ (**a**, **b**, **e**, **f**) and $n = 4$ (**c**, **d**, **g**, **h**)

confirm Eq. (5.31), which is independent of the polar angle φ. Maximal OAM densities, normalized to maximal intensity, are consistent with Eq. (5.33), i.e. normalized OAM is greater than m. The computed OAM density distributions of cylindrically polarized vortices are consistent with Eq. (5.44). According to [69], this expression has two overwhelming terms: the first and the second one. If $m > n$, i.e. if the vortex charge exceeds the polarization order, then the OAM density should

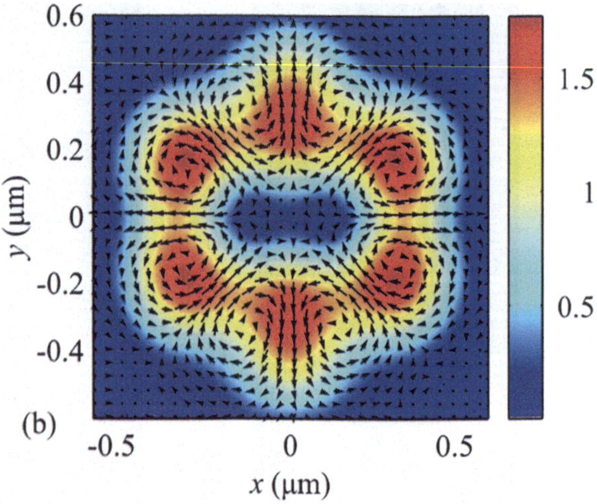

Fig. 5.5 Transverse Poynting vector in the light field depicted in Fig. 5.4a, b

be everywhere positive (Fig. 5.6h). If $m < n$, i.e. if the polarization order exceeds the vortex charge, then the OAM density distribution should have the negative inner ring, due to the second term in Eq. (5.44), surrounded by the positive outer ring, due to the first term in Eq. (5.44) Fig. 5.6f confirms this distribution.

The particles captured on the light ring is shown experimentally in [61, 62] to rotate in a circle counterclockwise ($\sigma = 1$) and clockwise ($\sigma = -1$) in the absence of an optical vortex in the sharp focus of the axicon exp(iar). Here a is the axicon parameter, with circular polarization $(1, i\sigma)^{\mathrm{T}}$, T is the transposition sign. This phenomenon is explained in [61, 62] by the fact that the paraxial OAM depends on the polarization state, and therefore when the rotation direction of the circular polarization vector changes, the particles rotation direction in a circle also changes. But this is not so. The particles in the sharp focus rotate in a circle due to the spin-orbit effect. Indeed, it follows from (5.29) that at $m = 0$ we obtain:

$$j_{1z,m=0} = \begin{cases} 2\left(I_{2,2}^2 + I_{1,1}^2\right), & \sigma = 1, \\ -2\left(I_{2,2}^2 + I_{1,1}^2\right), & \sigma = -1. \end{cases} \tag{5.49}$$

Figure 5.7 shows two identical ring intensity distributions (a, c) in the sharp focus (NA = 0.95) of an ideal spherical lens with the initial amplitude $B(\theta) = \exp(ia\tan\theta)$. This equation is obtained from the relation $r = f\tan\theta$ for the radial variable r on a spherical front converging to the focus, where f is the focal length, θ is the inclination angle to the optical axis of the rays converging to the geometric focus point. And the OAM density distribution on the intensity ring: for the right-handed circular polarization (b) and the left-handed circular polarization (d). Numerical simulation parameters are as follows: wavelength $\lambda = 532$ nm is used, the focal length is $f = 20\lambda$, the axicon parameter $a = 10$, the calculation region size is 8×8 μm.

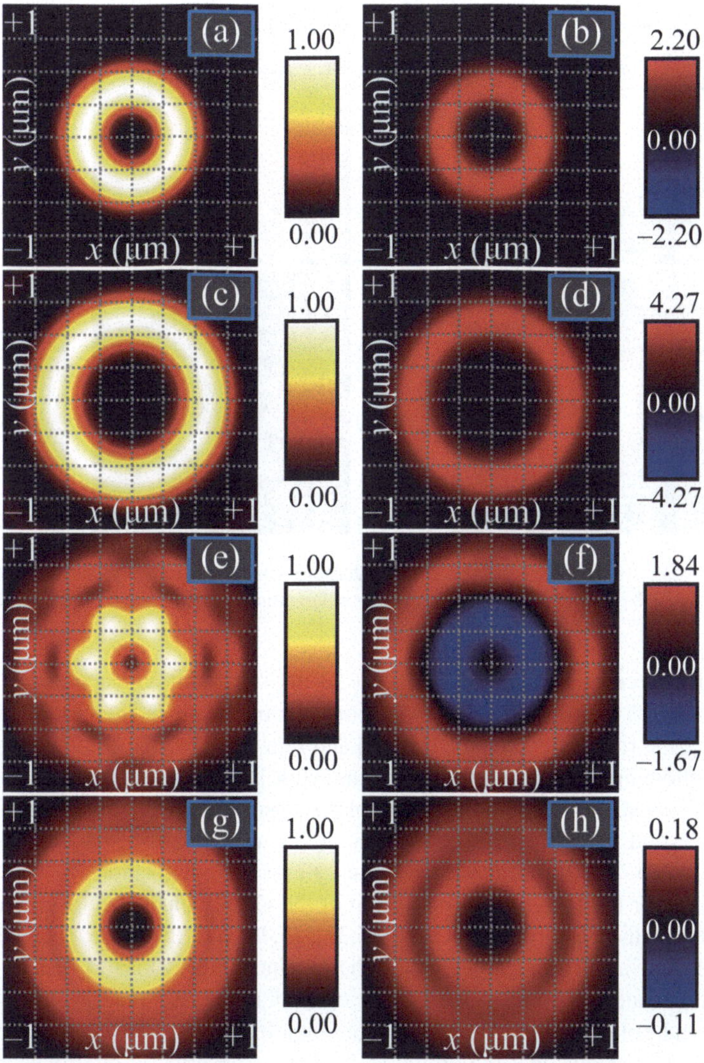

Fig. 5.6 Distributions of intensity (**a**, **c**, **e**, **g**) and density of longitudinal OAM component (**b**, **d**, **f**, **h**) in the sharp focus of circularly (**a–d**) (5.26) and cylindrically (**e–h**) (5.42) polarized optical vortices at different vortex and polarization orders: $m = 2$ (**a**, **b**, **e**, **f**), $m = 4$ (**c**, **d**, **g**, **h**), $n = 4$ (e, f), $n = 2$ (g, h)

Due to spin-orbit conversion the OAM density is seen from Fig. 5.7 to change sign when the direction of rotation of the polarization vector changes. An axicon is seen from Fig. 5.7 to form an intensity ring in the sharp focus. And along this ring, due to the spin-orbit conversion, light with right-handed circular polarization forms an azimuthal energy flow rotating counterclockwise around the ring. In this case, the OAM density on the ring is positive (Eq.(5.49) and Fig. 5.7b). And if the light has

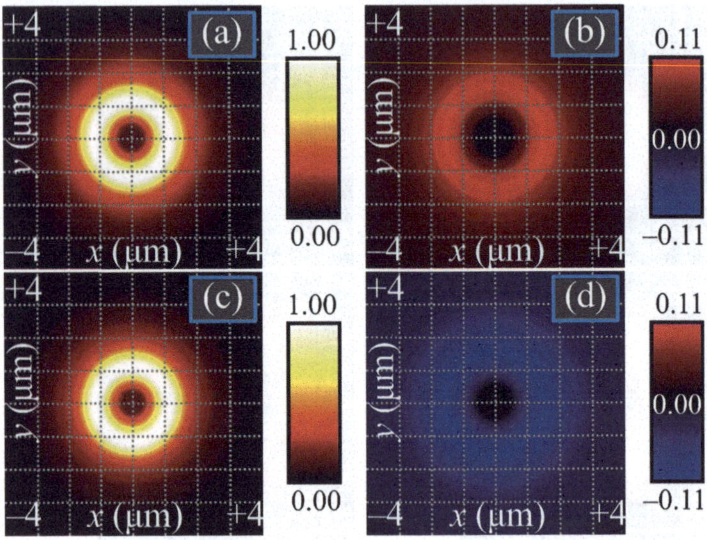

Fig. 5.7 Intensity distributions (**a**, **c**) and OAM densities (**b**, **d**) in the sharp focus of the field with right-handed (**a**, **b**) and left-handed (**c**, **d**) circular polarization

left-handed circular polarization, then due to the spin-orbit conversion, an energy flow rotating clockwise is formed in the focus along the ring. Therefore, the OAM density on the ring is negative (Eq. (5.49) and Fig. 5.7d).

The following results are obtained in the work [50]. For four initial vector light fields: an optical vortex with elliptical polarization (5.26), a superposition of light with linear and cylindrical polarizations (5.35), an optical vortex with cylindrical polarization (5.42) and light with circular-cylindrical polarization (5.46), explicit expressions for the longitudinal projection of the OAM density vector in the sharp focus are obtained: expressions (5.29), (5.37), (5.44) and (5.48), respectively. The OAM density in the focus is seen from the obtained expressions to depend on the polarization state of the initial field: on the order of cylindrical polarization n and on the elliptical polarization parameter α. From the obtained expressions for the OAM density it also follows that the orbital Hall effect is observed in the focus in some cases. In different areas of the focal plane the OAM density has a different sign (Fig. 5.4). Note that in the initial plane, fields (5.35), (5.46) and the field shown in Fig. 5.4 are without an optical vortex, but due to the local spin-orbit effect, regions are formed at the focus in which the energy flow (including the canonical one) rotates in different directions. Interestingly, spin-orbit coupling occurs in the case of sharp focusing of spatiotemporal optical vortices with circular [73] and radial [74] polarizations. The tilt of a 2D spatiotemporal optical vortex allows to form a transverse OAM with an arbitrary orientation at the focus [73]. 3D spatiotemporal optical vortices [74] have two orthogonal transverse OAMs, and the polarization structure near the focus has a toroidal topology.

5.3 Hybrid Orbital Angular Momentum of a Vector Light Beam

It is well known that vortex laser beams (for instance, Laguerre-Gaussian or Bessel-Gaussian beams) have helical wave front with a phase singularity (uncertainty) on the optical axis, as well as nonzero topological charge [75] and orbital angular momentum (OAM) [3]. Vector beams with inhomogeneous polarization (for example, radial or azimuthal) can also have the polarization singularities, characterized by polarization singularity indices [76, 77]. Thus, a questions arise: can the polarization vortices (for instance, cylindrical vector beams [78]) have some polarization OAM (pOAM) and which physical sense has the pOAM? Commonly used orbital angular momentum of a vortex field indicates the azimuthal velocity of the phase change and manifests itself in ability of a light beam with the OAM to rotate a particle along a circular path [79]. The same way, we can derive the pOAM quantity that indicates the azimuthal velocity of changing tilt of linear polarization vector in the beam section. In practice, such a polarization vortex can perform a mass transport [80] in a polarization-sensitive medium [81] and generate a spiral-shaped surface relief [82]. For instance, in work [82], spiral mass transport in an azopolymer thin film (chalcogenide glass) is done by a vector laser beam with a direction of the linear polarization vector linearly changing with the azimuthal angle. Thus, the light beam with azimuthally changing linear polarization vector «forces» the azopolymer molecules to shift along a circle. Therefore, it can be supposed that such polarization vortex has the polarization OAM.

It is also known that light fields with elliptic and circular polarization have the spin [1] and spin angular momentum [2] that allows them rotating absorbing particles around the center of mass [83]. The scattering force of light acting onto a particle can depend on the rotor (curl) of the spin angular momentum vector [84, 85]. Three-dimensional polarization distribution of a laser beam is employed for accelerating charged particles [86–88]. Cylindrical vector beams with inhomogeneous linear polarization, that do not have the spin, are also used for particles manipulation [89–91]. With inhomogeneous polarization of light, many optical phenomena are related, such as spin and orbital Hall effect [92, 93], polarization Möbius stripes [26], optical wheels [21, 22], spin-orbital conversion [8, 9]. For example, due to the spin-orbital conversion at the tight focus, an initial Gaussian beam with circular polarization generates in the focus an azimuthal energy flow that can rotate particles along a circle [94]. Due to the orbital Hall effect in the tight focus of a vector light field with inhomogeneous linear polarization, closed energy flows are locally generated, that also can rotate particles around a circle [95]. The reverse energy flux in the focus of vector fields as related with the rotor of the spin angular momentum [96].

For this reason, we find it interesting to consider an OAM for light fields with inhomogeneous polarization. In this work, we introduce and investigate two novel OAMs: polarization OAM, proportional to the azimuthal velocity of changing tilt of the polarization vector, and hybrid OAM, proportional to the azimuthal velocity of changing polarization ellipticity. These specified OAMs are compared with the conventional OAM, applied for scalar and vector vortex beams.

There is a physical reason why two characteristics of the scalar vortex field are considered separately: topological charge (TC) and the OAM. The TC characterizes azimuthal velocity of phase change of the wave front, i.e. TC indicates how many phase jumps by 2π has the wave front when passing along a closed contour around the optical axis. The OAM characterizes azimuthal change of the wavevector direction of photons, which is orthogonal to the wave front. Due to the OAM, photons move along straight lines, which do not intersect with the optical axis and are not parallel to it. Thus, when photons with OAM are absorbed by a particle, they transfer momentum to it and rotate the particle along a circular path. Similarly, for inhomogeneous vector fields, also two characteristics should be considered: polarization singularity index and polarization OAM. The polarization singularity index indicates the number of full turnovers of linear polarization vectors when passing along a closed contour around the beam optical axis in the beam transverse section. The OAM describes azimuthal velocity of change of linear polarization vector direction of each photon in the beam section. When such photons are absorbed by polarization-sensitive medium, they generate gradient of medium molecules concentration (molecules are stretched along different directions) and cause mass transfer (molecules motions, including along a circle).

5.3.1 Topological Charge of a Scalar Vortex Field and Polarization Singularity Index of a Vector Field

Here we consider paraxial vector coherent light fields (beams) with polarization singularities in a transverse section. Well known are vector beams with inhomogeneous linear polarization, that can have points with indefinite direction of linear polarization vector. Such points are called V-points. If a vector beam has a point with inhomogeneous elliptic polarization, then such a point with circular polarization is a polarization singularity point (C-point), since the direction of the major axis of the polarization ellipse is indefinite in this point. Such polarization singularity points are well studied [76, 77]. To characterize them, polarization indices are used, such as the Poincaré-Hopf index [97] and the Stokes index [98]. For formal description of these indices, we will follow the work [99], where the polarization singularity indices are defined similarly to the topological charge (TC) of scalar vortex beams. Thus, we will treat the TC of scalar optical vortices according to the M.V. Berry's formula [75]:

$$TC = \frac{1}{2\pi} \lim_{r \to \infty} \int_0^{2\pi} \frac{\partial}{\partial \varphi} [\arg E(r, \varphi)] d\varphi, \tag{5.50}$$

where $E(r, \varphi)$ is the complex amplitude of a scalar light field that depends on the transverse polar coordinates, longitudinal axis z is directed along the beam optical axis (propagation axis), arg is the argument of a complex number (phase of the light field), lim is the limit when the radial polar coordinate r tends to infinity. If a scalar light field has the complex amplitude

$$E(r, \varphi) = A(r) \exp(in\varphi), \tag{5.51}$$

then, substituting Eq. (5.51) into Eq. (5.50), we get that $TC = n$. Vector fields in the initial plane are described only by transverse components of the electric field strength vector (Jones vector):

$$\mathbf{E}(r, \varphi) = \begin{pmatrix} E_x(r, \varphi) \\ E_y(r, \varphi) \end{pmatrix}. \tag{5.52}$$

If the functions E_x and E_y are real-valued then the vector beam has in its section inhomogeneous linear polarization that can change the direction of the polarization vector in each point. Then, we can formally write a scalar complex amplitude of the form

$$E_c(r, \varphi) = E_x(r, \varphi) + iE_y(r, \varphi), \tag{5.53}$$

for which the formula (5.50) can be applied. In this case, the Poincaré-Hopf polarization singularity index [97] for the vector beam (5.52) is equal to the following expression:

$$\eta = \frac{1}{2\pi} \lim_{r \to \infty} \int_0^{2\pi} \frac{\partial}{\partial \varphi} [\arg E_c(r, \varphi)] d\varphi. \tag{5.54}$$

The index (5.54) indicates the integer number of turns by the 2π angle, made by the linear polarization vector in the section of the beam (5.52), when passing along a large-radius circle with the center on the optical axis. In practice, the circle radius should be chosen so that all polarization singularity points would be within this circle. If the functions E_x and E_y in Eq. (5.52) are complex-valued then the index (5.54) cannot be used since the complex amplitude (5.53) does not have the required structure. In this case, the Stokes index [98] is applied. For the field (5.52), the real components of the Stokes vector are given by:

$$\begin{aligned} S_1 &= |E_x|^2 - |E_y|^2, \\ S_2 &= 2\,\mathrm{Re}\left(E_x^* E_y\right), \\ S_3 &= 2\mathrm{Im}\left(E_x^* E_y\right). \end{aligned} \tag{5.55}$$

From the first two real Stokes components, scalar Stokes complex amplitude can be obtained:

$$S_c = S_1 + iS_2. \tag{5.56}$$

Using the M.V. Berry's formula (5.50) and the complex Stokes field (5.56), the Stokes polarization singularity index for the vector fields (5.52) can be defined as

$$\sigma = \frac{1}{2\pi} \lim_{r\to\infty} \int_0^{2\pi} \frac{\partial}{\partial\varphi} [\arg S_c(r,\varphi)] d\varphi. \tag{5.57}$$

It can be shown that the Stokes index is two times greater than the Poincaré-Hopf index.

5.3.2 *Orbital Angular Momentum of Scalar and Vector Fields*

Now we return back to the scalar vortex beams that are characterized not only by the TC (5.50), but also by the orbital angular momentum (OAM) [3]:

$$J_z = \mathrm{Im} \int_0^{\infty} \int_0^{2\pi} E^*(r,\varphi) \frac{\partial}{\partial\varphi} E(r,\varphi) r dr d\varphi. \tag{5.58}$$

Comparison of Eqs. (5.50) and (5.58) indicates that the TC and the OAM are different quantities, characterizing the vortex beams, since the TC (5.50) is defined only via the phase of the light field, whereas the expression for the OAM (5.58) contains both the phase and the amplitude. For vector light fields, similarly to Eq. (5.58), the vector OAM can be defined up to a constant factor $1/(8\pi\omega)$ with ω being the cyclic frequency of monochromatic light, as [67]:

$$\mathbf{J} = \mathrm{Im} \int_0^{\infty} \int_0^{2\pi} \sum_{p=x,y,z} E_p^* (\mathbf{r} \times \nabla) E_p r dr d\varphi, \tag{5.59}$$

where $\mathbf{r} = (x, y, z)$ is the position vector of a point in the beam section. However, for paraxial vector fields, the longitudinal component of the electric field strength vector can be neglected and, thus, only the longitudinal component of the vector (5.59) is significant. The longitudinal component of the OAM vector (5.59) for the vector field (5.52) is given by

$$J_{v,z} = \mathrm{Im} \int_0^{\infty} \int_0^{2\pi} \left[E_x^*(r,\varphi) \frac{\partial}{\partial \varphi} E_x(r,\varphi) + E_y^*(r,\varphi) \frac{\partial}{\partial \varphi} E_y(r,\varphi) \right] r dr d\varphi. \quad (5.60)$$

Physically, the vector OAM (5.60) is the average azimuthal velocity of changing phase of each of the transverse field components. For instance, for an initial vector vortex field written as

$$\mathbf{E}_1(r,\varphi) = \frac{A(r)}{\sqrt{2}} \begin{pmatrix} \exp(in\varphi) \\ \exp(im\varphi) \end{pmatrix}, \quad (5.61)$$

the OAM (5.61) is equal to

$$J_{v,z} = \left(\frac{n+m}{2} \right) W, \qquad W = 2\pi \int_0^{\infty} \int_0^{2\pi} A^2(r) r dr. \quad (5.62)$$

In Eq. (5.62), W stands for the beam power. As seen from Eq. (5.62), the OAM is proportional to the arithmetic mean of the topological charges of the vortices in each component of the vector field (5.61). The OAM (5.60) of a vector field with arbitrary inhomogeneous linear polarization, when the functions E_x and E_y in Eq. (5.52) are real-valued, is zero. Thus, whereas the indices (5.54) and (5.57) allow characterization a polarization singularity, the OAM (5.60) cannot characterize polarization singularities of a vector field. Rather, it allows only characterization of phase singularities of the electric field transverse components. For this reason, we should find analogues to the OAM (5.60), capable of characterizing the polarization vortices. Now we consider several candidates to the OAM of the polarization vortices. The vector field (5.52) can be written via the real and imaginary parts of the components E_x and E_y:

$$\mathbf{E}(r,\varphi) = \begin{pmatrix} E_x(r,\varphi) \\ E_y(r,\varphi) \end{pmatrix} = \begin{pmatrix} A_x + iB_x \\ A_y + iB_y \end{pmatrix}. \quad (5.63)$$

Using these designations, the OAM (5.60) is equal to

$$J_{v,z} = \int_0^{\infty} \int_0^{2\pi} \left[A_x \frac{\partial}{\partial \varphi} B_x - B_x \frac{\partial}{\partial \varphi} A_x + A_y \frac{\partial}{\partial \varphi} B_y - B_y \frac{\partial}{\partial \varphi} A_y \right] r dr d\varphi. \tag{5.64}$$

According to Eq. (5.64), the conventional vectorial OAM, applied to a vector field with inhomogeneous linear polarization, i.e. when the functions E_x and E_y in Eq. (5.63) are real-valued, is zero.

5.3.3 Orbital Angular Momentum of an Inhomogeneous Vector Field with Local Linear Polarization

Thus, we introduce here another OAM definition, which is different to Eqs. (5.58) and (5.64) and can characterize vector beams with inhomogeneous polarization. From now on, we call it a polarization OAM (pOAM) and define it as follows:

$$\begin{aligned} J_{p,z} = & \\ \mathrm{Re} & \int_0^{\infty} \int_0^{2\pi} \left[E_x^*(r,\varphi) \frac{\partial}{\partial \varphi} E_y(r,\varphi) - E_y^*(r,\varphi) \frac{\partial}{\partial \varphi} E_x(r,\varphi) \right] r dr d\varphi \\ = & \int_0^{\infty} \int_0^{2\pi} \left(A_x \frac{\partial}{\partial \varphi} A_y - A_y \frac{\partial}{\partial \varphi} A_x + B_x \frac{\partial}{\partial \varphi} B_y - B_y \frac{\partial}{\partial \varphi} B_x \right) r dr d\varphi. \end{aligned} \tag{5.65}$$

As seen from Eq. (5.65), real and imaginary parts in the integrand are separated, and the OAM (5.65) is thus nonzero, when the functions E_x and E_y in Eq. (5.63) are real-valued. We can verify it by an example. As an initial field, we consider a cylindrical vector beam with the Jones vector

$$\mathbf{E}_2(r,\varphi) = A(r) \begin{pmatrix} \cos(n\varphi) \\ \sin(n\varphi) \end{pmatrix}. \tag{5.66}$$

Substituting Eq. (5.66) into Eq. (5.65), we get $J_{p,z} = nW$, whereas applying the OAM expression (5.64) to the field (5.66) yields zero: $J_{v,z} = 0$. We note that for the cylindrical vector beam (5.66), polarization singularity index (5.54) is n, while the Stokes index equals $2n$. The index (5.54) for the field (5.66) indicates that upon passing in the beam section along a closed contour around the optical axis, linear polarization vectors make n full turnovers. Thus, the pOAM (5.65) describes the average azimuthal velocity of rotation of linear polarization vectors in a section of an inhomogeneous vector beam. We note also that the polarization OAM (5.65), applied to a vortex beam with circular polarization

$$\mathbf{E}_3(r,\varphi) = A(r)\frac{\exp(in\varphi)}{\sqrt{2}}\begin{pmatrix} 1 \\ -i \end{pmatrix}, \tag{5.67}$$

yields the same result, $J_{p,z} = nW$. Thus, the polarization OAM indicates the average azimuthal velocity of phase change of a circularly polarized vortex field (5.67), and the average azimuthal velocity of rotation of linear polarization vectors of an inhomogeneously polarized beam (5.66), although for linearly polarized optical vortex the pOAM is zero.

5.3.4 *Orbital Angular Momentum of a Hybrid Vector Field with Elliptic Polarization*

The pOAM definition (5.65) can be modified if instead of the real part of a complex number, the imaginary part is taken, and if the difference of two terms is replaced by sum. Then we obtain a new definition of the polarization OAM. For distinguishing it from the definition (5.65), we call this OAM the hybrid polarization OAM (hOAM):

$$\begin{aligned} J_{h,z} &= \\ &\operatorname{Im}\int_0^\infty\int_0^{2\pi}\left[E_x^*(r,\varphi)\frac{\partial}{\partial\varphi}E_y(r,\varphi) + E_y^*(r,\varphi)\frac{\partial}{\partial\varphi}E_x(r,\varphi)\right]rdrd\varphi \\ &= \int_0^\infty\int_0^{2\pi}\left(A_x\frac{\partial}{\partial\varphi}B_y + A_y\frac{\partial}{\partial\varphi}B_x - B_x\frac{\partial}{\partial\varphi}A_y - B_y\frac{\partial}{\partial\varphi}A_x\right)rdrd\varphi. \end{aligned} \tag{5.68}$$

If the field (5.66) is substituted into the hOAM (5.68) then we get zero since the imaginary part of the integral is taken in Eq. (5.68) whereas the integrand is real-valued as is the field (5.66). Now we consider a hybrid vector field, which combines properties of both a cylindrical vector field (5.66) and of circularly polarized field:

$$\mathbf{E}_4(r,\varphi) = A(r)\begin{pmatrix} \cos(n\varphi) \\ i\sin(n\varphi) \end{pmatrix}. \tag{5.69}$$

With increasing azimuthal angle, polarization of the vector field (5.69) changes from linear to elliptic and circular, and, further, changes again from elliptic to linear, repeating periodically. Substituting Eq. (5.69) into Eq. (5.68), we get $J_{h,z} = nW$. Thus, the hybrid OAM (5.68) indicates the azimuthal velocity of changing ellipticity degree of light beams with inhomogeneous elliptic polarization. Comparison of the conventional OAM for a vector field with the hybrid OAM (5.68) reveals that the OAM (5.68) is nonzero if the components of the beam (5.63) are pure imaginary or pure real, like those of the beam (5.69). For the conventional OAM (5.64) to be nonzero, it is necessary that both components of the initial field had both real and imaginary parts. As an example, we consider a vortex beam with the topological charge m and with nth-order cylindrical polarization:

$$\mathbf{E}_5(r,\varphi) = A(r)\exp(im\varphi)\begin{pmatrix} \cos(n\varphi) \\ \sin(n\varphi) \end{pmatrix}. \tag{5.70}$$

Substituting Eq. (5.70) into Eq. (5.64), we obtain that the conventional OAM of the beam (5.70) is equal to $J_{v,z} = mW$. If Eq. (5.70) is substituted into Eq. (5.65),we derive that the polarization OAM of the beam (5.70) equals $J_{p,z} = nW$. The hybrid OAM of the beam (5.70) is zero. Similarly, we consider here a vortex beam with hybrid polarization:

$$\mathbf{E}_6(r,\varphi) = A(r)\exp(im\varphi)\begin{pmatrix} \cos(n\varphi) \\ i\sin(n\varphi) \end{pmatrix}. \tag{5.71}$$

It is easy to prove for the beam (5.71) that $J_{v,z} = mW$, $J_{p,z} = 0$, $J_{h,z} = nW$. We note that all three OAMs [Eqs. (5.60), (5.65) and (5.68)] are propagation invariant and do not changed their value upon free space propagation of a light field.

5.3.5 *Numerical Simulation*

To confirm the theoretical values of the three OAM types, we made numerical computations. As a rotationally symmetrical optical vortex (5.51), we used the Laguerre-Gaussian beams in all calculations.

Intensity and polarization distributions of the light beams are illustrated in Fig. 5.8. For all these patterns, we computed the vector OAM $J_{\mathrm{v},z}$ (11), polarization OAM $J_{\mathrm{p},z}$ (5.65), and hybrid OAM $J_{\mathrm{h},z}$ (19), normalized to the beam power: 6.998, 6.998, 6.328×10^{-22} [Fig. 5.8a]; 6.998, −6.998, 6.328×10^{-22} [Fig. 5.8b]; 6.998, 0, 0 [Fig. 5.8c]; 6.998, 6.328×10^{-22}, 6.998 [Fig. 5.8d]; 0, 2.995, 0 [Fig. 5.8e]; 0, 0, 2.995 [Fig. 5.8f]; 6.997, 2.998, 2.225×10^{-18} [Fig. 5.8g]; 6.997, 2.225×10^{-18}, 2.998 [Fig. 5.8h].

Further, we computed all three OAMs for the two-index light field (5.61). In this case, the Laguerre-Gaussian beams cannot be chosen as the envelope functions, since, in contrast to the Poincaré beams, the beam (5.61) is composed of beams, whose topological charge is different not only by the sign, but also by its absolute value. Therefore, we consider here fields (5.61) with each component being a simple Gaussian optical vortex, i.e. $A(r) = \exp[-(r/w)^2$ with w being the waist radius.

Shown in Fig. 5.9 are the intensity and polarization distributions, as well as the densities of the polarization and hybrid OAM for three two-index beams (5.61) with different indices, when they both are odd [Fig. 5.9a–g], both are even [Fig. 5.9b–h], and are of different parity [Fig. 5.9c–i].

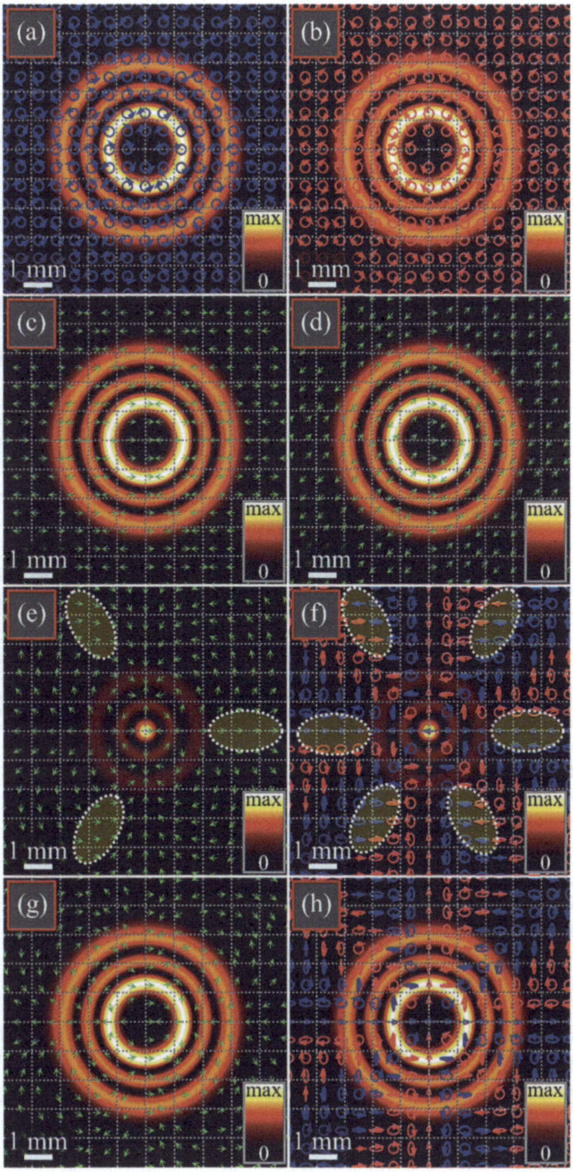

Fig. 5.8 Intensity and polarization distributions in the initial plane of vector beams for the following parameters: waist radius $w_0 = 1$ mm, lower (radial) index of the Laguerre-Gaussian beam $p = 2$, upper (azimuthal) index is $m = 0$ (**e**,**f**) and $m = 7$ (**a–d**,**g**,**h**), order of cylindrical/hybrid polarization (**e–h**) $n = 3$, polarization is left circular (**a**), right circular (**b**), linear horizontal (**c**), linear by 45° (**d**), cylindrical without a vortex (**e**), hybrid without a vortex (**f**), cylindrical with a vortex (**g**), hybrid with a vortex (**h**). Green color (**c**,**d**, **e**,**g**) illustrates polarization directions of beams with inhomogeneous linear polarization. Red and blue colors (**a**,**b**,**f**,**h**) denote respectively right and left elliptic polarization. Dashed ellipses (**e**,**f**) show areas with horizontal linear polarization

According to Eq. (5.61), if the beam orders are of the same parity, the Jones vector on the horizontal axis x ($\varphi = 0$, $\varphi = \pi$) and on the vertical axis y ($\varphi = \pm\pi/2$) consists of real-valued components and the field has linear polarization, directed by the angle 45° or 135° to the horizontal axis. If orders are of different parity, polarization on the horizontal axis is linear, but on the vertical axis it is circular. It is confirmed in Fig. 5.9.

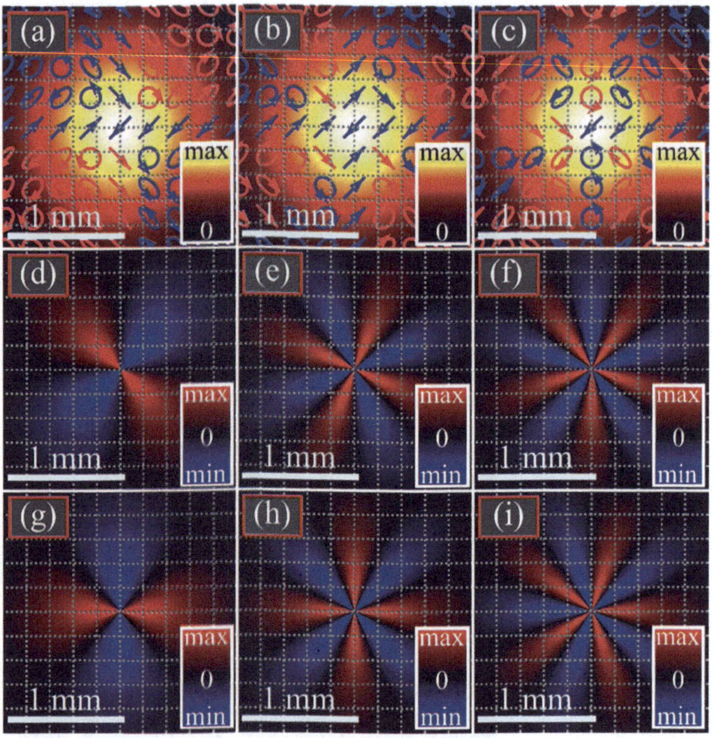

Fig. 5.9 Intensity and polarization (**a–c**) distributions, as well as polarization OAM (**d–f**) and hybrid OAM (**g–i**) densities of three two-index beams (5.61) with different values of the indices n and m. For computations, the following parameters were used: waist radius of the Gaussian beam $w = 1$ mm, beam order along the components E_x and E_y: $(n, m) = (3, 5)$ (**a**), $(n, m) = (4, 8)$ (**b**), $(n, m) = (3, 8)$ (**c**). Red and blue color (**a–c**) denotes respectively right and left elliptic polarization. Red and blue color (**d–i**) denotes areas with respectively positive and negative density of the polarization or hybrid OAM

Computations also confirm the values of the normalized vectorial (5.60), polarization (5.65) and hybrid (5.68) OAM: $(n + m)/2$, 0, 0. For Fig. 5.9a,d,g, the following values were obtained: $3.999 \approx 4$, $1.647 \times 10^{-17} \approx 0$, $-3.029 \times 10^{-19} \approx 0$. For Fig. 5.9b,e,h, the computed values are $5.997 \approx 6$, $-1.221 \times 10^{-16} \approx 0$, -0.231. For Fig. 5.9c,f,i, we obtained $5.497 \approx 5.5$, $3.752 \times 10^{-17} \approx 0$, $7.228 \times 10^{-17} \approx 0$.

5.3.6 *Experiment*

In this experiment, whose schematic setup is depicted in Fig. 5.10, a cylindrical vector beam (5.66) with $n = 2$ was also generated by using a vector wave plate VWP (Thorlabs WPV10–633) and by a Gaussian laser beam with the wavelength of 633 nm. Cylindrical vector beam sequentially passed through a quarter-wave plate

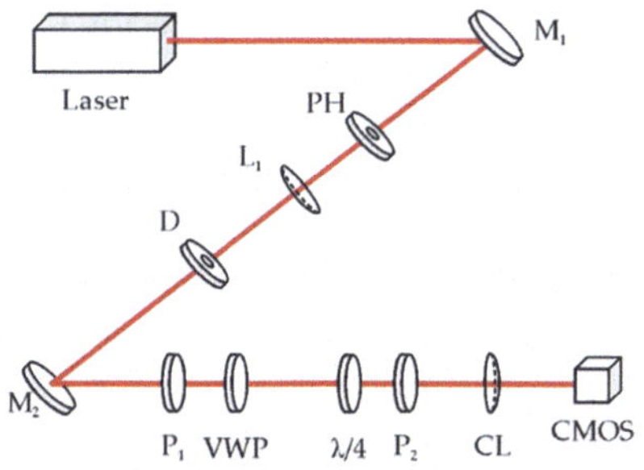

Fig. 5.10 Schematic experimental setup: Laser—laser with the wavelength of 633 nm, M_1 and M_2—mirrors, PH, D—pinhole and a diaphragm for filtering the noise in the beam, L_1—spherical lens, P_1—polarizer, VWP—q-plate (q = 1), λ/4—quarter-wave plate, P_2—polarizer inclined by the angle of 45°, CL—cylindrical lens, CMOS—CCD-camera

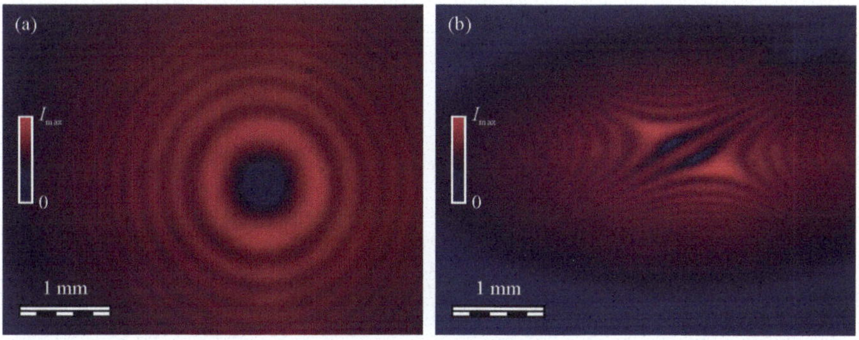

Fig. 5.11 Image of the phase vortex (**a**), and intensity distribution in the focus of the cylindrical lens (**b**)

and a polarizer P_2, oriented by an angle of 45° with respect to the quarter-wave plate. After such sequential passing through the wave plate and the polarizer, vector beam was converted into an optical vortex. An image of the generated vortex is given in Fig. 5.11a. For determining the vortex topological charge, generated beam passed through a cylindrical lens CL. Intensity distribution in the lens focus is illustrated in Fig. 5.11b. As seen in Fig. 5.11b, intensity in the center of the focal plane contains a light spot with three maxima and two minima.

Therefore, according to [100], the TC of the vortex in Fig. 5.11a equals two, and the pOAM of the beam from Fig. 5.11b is thus also equal to two: $J_{p,z}/W = n = 2$. In a general case, for measuring the OAMs (5.60), (5.65) and (5.68), separate field components E_x and E_y should be obtained by polarizers. Then, using a reference beam, two interferograms should be generated and registered, from which amplitudes and phases of the fields E_x and E_y should be obtained. Further, the OAMs should be calculated by Eqs. (5.60), (5.65) and (5.68).

In this section, similarly to the well-known vector OAM, characterizing azimuthal phase change of the transverse components of the electric vector, we have considered a polarization OAM, characterizing azimuthal change of direction of the instant linear polarization vector, and a hybrid OAM, which also indicates azimuthal change of ellipticity degree of the polarization ellipse in a section of hybrid vector field. Values of these OAMs have been obtained for several examples of vector fields. For instance, conventional and polarization OAM are equal to each other for a circularly polarized optical vortex. Conventional and hybrid OAM are nonzero for an optical vortex with hybrid polarization. Vector and hybrid OAM are extrinsic OAMs and they are not quantities defined for single photons in a light beam. Instead, they describe the whole beam and characterize mass transport of molecules of a polarization-sensitive matter. Two ways have been proposed for experimental measuring the polarization OAM of a vector beam. The experimental results are consistent with the theory [101].

References

1. J.H. Poynting, Proc. R. Soc. London. Ser. A, Contain. Pap. a Math. Phys. Character **82**, 560 (1909)
2. R.A. Beth, Phys. Rev. **50**, 115 (1936)
3. L. Allen, M.W. Beijersbergen, R.J.C. Spreeuw, J.P. Woerdman, Phys. Rev. A **45**, 8185 (1992)
4. N. Bokor, Y. Iketaki, T. Watanabe, M. Fujii, Opt. Express **13**, 10440 (2005)
5. Z. Bomzon, M. Gu, J. Shamir, Appl. Phys. Lett. **89**, 241104 (2006)
6. B. Richards, E. Wolf, Proc. R. Soc. A Math. Phys. Eng. Sci. **253**, 358 (1959)
7. Z. Bomzon, M. Gu, Opt. Lett. **32**, 3017 (2007)
8. Y. Zhao, J.S. Edgar, G.D.M. Jeffries, D. McGloin, D.T. Chiu, Phys. Rev. Lett. **99**, 073901 (2007)
9. T.A. Nieminen, A.B. Stilgoe, N.R. Heckenberg, H. Rubinsztein-Dunlop, J. Opt. A Pure Appl. Opt. **10**, 115005 (2008)
10. Y. Zhao, D. Shapiro, D. Mcgloin, D.T. Chiu, S. Marchesini, Opt. Express **17**, 23316 (2009)
11. Y. Gorodetski, A. Niv, V. Kleiner, E. Hasman, Phys. Rev. Lett. **101**, 043903 (2008)
12. P.B. Monteiro, P.A.M. Neto, H.M. Nussenzveig, Phys. Rev. A **79**, 033830 (2009)
13. O.G. Rodríguez-Herrera, D. Lara, K.Y. Bliokh, E.A. Ostrovskaya, C. Dainty, Phys. Rev. Lett. **104**, 253601 (2010)
14. M.R. Foreman, P. Török, New J. Phys. **13**, 063041 (2011)
15. K.Y. Bliokh, E.A. Ostrovskaya, M.A. Alonso, O.G. Rodríguez-Herrera, D. Lara, C. Dainty, Opt. Express **19**, 26132 (2011)
16. H. Li, C. Ma, J. Wang, M. Tang, X. Li, Opt. Express **29**, 39419 (2021)
17. A.V. Arzola, L. Chvátal, P. Jákl, P. Zemánek, Sci. Rep. **9**, 4127 (2019)
18. J.-X. Guo, W.-Y. Wang, T.-Y. Cheng, J.-Q. Lü, Front. Phys. **10** (2022)
19. Y. Wu, P. Yu, Y. Liu, Z. Wang, Y. Li, L. Gong, J. Light. Technol. **41**, 2252 (2023)
20. K.Y. Bliokh, F.J. Rodríguez-Fortuño, F. Nori, A.V. Zayats, Nat. Photonics **9**, 796 (2015)
21. O. Graydon, Nat. Photonics **7**, 672 (2013)
22. A. Aiello, P. Banzer, M. Neugebauer, G. Leuchs, Nat. Photonics **9**, 789 (2015)
23. W. Miao, X. Pang, W. Liu, IEEE Photonics J. **12**, 1 (2020)
24. J. Berškys, S. Orlov, *2021 Conf. Lasers Electro-Optics Eur. Eur. Quantum Electron. Conf.* (Optica Publishing Group, 2021), p. eg_p_10
25. E.J. Galvez, I. Dutta, K. Beach, J.J. Zeosky, J.A. Jones, B. Khajavi, Sci. Rep. **7**, 13653 (2017)

26. T. Bauer, P. Banzer, E. Karimi, S. Orlov, A. Rubano, L. Marrucci, E. Santamato, R.W. Boyd, G. Leuchs, Science. (80-.) **347**, 964 (2015)
27. O.V. Angelsky, I.I. Mokhun, A.Y. Bekshaev, C.Y. Zenkova, J. Zheng, Front. Phys. **11**, 1 (2023)
28. C. Wan, Q. Zhan, Opt. Express **27**, 11516 (2019)
29. K.Y. Bliokh, M.A. Alonso, D. Sugic, M. Perrin, F. Nori, E. Brasselet, Phys. Fluids **33**, 077122 (2021)
30. I. Freund, Opt. Lett. **45**, 3333 (2020)
31. W. Shu, C. Lin, J. Wu, S. Chen, X. Ling, X. Zhou, H. Luo, S. Wen, Phys. Rev. A **101**, 23819 (2020)
32. M. Zhang, H. Ren, X. Ouyang, M. Jiang, Y. Lu, Y. Hu, S. Fu, Z. Li, Z. Chen, B.-O. Guan, Y. Cao, X. Li, ACS Photonics **8**, 1169 (2021)
33. W. Wang, R. Zhao, Q. Kang, R. Wang, X. Liu, T. Liu, S.-W. Fan, Z. Guo, Appl. Opt. **62**, 8159 (2023)
34. M. Neugebauer, P. Banzer, T. Bauer, N. Lindlein, A. Aiello, C. Marquardt, J. Korger, G. Leuchs, in Conf., *Lasers Electro-Optics 2012* (OSA, Washington, DC, 2012), p. QW1E.4
35. K. Zhao, Z. Zhang, H. Zang, J. Du, Y. Lu, P. Wang, Opt. Lett. **46**, 1896 (2021)
36. D.O. Ignatyeva, C.S. Davies, D.A. Sylgacheva, A. Tsukamoto, H. Yoshikawa, P.O. Kapralov, A. Kirilyuk, V.I. Belotelov, A.V. Kimel, Nat. Commun. **10**, 4786 (2019)
37. F. Hendriks, R.R. Rojas-Lopez, B. Koopmans, M.H.D. Guimarães, Nat. Commun. **15**, 1298 (2024)
38. J. Zhou, H. Ma, Y. Zhang, S. Zhang, C. Min, X. Yuan, Opt. Lett. **47**, 1494 (2022)
39. Y. Wu, X. Hu, Y. Li, R. Chen, Photonics **10**, 820 (2023)
40. H. Li, C. Wang, M. Tang, X. Li, Opt. Express **28**, 18607 (2020)
41. F.G. Mitri, J. Opt. Soc. Am. A **33**, 1661 (2016)
42. Y. Shen, Q. Zhang, P. Shi, L. Du, X. Yuan, A.V. Zayats, Nat. Photonics **18**, 15 (2024)
43. Y. Zeng, Y. Yu, X. Shen, J. Chen, Q. Zhan, Nano **13**, 251 (2024)
44. S.M. Barnett, C. Cisowski, A. McWilliams, F.C. Speirits, Z. Ye, J.B. Goette, S. Franke-Arnold, in *Act. Photonic Platforms 2023*, ed. by G.S. Subramania, S. Foteinopoulou, (SPIE, 2023), p. 57
45. V.V. Kotlyar, A.A. Kovalev, A.M. Telegin, Photonics **10**, 160 (2023)
46. V.V. Kotlyar, S.S. Stafeev, V.D. Zaitsev, A.M. Telegin, E.S. Kozlova, Appl. Sci. **13**, 8361 (2023)
47. B. Ghosh, A. Daniel, B. Gorzkowski, A.Y. Bekshaev, R. Lapkiewicz, K.Y. Bliokh, J. Opt. Soc. Am. B **41**, 1276 (2024)
48. A.A. Kovalev, V.V. Kotlyar, Micromachines **14**, 494 (2023)
49. V.V. Kotlyar, A.A. Kovalev, S.S. Stafeev, A.M. Telegin, J. Opt. Soc. Am. A **42**, 52 (2025)
50. A.A. Kovalev, V.V. Kotlyar, S.S. Stafeev, J. Opt. **27**, 45604 (2025)
51. S.S. Stafeev, V.D. Zaitsev, V.V. Kotlyar, Opt. Commun. **583**, 131803 (2025)
52. A. Forbes, M. de Oliveira, M.R. Dennis, Nat. Photonics **15**, 253 (2021)
53. C. He, Y. Shen, A. Forbes, Light Sci. Appl. **11**, 205 (2022)
54. S.-Y. Huang, G.-L. Zhang, Q. Wang, M. Wang, C. Tu, Y. Li, H.-T. Wang, Optica **8**, 1231 (2021)
55. Y. Shen, X. Wang, Z. Xie, C. Min, X. Fu, Q. Liu, M. Gong, X. Yuan, Light Sci. Appl. **8**, 90 (2019)
56. J. Wang, Photonics Res. **4**, B14 (2016)
57. C.J. Bustamante, Y.R. Chemla, S. Liu, M.D. Wang, Nat. Rev. Methods Prim. **1**, 25 (2021)
58. S. Slussarenko, D.J. Joch, N. Tischler, F. Ghafari, L.K. Shalm, V.B. Verma, S.W. Nam, G.J. Pryde, npj Quantum Inf. **8**, 20 (2022)
59. B. Zhang, S. Daly, C. Zhu, M.O. Lenz, L.E. Weiss, L.-M. Needham, R. Peters, S.F. Lee, K. O'Holleran, Optica **11**, 1519 (2024)
60. H. Kawaguchi, R. Yasuhara, H. Yang, C. Hori, R. Miyagawa, K. Sugioka, M. Ota, H. Uehara, Opt. Mater. Express **14**, 424 (2024)

61. L. Allen, M.J. Padgett, Opt. Commun. **184**, 67 (2000)
62. X.-L. Wang, J. Chen, Y. Li, J. Ding, C.-S. Guo, H.-T. Wang, Phys. Rev. Lett. **105**, 253602 (2010)
63. V.V. Kotlyar, A.A. Kovalev, A.P. Porfirev, Phys. Rev. A **97**, 053833 (2018)
64. A.A. Kovalev, V.V. Kotlyar, S.S. Stafeev, Micromachines **14**, 777 (2023)
65. K.Y. Bliokh, F. Nori, Phys. Rep. **592**, 1 (2015)
66. K.Y. Bliokh, M.A. Alonso, E.A. Ostrovskaya, A. Aiello, Phys. Rev. A **82**, 063825 (2010)
67. L. Fang, J. Wang, Opt. Express **25**, 23364 (2017)
68. A.A. Kovalev, S.S. Stafeev, A.M. Telegin, Comput. Opt. **49**, 933 (2025)
69. A.A. Kovalev, V.V. Kotlyar, Comput. Opt. **47**, 533 (2023)
70. V.V. Kotlyar, S.S. Stafeev, A.G. Nalimov, Photonics **8**, 227 (2021)
71. V.V. Kotlyar, S.S. Stafeev, A.M. Telegin, Optik (Stuttg). **287**, 171103 (2023)
72. V.V. Kotlyar, S.S. Stafeev, E.S. Kozlova, A.G. Nalimov, Sensors **21**, 6424 (2021)
73. J. Chen, L. Yu, C. Wan, Q. Zhan, ACS Photonics **9**, 793 (2022)
74. J. Chen, P. Zheng, Q. Zhan, Opt. Express **30**, 46666 (2022)
75. M.V. Berry, J. Opt. A Pure Appl. Opt. **6**, 259 (2004)
76. M.V. Berry, J. Opt. A Pure Appl. Opt. **6**, 675 (2004)
77. I. Freund, Opt. Commun. **201**, 251 (2002)
78. Q. Zhan, Adv. Opt. Photon. **1**, 1 (2009)
79. H. He, M.E.J. Friese, N.R. Heckenberg, H. Rubinsztein-Dunlop, Phys. Rev. Lett. **75**, 826 (1995)
80. A. Ambrosio, P. Maddalena, L. Marrucci, Phys. Rev. Lett. **110**, 146102 (2013)
81. Y. Zhai, L. Cao, Y. Liu, X. Tan, Materials (Basel). **13**, 5562 (2020)
82. A. Porfirev, S. Khonina, D. Porfirev, N. Ivliev, Appl. Opt. **63**, 3779 (2024)
83. M.E.J. Friese, J. Enger, H. Rubinsztein-Dunlop, N.R. Heckenberg, Phys. Rev. A **54**, 1593 (1996)
84. Y. Zhao, L. Zhou, X. Jiang, L. Zhu, Q. Shi, Nano **14**, 691 (2024)
85. S. Albaladejo, M.I. Marqués, M. Laroche, J.J. Sáenz, Phys. Rev. Lett. **102**, 113602 (2009)
86. N.B. Baranova, B.Y. Zeldovich, M.O. Scully, JETP **78**, 249 (1994)
87. V.H. Mellado, S. Hacyan, R. Jauregui, Laser Part. Beams **24**, 559 (2006)
88. K. Polak, J.-C. Gayde, M. Sulc, J. Phys. Conf. Ser. **2687**, 042003 (2024)
89. O.M. Maragò, P.H. Jones, P.G. Gucciardi, G. Volpe, A.C. Ferrari, Nat. Nanotechnol. **8**, 807 (2013)
90. P. Meng, Z. Man, A.P. Konijnenberg, H.P. Urbach, Opt. Express **27**, 35336 (2019)
91. R. Yang, R. Li, S. Qin, C. Ding, F.G. Mitri, J. Opt. **19**, 025602 (2017)
92. C. Leyder, M. Romanelli, J.P. Karr, E. Giacobino, T.C.H. Liew, M.M. Glazov, A.V. Kavokin, G. Malpuech, A. Bramati, Nat. Phys. **3**, 628 (2007)
93. V.V. Kotlyar, S.S. Stafeev, V.D. Zaitsev, A.A. Kovalev, Phys. Lett. A **458**, 128596 (2023)
94. V.V. Kotlyar, A.G. Nalimov, S.S. Stafeev, J. Opt. Soc. Am. B **36**, 2850 (2019)
95. V.V. Kotlyar, S.S. Stafeev, E.S. Kozlova, M.A. Butt, Photonics **9**, 970 (2022)
96. V.V. Kotlyar, S.S. Stafeev, Optik (Stuttg). **245**, 167703 (2021)
97. S. Bansal, P. Senthilkumaran, Opt. Lasers Eng. **160**, 107295 (2023)
98. R. Ruchi, P. Senthilkumaran, S.K. Pal, Int. J. Opt. **1** (2020, 2020)
99. V.V. Kotlyar, A.A. Kovalev, V.D. Zaitsev, Photonics **9**, 298 (2022)
100. V.V. Kotlyar, A.A. Kovalev, A.P. Porfirev, Appl. Opt. **56**, 4095 (2017)
101. V.V. Kotlyar, A.A. Kovalev, A.M. Telegin, S.S. Stafeev, Opt. Mem. Neural Networks **34**, S292 (2025)

Chapter 6
Features of Vector Beams in Sharp Focus

6.1 Helicity Density of Higher-Order Poincaré Beams in Tight Focus

Currently, researchers all over the world are actively studying beams with non-uniform cross-sectional polarization, i.e., polarization-inhomogeneous beams. These beams include a wide class of beams: cylindrical vector beams [1], beams with hybrid polarization [2, 3], beams with multiple singularity centers [4, 5], etc. The reason for this interest is in numerous applications, for example in optical communications [6, 7], micromanipulations [8, 9], or material processing [10]. Poincaré beams [11, 12] are polarization-inhomogeneous beams with radial symmetry. Special cases of Poincaré beams are beams with radial and azimuthal polarizations [1], and optical vortices (Gauss-Laguerre modes) with circular polarization [13, 14]. In [13], the focusing of Poincaré beams was investigated theoretically and numerically using the Richards-Wolf formalism. It was shown that Poincaré beams demonstrate a set of unique phenomena at the focus, such as the reverse energy flow [15, 16] and spin-orbit conversion [17]. For some special cases of Poincaré beams, polarization Möbius strips can be observed at the focus [18]. Poincaré beams can be generated interferometrically [14], using a spatial light modulator [19, 20] or metasurfaces [21].

Another actively studied beam characteristic is chirality [22], which has recently also attracted the attention of researchers. Chirality coincides with helicity [23], it differs in multiplier and sign. Chirality of Poincaré beams was previously investigated by M. Babiker et al. in [23]. In particular, it was shown [23] that chirality on the optical axis is observed for beam of orders $n = 0$ and $n = 1$, moreover, for $n = 1$ it is maximal. M. Babiker et al. called this effect superchirality. Later K. Forbes criticized the effect of superchirality, pointing out the parameters in [24] are non-physical. In the ensuing discussion [25] M. Babiker et al. responded to K. Forbes' criticism points.

V. Kotlyar et al., *Inverse Canonical Energy Flow in Laser Beams*,
https://doi.org/10.1007/978-3-032-27289-8_6

In this section, the helicity density of higher-order Poincaré beams was investigated using Richards-Wolf formalism under tight focusing conditions. It was shown that the helicity density for higher-order Poincaré beams always has radial symmetry and does not depend on the azimuthal angle of the beam section. The absolute value of helicity density is maximum when the polar angle of the beam is zero or π (this case corresponds to optical vortex with circular polarization), and it is absent when the polar angle is π/2 (this case corresponds to cylindrical vector beam). Non-zero values of helicity density on the optical axis are observed for beam orders 0, 1, and 2. If the polar angle of the beam is zero, the intensity of higher-order Poincaré beams coincides with the helicity density up to a factor. It was theoretically and numerically shown that for ideal case (the numerical aperture tends to unity, the lens is limited by a thin annular aperture) helicity density for the beam of order $n = 1$ is larger on the optical axis, then for the beams of order $n = 0$ or $n = 2$; however, for standard aplanatic microobjectives (NA = 0.95) helicity density on the optical axis is larger for the beam of order $n = 0$. All results obtained for helicity could be adapted to chirality. In contrast to the papers [23–25], the use of the Richards-Wolf formalism allows us to calculate all six components of the electric and magnetic fields at the focus and takes into account vector effects, which especially have place in focal spots of lenses with large numerical apertures. Only taking into account vector effects in focus allows us to observe the case when the helicity density for the beam of order $n = 1$ is larger on the optical axis, then for the beams of order $n = 0$ or $n = 2$.

6.1.1 Theoretical Background

The investigated higher-order Poincaré beams have the form [11, 12]:

$$\mathbf{E}_P(\alpha, \varphi) = \frac{A(\alpha)}{\sqrt{2}} \begin{pmatrix} ae^{-in\varphi} + be^{in\varphi} \\ iae^{-in\varphi} - ibe^{in\varphi} \end{pmatrix} \tag{6.1}$$

where $a = \cos(\theta/2)e^{-i\psi/2}$, $b = \sin(\theta/2)e^{i\psi/2}$; θ and ψ polar and azimuthal angles that define the position of polarization on the generalized Poincaré sphere (Fig. 6.1a), φ is azimuthal angle in the plane perpendicular to the optical axis (Fig. 6.1b). $A(r)$ is the amplitude of the initial field, depending only on the polar angle α (Fig. 6.1b). In general, a and b satisfy the following conditions [26]:

$$\begin{aligned} a^*b - b^*a &= i\sin(\theta)\sin(\psi) \\ b^2 - a^2 &= \cos(\theta) \\ a^2 + b^2 &= 1 \end{aligned} \tag{6.2}$$

In this section we used the Richards-Wolf formalism [27, 28]:

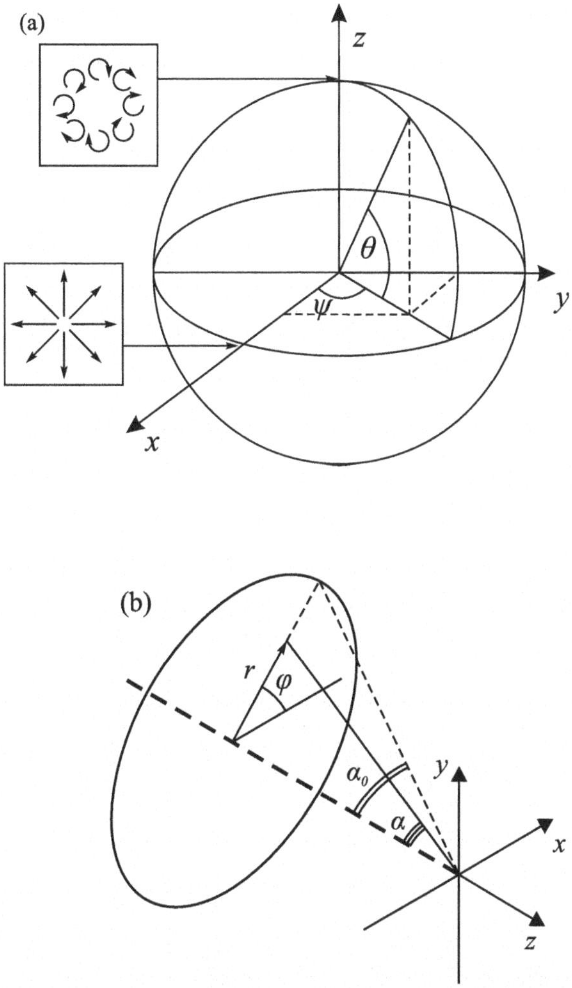

Fig. 6.1 Poincaré sphere (**a**) and schematic image of high numerical aperture lens (**b**)

$$\mathbf{U}(\rho, \psi, z) = -\frac{if}{\lambda} \int_0^{\alpha_{\max}} \int_0^{2\pi} B(\alpha, \varphi) T(\alpha) \mathbf{P}(\alpha, \varphi) \times \\ \times \exp\{ik[\rho \sin\alpha \cos(\varphi - \psi) + z \cos\alpha]\} \sin\alpha d\alpha d\varphi \tag{6.3}$$

where $\mathbf{U}(\rho, \psi, z)$ is electric or magnetic field in the area of focal point, $B(\alpha, \varphi)$ is electric or magnetic incident field (α is the polar angle, φ is the azimuthal angle). $T(\alpha)$ is lens' apodization function, f is the focal length, $k = 2\pi/\lambda$ is the wavenumber, λ is the wavelength, α_{max} is the maximum polar angle determined by the numerical aperture of the lens ($NA = \sin \alpha_{max}$). Figure 6.1b schematically shows the high numerical aperture lens. $\mathbf{P}(\alpha, \varphi)$ is the polarization vector, for the electric and magnetic fields:

$$\mathbf{P}(\alpha,\phi)=\begin{bmatrix}\sin^2\phi+\cos^2\phi\cos\alpha & \sin\phi\cos\phi(\cos\alpha-1) & \cos\phi\sin\alpha\\ \sin\phi\cos\phi(\cos\alpha-1) & \cos^2\phi+\sin^2\phi\cos\alpha & \sin\phi\sin\alpha\\ -\sin\alpha\cos\phi & -\sin\alpha\sin\phi & \cos\alpha\end{bmatrix}\times\begin{bmatrix}U_{x,in}\\ U_{y,in}\\ U_{z,in}\end{bmatrix} \tag{6.4}$$

where $U_{x,in}$, $U_{y,in}$ and $U_{z,in}$ are the x-, y- and z-components of the incident field. In our investigation the transverse components of the incident beam follow from Eq. (6.1), the longitudinal component of the incident field in the initial plane was assumed to be zero $E_{z,in}=0$.

In [13], using the Richards-Wolf formalism, it was obtained formulas for calculating all components of the electric and magnetic field at the focus of the Poincaré beam:

$$\begin{aligned}
E_x&=\frac{i^{n-1}}{\sqrt{2}}\left[\left(be^{in\varphi}+ae^{-in\varphi}\right)I_{0,n}+\left(be^{i(n-2)\varphi}+ae^{-i(n-2)\varphi}\right)I_{2,n-2}\right],\\
E_y&=\frac{i^{n}}{\sqrt{2}}\left[\left(ae^{-in\varphi}-be^{in\varphi}\right)I_{0,n}-\left(ae^{-i(n-2)\varphi}-be^{i(n-2)\varphi}\right)I_{2,n-2}\right],\\
E_z&=\sqrt{2}i^n\left(be^{i(n-1)\varphi}+ae^{-i(n-1)\varphi}\right)I_{1,n-1},\\
H_x&=\frac{i^{n}}{\sqrt{2}}\left[\left(be^{in\varphi}-ae^{-in\varphi}\right)I_{0,n}+\left(be^{i(n-2)\varphi}-ae^{-i(n-2)\varphi}\right)I_{2,n-2}\right],\\
H_y&=\frac{i^{n-1}}{\sqrt{2}}\left[\left(be^{in\varphi}+ae^{-in\varphi}\right)I_{0,n}-\left(be^{i(n-2)\varphi}+ae^{-i(n-2)\varphi}\right)I_{2,n-2}\right],\\
H_z&=\sqrt{2}i^{n+1}\left(be^{i(n-1)\varphi}-ae^{-i(n-1)\varphi}\right)I_{1,n-1}.
\end{aligned} \tag{6.5}$$

where

$$I_{\nu,\mu}=\left(\frac{4\pi f}{\lambda}\right)\int\limits_0^{\alpha_0}\sin^{\nu+1}\left(\frac{\alpha}{2}\right)\cos^{3-\nu}\left(\frac{\alpha}{2}\right)\times \times A(\alpha)e^{ikz\cos\alpha}J_\mu(x)d\alpha, \tag{6.6}$$

where $J_\mu(x)$ is Bessel function of order μ, $x = kr\sin\theta$, $A(\alpha)$ is radially symmetric real function of beam amplitude (6.1).

We will further consider the distribution of the components of the electric and magnetic fields at the focus at $z=0$, for this case the integral (6.6) takes real values. The helicity density at the focus will be calculated using the formula [29]:

$$\eta = -\frac{\sqrt{\varepsilon_0 \mu_0}}{2\omega} \mathrm{Im}(\mathbf{E}^* \cdot \mathbf{H}) \tag{6.7}$$

It should be noted that helicity could be calculated by integral of the helicity density (6.7) over the x-y plane [26]. Helicity density (6.7) differs from chirality density only by a multiplier for monochromatic beams [30]. Below we will investigate helicity density. We omit the multiplier standing in front of the imaginary part in expression (6.7), but we will take into account the sign. Substituting Eqs. (6.5) into (6.7), we obtain the following expression for the helicity density of Poincaré beams in the focal plane ($z = 0$):

$$\eta = \cos\theta\left(I_{0,n}^2 + I_{2,n-2}^2 + 2I_{1,n-1}^2\right) \tag{6.8}$$

From Eq. (6.8) it follows that, first, the shape of helicity density in transverse (focal) plane always has radial (axial) symmetry—it does not depend from azimuthal angle φ (Fig. 6.1b). Second, helicity density depends on the polar angle θ of the Poincaré beam (6.1); however, it does not depend on the azimuthal angle ψ of the beam (Fig. 6.1a). This fact was previously pointed out by M. Babiker et al. [23, 26] in scalar approximation. Third, the helicity density is maximum at $\theta = 0$ and is minimum when $\theta = \pi$. This fact was also previously pointed out by M. Babiker et al. [23, 26] in scalar approximation. In this paper we confirm it in vectorial approximation. It should be noted that angle $\theta = 0$ corresponds to optical vortex with a topological charge $-n$ and right-hand circular polarization $(\mathbf{x} + i\mathbf{y})/\sqrt{2}$. I.e. maximal helicity density corresponds to not Poincaré beam, but to simple scalar beam, which pointed out K. Forbes in the discussion [23–25]. Moreover, this maximum helicity for circularly polarized vortex beams was demonstrated by K. Forbes and G. Jones in [31, 32]. Fourth, helicity density is absent at $\theta = \pi/2$ (this angle, in particular, corresponds to radial and azimuthal polarizations [1], as well as high-order inhomogeneous cylindrical polarization [33]). This conclusion is confirmed by the results of work [34]. Fifth, for beam orders $n = 0$, 1 and 2, non-zero helicity density values on the optical axis will be observed (the second index in integral (6.6) corresponds to the order of the Bessel function under the integral).

By integrating over the beam cross section, the helicity density can be obtained at the beam focus (6.1):

$$\begin{aligned}
&\overline{\eta} = \cos\theta \int_0^\infty \int_0^{2\pi} \left(I_{0,n}^2 + I_{2,n-2}^2 + 2I_{1,n-1}^2\right) r dr d\varphi = \\
&\cos\theta(W_0 + W_2 + 2W_1) = W\cos\theta, \\
&W_0 = 2\pi \int_0^\infty I_{0,n}^2(r) r dr, \\
&W_2 = 2\pi \int_0^\infty I_{2,n-2}^2(r) r dr, \\
&W_1 = 2\pi \int_0^\infty I_{1,n-1}^2(r) r dr.
\end{aligned} \tag{6.9}$$

In (6.9) W is the total power of the beam (6.1). Equation (6.9) is obtained on the basis of the following integral:

$$\begin{aligned}
W_\nu &= 2\pi \int_0^\infty \left|I_{\nu,\mu}(r)\right|^2 r dr = \\
&= 4\pi f^2 \int_0^{\alpha_0} \sin^{2\nu+1}\left(\frac{\alpha}{2}\right) \cos^{5-2\nu}\left(\frac{\alpha}{2}\right) |A(\alpha)|^2 d\alpha.
\end{aligned} \tag{6.10}$$

In work [13] an expression for calculating the intensity of Poincaré beams was found, which had the form:

$$\begin{aligned}
&I(r,\varphi) = |E_x|^2 + \left|E_y\right|^2 + |E_z|^2 = \\
&= I_{0,n}^2 + I_{2,n-2}^2 + 2I_{1,n-1}^2 + \\
&+ 2\sin\theta\cos(2(n-1)\varphi + \psi)\left(I_{0,n}I_{2,n-2} + 2I_{1,n-1}^2\right).
\end{aligned} \tag{6.11}$$

From (6.11) it follows that in the general case the intensity has central symmetry, and at $n = 1$ or $\theta = 0$—also axial symmetry. From the comparison of (6.8) and (6.11) it is evident that at $\theta = 0$ the intensity of Poincaré beams coincides with the helicity density up to a factor. By integrating the intensity at the focus over the beam cross-section, we obtain the total beam power (6.1):

$$\begin{aligned} W = & \int_0^{\infty}\int_0^{2\pi} I(r,\varphi) r dr d\varphi = \\ & \int_0^{\infty}\int_0^{2\pi} \left(I_{0,n}^2 + I_{2,n-2}^2 + 2I_{1,n-1}^2\right) r dr d\varphi + \\ & +2\sin\theta \int_0^{\infty}\int_0^{2\pi} \cos(2(n-1)\varphi + \psi)\times \\ & \left(I_{0,n}I_{2,n-2} + 2I_{1,n-1}^2\right) r dr d\varphi = \\ & W_0 + W_2 + 2W_1. \end{aligned} \tag{6.12}$$

In (6.12) the last integral is zero, since the integration of a periodic function over the angle φ over its period is equal to zero. Comparing (6.9) and (6.12) it is evident that the helicity of beam (6.1) at the focus η is equal to the beam power *W* multiplied by cosθ.

As already noted, non-zero values of the on-axis helicity density are observed for beam orders $n = 0$, 1 and 2. In the discussion by M. Babiker [23, 25] and K. Forbes [24], special attention was paid to the case when the on-axis chirality for a beam of order $n = 1$ is greater than for a beam of order $n = 0$.

Let us assume that we are considering the focusing of a Poincaré beam (6.1) by a wide-aperture lens in front of which an annular diaphragm is placed, then the input beam will have the form

$$A(\alpha) = \delta(\alpha_0) \tag{6.13}$$

We will also assume that the numerical aperture of the lens tends to unity:

$$NA \to 1, \alpha_0 \to \frac{\pi}{2}, \tag{6.14}$$

This situation is physically unrealizable in full; however, it can be approached using flat diffractive lenses [35] or metalenses [36] limited by annular diaphragms. Then the integral (6.6) can be transformed to the form

$$I_{\nu,\mu} = \left(\frac{4\pi f}{\lambda}\right) \sin^{\nu+1}\left(\frac{\pi}{4}\right) \cos^{3-\nu}\left(\frac{\pi}{4}\right) J_\mu(kr) \tag{6.15}$$

Then helicity density will be written as

$$\eta = C\left(J_n^2(kr) + J_{n-2}^2(kr) + 2J_{n-1}^2(kr)\right), \tag{6.16}$$

where

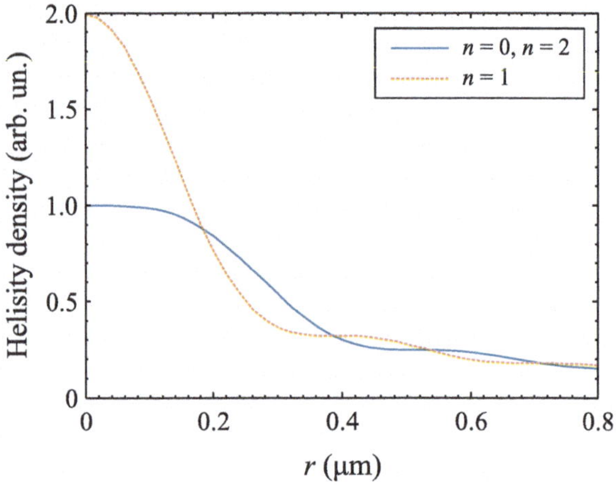

Fig. 6.2 Helicity density in tight focus (the numerical aperture is equal to unity, the lens is limited by a thin annular aperture) calculated using Eqs. (6.18)–(6.19)

$$C = \frac{4\pi^2 f^2}{\lambda^2} \cos\theta T^2\left(\frac{\pi}{2}\right). \tag{6.17}$$

Then helicity density (6.16) for beam orders $n = 0$ and $n = 2$ will take the form

$$\eta = C\left(J_2^2(kr) + J_0^2(kr) + 2J_1^2(kr)\right) \tag{6.18}$$

For the beam of order $n = 1$ the Eq. (6.16) is equal to:

$$\eta = 2C\left(J_0^2(kr) + J_1^2(kr)\right) \tag{6.19}$$

Only the zeroth order of the Bessel function in (6.18) and (6.19) is non-zero on the optical axis ($r = 0$), i.e. only one term with $J_0(0)$ remains in (6.18) and (6.19), and in (6.19) it is twice larger then in (6.18). Thus, in this specific case, when the lens is limited by an annular aperture, and the aperture angle tends to $\pi/2$ for beam order $n = 1$, the helicity density as well as chirality on the axis can be maximal—twice the helicity density for the beams of orders $n = 0$ and $n = 2$. Figure 6.2 shows the helicity density distribution at the focus for this case (6.13), (6.14) ($\lambda = 0.633$ μm).

6.1.2 *Numerical Simulation*

Below are shown the results of numerical simulation of the tight focusing of the Poincaré beam (1) with the wavelength $\lambda = 633$ nm and the flat wave front by the aplanatic objective with the numerical aperture NA = 0.95 (this numerical aperture corresponds to the commercially available objectives). For some particular cases of the beam, the helicity density distribution is shown in the figures below. In particular, Fig. 6.3 shows the result of focusing the Poincaré beam with the parameters

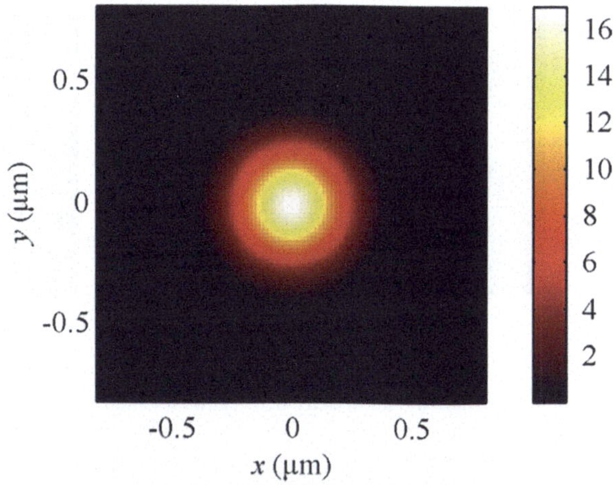

Fig. 6.3 Intensity for the Poincaré beam (6.1) with parameters $n = 0$, $\psi = 0$, $\theta = 0$. For these parameters helicity density coincides with intensity

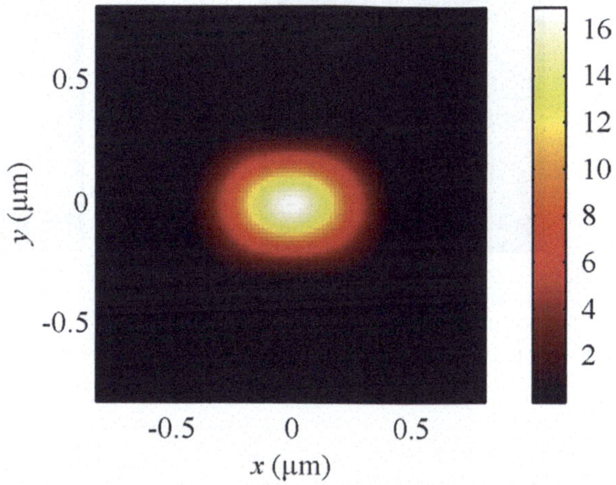

Fig. 6.4 Intensity for the Poincaré beam (6.1) with parameters $n = 0$, $\psi = 0$, $\theta = \pi/2$. For these parameters there is no helicity density

$n = 0$, $\psi = 0$, $\theta = 0$. In this case, the intensity coincides with the helicity density. Figure 6.4 shows the intensity distribution when focusing the beam with the parameters $n = 0$, $\psi = 0$, $\theta = \pi/2$; helicity density is absent with these parameters. Figure 6.5 shows the result of focusing the beam (1) for $n = 0$, $\psi = \pi/4$, $\theta = \pi/4$. The intensity and helicity density with these parameters do not coincide. Helicity density has a radially symmetrical shape, while the intensity has the elliptical shape. In this case, the presence of the factor $\cos(2(n\text{-}1)\varphi + \psi)$ in formula (6.11) is responsible for the elliptical shape and rotation of the spot. Figures 6.7 show the effect of the topological charge of the beam vortex n on the helicity density.

From Fig. 6.7 it follows that the helicity density has non-zero values on the axis if the beam order n is equal to zero, one, or two, and it is equal to zero for other orders. This simulation is in agreement with Eq. (6.8). From Fig. 6.3, 6.6, 6.7 it also follows that when using standard wide-aperture objectives (NA = 0.95) the helicity density

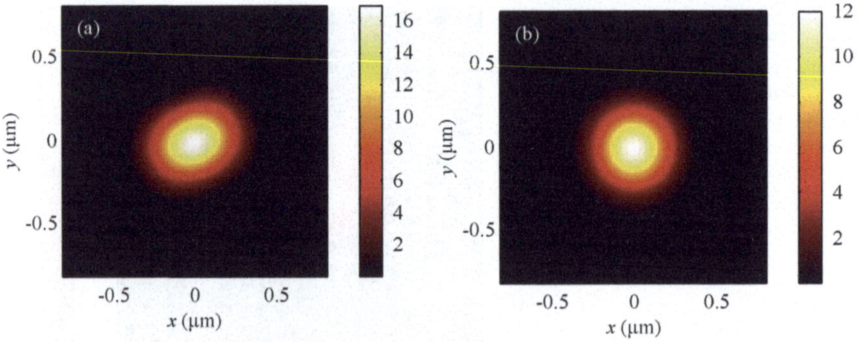

Fig. 6.5 Intensity (**a**) and helicity density (**b**) for the Poincaré beam (6.1) with parameters $n = 0$, $\psi = \pi/4$, $\theta = \pi/4$

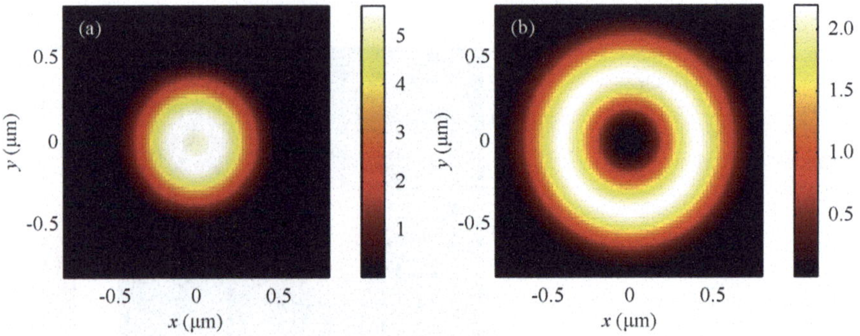

Fig. 6.6 Helicity density for Poincaré beams (1) with parameters n = 1 (**a**) and n = 3 (**b**) (ψ = 0, θ = 0)

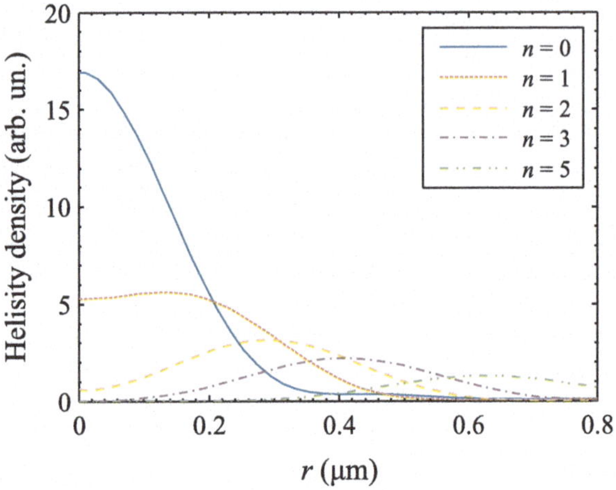

Fig. 6.7 Helicity density cross-sections for different beam orders ($\psi = 0$, $\theta = 0$)

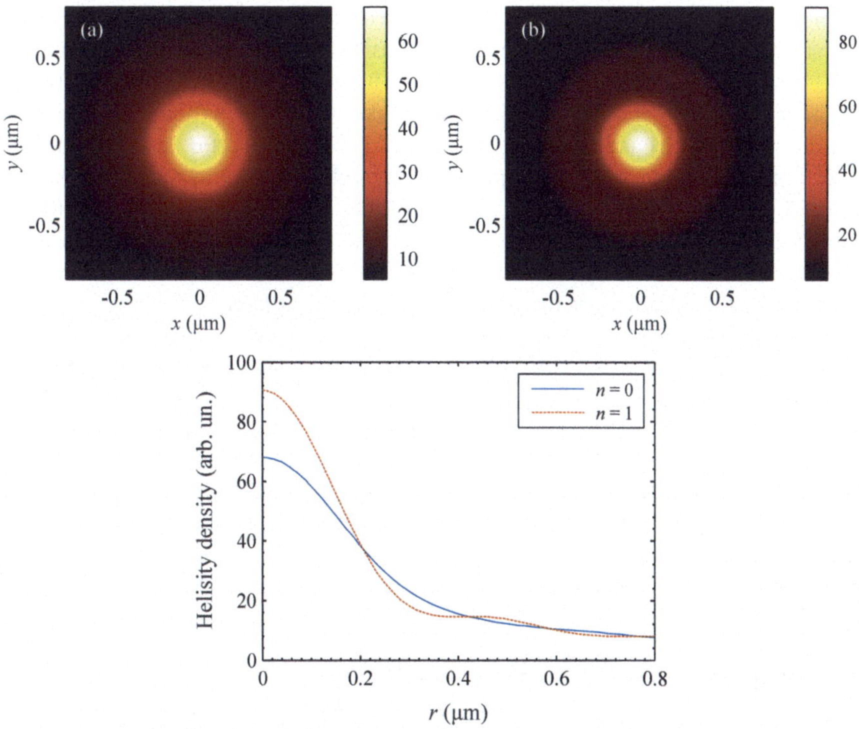

Fig. 6.8 Distribution of helicity density (**a**,**b**) and its cross-sections (**c**) when focusing Poincaré beams ($\psi = 0$, $\theta = 0$) of orders $n = 0$ (**a**) and $n = 1$ (**b**) by diffractive lens (NA = 0.99), limited by an annular aperture $\alpha_{min} = 0.9 \times \arcsin(\text{NA})$

on the axis is maximum for the zeroth beam order is larger for the beam of order $n = 1$. From Fig. 6.7 it follows that further increase of n leads to widening of the ring (for example, at $n = 12$ the radius of the ring in the simulation was equal to 1.6 μm or 2.5λ, at $n = 20$ it equal to was 2.4 μm or 3.8λ). The analytical equation (6.8) is valid for any n, however, it should be noted that Richards-Wolf integrals are intended for calculating the light field in the vicinity of tight focus and with increasing distance to the focus their accuracy decreases, therefore, calculation of higher modes with ring radii large enough, may carry potential errors. Moreover, the experimental focusing of vortices with large n to the small area is challenging [37, 38]. Taking the effects of large n into account is beyond the scope of this article.

Now let consider tight focusing by a diffractive lens with a numerical aperture of NA = 0.99, limited by an annular aperture with a minimum angle of $\alpha_{min} = 0.9 \times \arcsin(\text{NA})$. Figure 6.8 shows helicity density distribution for the beams of orders $n = 0$ and $n = 1$ ($\psi = 0$, $\theta = 0$). Figure 6.8 shows that in this very specific case helicity density on the optical axis for the beam of order $n = 1$ is larger than helicity density for the beam of order $n = 0$.

In this section, the helicity in tightly focused higher-order Poincaré beams was investigated theoretically and numerically using the Richards-Wolf formalism [41]. It was shown that the helicity density for higher-order Poincaré beams always has radial symmetry. The helicity density is maximum at $\theta = 0$ (optical vortex with circular polarization) and is absent at $\theta = \pi/2$ (cylindrical vector beam). Non-zero values of helicity density on the optical axis are observed for beam orders $n = 0$, 1, and 2. At $\theta = 0$, the intensity of higher-order Poincaré beams coincides with the helicity density up to a factor. It was theoretically and numerically shown that for ideal case (the numerical aperture tends to unity, the lens is limited by a thin annular aperture) helicity density for the beam of order $n = 1$ is larger on the optical axis, then for the beams of order $n = 0$ or $n = 2$; however, for standard aplanatic microobjectives (NA = 0.95) helicity density on the optical axis is larger for the beam of order $n = 0$. Our results are consistent with the results previously obtained by M. Babiker et al. in [26]; however, in our research it was taken into account the vectorial effects which have place in the focus of a lens with high numerical aperture. All results obtained for helicity could be adapted to chirality. In the future investigations the results of this paper could be generalized to the case of hybrid-order Poincaré beams [39, 40].

6.2 Skyrmion Based on Poincaré Beams

In optics, recent years have seen the arrival of new publications dealing with a new object of study, namely, skyrmions, or optical quasiparticles. Such vector fields are called quasiparticles because they abruptly decay with increasing distance from their center. In microparticle physics, the skyrmions have been known since 1961 [42], being named after T.H.R. Skyrme who was the first to propose them. An optical skyrmion is constructed as follows. First, the original 2D Jones vector is chosen for a vector field of Poincare-beam type. Next, the normalized Stokes vector is calculated, forming a basis for constructing a 3D skyrmion field [43–46]. The polarization pattern over the skyrmion cross-section can be either Neel type or Bloch type (vortical distribution of the polarization vector) [47]. Unlike other vector fields with V-point, C-point (or other) polarization singularities, which serve as centers of topological charge distributions shaping lemon-, star-, monster-type and other patterns, the skyrmion vector field has just positive Cartesian projections, with no negative projections. We note that the vector at the skyrmion center and peripheral vectors are mutually orthogonal. While at the center the vector is directed along the optical axis, on the periphery the skyrmion vectors are found in the orthogonal plane.

However, all skyrmion types share a common topo-logical characteristic similar to the topological charge (winding number), or the polarization singularity index of any vector field, and called the skyrmion number [47–49]. The skyrmion number shows how many circles the polarization state makes on the Poincare sphere on a full circle at the beam cross-section. Just like the topological charge, the skyrmion number conserves upon the paraxial beam propagation. For Poincare beams built as superposition of two Laguerre-Gaussian beams it has been shown [49] that the skyrmion number is given by the difference of two topological charges of the constituent beams in the Poincare beam. So far, no general proof of the conservation of the skyrmion number upon propagation has been offered. An alternative definition of the skyrmion number uses Stokes components [50] and is based on a contour integral over a closed path with respect to the gradient of phase of a complex Stokes field. This definition is not constructive.

In this section, by retrofitting the skyrmion number definition from Ref. [50], we propose a novel constructive expression for the skyrmion number based on an integral taken over an infinite-radius axis-centered circle with respect to the derivative of the complex Stokes field phase relative to the azimuthal angle. By retrofitting we mean that the familiar formula for calculating the skyrmion number is simplified, allowing the skyrmion number to be derived in an analytical form from an analytical relationship. Besides, the relationship for calculating the skyrmion number in [50] reveals that the skyrmion number depends on the radial variable. In the retrofitted formula proposed herein, the skyrmion number is an integer which is independent of the radial variable and conserves upon free-space propagation of the skyrmion.

In this work, using two source vector fields in the form of Poincare beams as an example, we show that the familiar definition of the skyrmion number [47, 48] and the new one lead to the same result. Besides, the partial skyrmion numbers calculated based on each projection of the skyrmion vector field are shown to be equal to each other and to a third of the full integer skyrmion number. It is numerically shown that upon paraxial propagation, both partial skyrmion numbers and the full skyrmion number are preserved. It is worth noting that one of two examples of the Poincare beams analyzed herein is not built as superposition of Laguerre-Gaussian beams.

6.2.1 Theoretical Background

For a source 2D vector light field whose Jones vector is given by

$$\mathbf{E} = \begin{pmatrix} E_x(r, \varphi) \\ E_y(r, \varphi) \end{pmatrix}, \tag{6.20}$$

where the complex functions $E_x(r, \varphi)$ and $E_y(r, \varphi)$ describe amplitudes of the transverse electric field projections in the polar coordinates (r, φ), a 3D vector field composed of three normalized Stokes components can be constructed in the form:

$$\begin{aligned} S_x &= \frac{|E_x|^2 - |E_y|^2}{|E_x|^2 + |E_y|^2}, \\ S_y &= \frac{2\,\mathrm{Re}\left(E_x^* E_y\right)}{|E_x|^2 + |E_y|^2}, \\ S_z &= \frac{2\mathrm{Im}\left(E_x^* E_y\right)}{|E_x|^2 + |E_y|^2}. \end{aligned} \tag{6.21}$$

Just as an optical vortex with complex amplitude $E(r, \varphi)$ is described by its topological charge [51].

$$TC = \lim_{r \to \infty} \frac{\mathrm{Im}}{2\pi} \int_0^{2\pi} \frac{\partial \arg(E(r,\varphi))}{\partial \varphi} d\varphi, \tag{6.22}$$

(where lim means that we take a limit of the function by tending r to infinity, Im is the imaginary part of the number), so is skyrmion (6.21) described by a skyrmion number:

$$NS = \frac{1}{4\pi} \iint_\sigma \mathbf{S} \cdot \left(\frac{\partial \mathbf{S}}{\partial x} \times \frac{\partial \mathbf{S}}{\partial y}\right) dxdy = \frac{1}{4\pi} \iint_\sigma \sum dxdy, \tag{6.23}$$

where σ is a 2D integration domain and $\times$ denotes the cross product. A skyrmion beam [47–49] is a 3D topological vector field $\mathbf{\Sigma} = (\Sigma_x, \Sigma_y, \Sigma_z)$ related with the Stokes vector $\mathbf{S} = (S_x, S_y, S_z)$, which, in turn, is related with the 2D field $\mathbf{E}$ in Eq. (6.20). Using the integrands in (6.23), projections of the skyrmion vector field $\mathbf{\Sigma}$ may be expressed by

$$\begin{aligned} \Sigma_x &= S_x \left(\frac{\partial S_y}{\partial x} \frac{\partial S_z}{\partial y} - \frac{\partial S_z}{\partial x} \frac{\partial S_y}{\partial y} \right), \\ \Sigma_y &= S_y \left(\frac{\partial S_z}{\partial x} \frac{\partial S_x}{\partial y} - \frac{\partial S_x}{\partial x} \frac{\partial S_z}{\partial y} \right), \\ \Sigma_z &= S_z \left(\frac{\partial S_x}{\partial x} \frac{\partial S_y}{\partial y} - \frac{\partial S_y}{\partial x} \frac{\partial S_x}{\partial y} \right). \end{aligned} \tag{6.24}$$

It can be shown [50] that the skyrmion field in (6.24) is transverse, i.e. the field divergence is zero:

$$\nabla \cdot \mathbf{\Sigma} = 0. \tag{6.25}$$

Within this framework, equality (6.25) is not possible to prove analytically because this requires the knowledge of the derivative of field (6.20) by the longitudinal coordinate z. Using a divergence theorem, we find that the skyrmion field flux through any closed surface $\mathbf{\Pi}$ equals zero:

$$\oint \mathbf{\Sigma} d\Pi = 0. \tag{6.26}$$

From (6.26) it follows that as skyrmion (6.24) is paraxially propagating in free space, skyrmion number (6.23) is preserved for any projection individually:

$$\iint_{R^2} \sum\nolimits_i (x, y, z) dxdy = \iint_{R^2} \sum\nolimits_i (x, y, 0) dxdy, \quad i = x, y, z. \tag{6.27}$$

Among other things, this conclusion is based on the fact that in (6.24), projections of the skyrmion vector field can vary cyclically. Because of this, the full skyrmion number NS is also preserved:

$$\begin{aligned} NS = &\frac{1}{4\pi} \iint_{R^2} \left[\sum\nolimits_x (x, y, z) + \sum\nolimits_y (x, y, z) \right. \\ &\left. + \sum\nolimits_z (x, y, z) \right] dxdy \\ = &\frac{1}{4\pi} \iint_{R^2} \left[\sum\nolimits_x (x, y, 0) + \sum\nolimits_y (x, y, 0) \right. \\ &\left. + \sum\nolimits_z (x, y, 0) \right] dxdy. \end{aligned} \tag{6.28}$$

6.2.2 *Skyrmion Field in the Source Plane for a Generalized Poincare Beam*

For the skyrmion field to be constructed based on source field (6.20), some conditions need to be satisfied. Actually, three Stokes vector projections in (6.21) can be constructed for any source field (6.20). But not any Stokes vector (6.21) is suited for constructing skyrmion field (6.24). Putting it more exactly, although an arbitrary skyrmion field can be expressed via the Stokes vector but the skyrmion number derived so will be zero. To clarify this assertion we will change from the Cartesian derivatives in (6.24) to the polar derivatives, which yields:

$$\begin{aligned}\Sigma_x &= \frac{S_x}{r}\left(\frac{\partial S_y}{\partial r}\frac{\partial S_z}{\partial \varphi} - \frac{\partial S_z}{\partial r}\frac{\partial S_y}{\partial \varphi}\right),\\ \Sigma_y &= \frac{S_y}{r}\left(\frac{\partial S_z}{\partial r}\frac{\partial S_x}{\partial \varphi} - \frac{\partial S_x}{\partial r}\frac{\partial S_z}{\partial \varphi}\right),\\ \Sigma_z &= \frac{S_z}{r}\left(\frac{\partial S_x}{\partial r}\frac{\partial S_y}{\partial \varphi} - \frac{\partial S_y}{\partial r}\frac{\partial S_x}{\partial \varphi}\right).\end{aligned} \tag{6.29}$$

As seen from (6.29), all projections of source field (6.20), which is independent of the radial variable, are zero because all derivatives by the radial variable are zero. Besides, if source field (6.20) is composed of real functions, then S_z in Eq. (6.21) is zero, as are its derivatives by φ in (6.29). Therefore, the entire skyrmion field in (6.29) equals zero. Hence, we infer that to obtain non-zero projections (6.29) of the skyrmion field, source field (6.20) needs to (i) depend on both polar coordinates, (ii) have a unit absolute value and (iii) be complex, while Stokes vector (2) has to have all non-zero projections with at least one projection being independent of the azimuthal angle.

It appears that satisfying the above-said conditions is a source Poincare beam vector field [52, 53], written in a special form:

$$\begin{aligned}\mathbf{E} &= \frac{1}{\sqrt{r^2+f^2}}\begin{pmatrix} fe^{-i(n\varphi+\varphi_0)} + re^{+i(n\varphi+\varphi_0)} \\ ife^{-i(n\varphi+\varphi_0)} - ire^{+i(n\varphi+\varphi_0)} \end{pmatrix}\\ &= \frac{1}{\sqrt{r^2+f^2}}\left[fe^{-i(n\varphi+\varphi_0)}\begin{pmatrix}1\\ i\end{pmatrix} + re^{+i(n\varphi+\varphi_0)}\begin{pmatrix}1\\ -i\end{pmatrix}\right],\end{aligned} \tag{6.30}$$

where f is the focal length of a spherical lens and φ_0 is a constant phase. Using (6.21), Stokes vector projections of source field (6.30) are expressed in the form:

$$\begin{aligned}S_x &= \frac{2rf}{r^2+f^2}\cos(2n\varphi + 2\varphi_0),\\ S_y &= \frac{2rf}{r^2+f^2}\sin(2n\varphi + 2\varphi_0),\\ S_z &= \frac{f^2-r^2}{r^2+f^2}.\end{aligned} \tag{6.31}$$

The longitudinal Stokes vector projection in (6.31) shows that field (6.30) has right handed elliptical polarization ($S_z > 0$) at $r < f$, left handed elliptical polarization ($S_z < 0$) at $r > f$, and linear polarization ($S_z = 0$) at $r = f$. On the optical axis ($r = 0$), there occurs a C-point, i. e. right handed circular polarization ($S_z = 1$). The sum of squares of the transverse Stokes vector projections is independent of the azimuthal angle φ and equals

$$S_x^2 + S_y^2 = \frac{4r^2 f^2}{\left(r^2 + f^2\right)^2}. \tag{6.32}$$

From (6.32), it is seen that when making an axis-centered arbitrary circle of radius r in the cross-section, the transverse Stokes vector projection is φ-independent but varies with radius r, increasing from zero at $r = 0$ to its maximum at $r = f$, before starting to decay to zero at infinity.

Substituting (6.31) into (6.29) yields projections of the skyrmion field for source field (6.30):

$$\begin{aligned} \Sigma_x &= \frac{32nr^2 f^4}{\left(r^2 + f^2\right)^4} \cos^2(2n\varphi + 2\varphi_0), \\ \Sigma_y &= \frac{32nr^2 f^4}{\left(r^2 + f^2\right)^4} \sin^2(2n\varphi + 2\varphi_0), \\ \Sigma_z &= \frac{8nf^2\left(f^2 - r^2\right)^2}{\left(r^2 + f^2\right)^4}. \end{aligned} \tag{6.33}$$

Summing up the transverse skyrmion field projections in (6.33) yields:

$$\sum\nolimits_x + \sum\nolimits_y = \frac{32nr^2 f^4}{\left(r^2 + f^2\right)^4}. \tag{6.34}$$

Thus, we have derived a function independent of the azimuthal angle φ. That is, when making an axis-centered, arbitrary-radius circle in the cross-section of beam (6.30), the sum of the transverse projections (6.33) conserves, although the modulus of the transverse skyrmion vector field in (6.33) does not conserve. As r tends to infinity, all skyrmion projections quickly tend to zero as r^{-4} and r^{-6}. Interestingly, all skyrmion projections in (6.33) are positive definite functions and have the same average value over the beam cross-section:

$$\begin{aligned} \int_0^\infty \int_0^{2\pi} \sum\nolimits_x (r, \varphi) r dr d\varphi &= \int_0^\infty \int_0^{2\pi} \sum\nolimits_y (r, \varphi) r dr d\varphi \\ = \int_0^\infty \int_0^{2\pi} \sum\nolimits_z (r, \varphi) r dr d\varphi &= \frac{8}{3}\pi n. \end{aligned} \tag{6.35}$$

Hence, for generalized Poincare beam (6.30), skyrmion number (6.23) equals

$$NS = \frac{1}{4\pi} \int_0^{\infty} \int_0^{2\pi} \left(\sum_x + \sum_y + \sum_z \right) r dr d\varphi = 2n. \tag{6.36}$$

From (6.36), it is seen that the skyrmion number is twice as large as topological charge n of an optical beam comprising a Poincare beam. In this case, the skyrmion number is seen to coincide with the Stokes index of the beam polarization singularity, which is twice as large as the Poincare-Hopf polarization index and equals n for the Poincare beam [53].

An alternative formula for calculating the skyrmion number was given in Ref. [50], although with no proof:

$$\begin{aligned} NS &= -\frac{1}{4\pi} \oint S_z \nabla \Phi d\mathbf{l}, \\ \Phi &= \arg\left(S_x + iS_y\right). \end{aligned} \tag{6.37}$$

In (6.37), the skyrmion number is calculated as an integral along a closed contour of the product of the third Stokes vector projection by the gradient of argument of the complex transverse Stokes field Φ. To describe the skyrmion topology in a proper way, an arbitrary integration contour in (6.37) should be replaced by a circle, the gradient of Φ by the derivative by the azimuthal angle φ, while also tending the circle radius to infinity so that all singularity points of vector field (6.31) are included. Then, instead of Eq. (6.37), the skyrmion number will be given by

$$NS = -\lim_{r \to \infty} \frac{\mathrm{Im}}{2\pi} \int_0^{2\pi} \frac{S_z}{S_x + iS_y} \frac{\partial}{\partial \varphi} \left(S_x + iS_y\right) d\varphi. \tag{6.38}$$

Note that in (6.37) the integral is divided by 4π, and in (6.38) the integral is divided by 2π. In (6.38), the circle lies in a plane transverse to the optical axis, therefore it is necessary to divide by the length of the circle of unit radius, and not by the area of the sphere of unit radius. When compared with (6.37), expression (6.38) for the skyrmion number is more constructive in many respects, being similar to the Berry formula [50] for calculating the topological charge of scalar optical vortices and to how the Poincare-Hopf index of vector light fields with polarization singularity points is calculated [54]. Using the Stokes field in (6.31), skyrmion number (6.30) calculated by the use of Eq. (6.38) is given by

$$
\begin{aligned}
NS &= -\lim_{r\to\infty}\frac{\mathrm{Im}}{2\pi}\int_0^{2\pi}\left(\frac{f^2-r^2}{f^2+r^2}\right) \\
&\quad \times \frac{\partial}{\partial\varphi}\left(\frac{2fre^{i(2n\varphi+2\varphi_0)}}{f^2+r^2}\right)\left(\frac{2fre^{i(2n\varphi+2\varphi_0)}}{f^2+r^2}\right)^{-1} d\varphi \\
&= 2n.
\end{aligned}
\tag{6.39}
$$

A comparison of the number of skyrmion (6.33) derived using Eq. (6.36) with that derived from new formula (6.38) proposed above shows that the two numbers coincide and equal $2n$. Hence, the skyrmion number can be calculated using just two first Stokes vector components, S_x and S_y (6.21). The third normalized Stokes vector component, S_z in (6.38) equals +1 or −1 at infinity. For these components to be measured, one needs to measure four intensity distributions at different angles of inclination of the polarizer axis to the Cartesian axis x: 0°, 90°, 45° and −45°. In this case, the Stokes vector components will be expressed via these intensity as follows: $S_x = I_0 - I_{90}$, $S_y = I_{45} - I_{-45}$. In the next step, the skyrmion number NS can be calculated using (6.38). We note that in Ref. [55], Stokes vector components were measured experimentally for different skyrmion types. However, owing to the complexity of Eq. (6.23), the numbers of the generated skyrmions were not calculated in Ref. [55]. Had the formula (6.38) been known to the authors of Ref. [55], the measurement of the skyrmion number could be rendered easier.

6.2.3 Skyrmion Field for Superposition of Two Optical Vortices with Orthogonal Linear Polarizations

Another example of a skyrmion field in the source plane we study below is analogous to Poincare beam (6.11), but instead of circularly polarized beams comprise two mutually orthogonal linear polarizations. We also assume the optical vortices to have different topological charges and intensity described by a power function similar to the single-ring Laguerre-Gaussian beams. Then, in the source plane, the Jones vector of such a beam takes the form:

$$
\begin{aligned}
\mathbf{E} &= \frac{1}{\sqrt{r^{2n}+r^{2m}}}\begin{pmatrix} r^n e^{i(n\varphi+\varphi_0)} \\ r^m e^{i(m\varphi+\varphi_0)} \end{pmatrix} \\
&= \frac{1}{\sqrt{r^{2n}+r^{2m}}}\left[r^n e^{i(n\varphi+\varphi_0)}\begin{pmatrix}1\\0\end{pmatrix} + r^m e^{i(m\varphi+\varphi_0)}\begin{pmatrix}0\\1\end{pmatrix}\right],
\end{aligned}
\tag{6.40}
$$

where n and m are positive integer numbers that define topological charges of the optical vortices. By analogy with Eq. (6.31), projections of the Stokes vector normalized to intensity for field (6.40) are given by

$$\begin{aligned} S_x &= \frac{r^{2n} - r^{2m}}{r^{2n} + r^{2m}}, \\ S_y &= \frac{2r^{n+m}}{r^{2n} + r^{2m}} \cos((m-n)\varphi), \\ S_z &= \frac{2r^{n+m}}{r^{2n} + r^{2m}} \sin((m-n)\varphi). \end{aligned} \tag{6.41}$$

Stokes field (6.41) is identical in form to the magnetization vector projections discussed in Ref. [43]. From a comparison of Eqs. (6.31) and (6.41), both skyrmions are seen to be structurally similar, with one Stokes vector projection being azimuthal-angle independent and two other Stokes vector projections depending on sin θ and cos θ, where θ is linearly dependent on the azimuthal angle φ. Therefore, the sum of the two projections squared is independent of the azimuthal angle, as is the third Stokes vector projection. The difference is that in this case, the longitudinal and transverse projections need to be added:

$$S_z^2 + S_y^2 = \frac{4r^{2(n+m)}}{\left(r^{2n} + r^{2m}\right)^2}. \tag{6.42}$$

Unlike field (6.30), which changes handedness with changing radius, the elliptic polarization of field (6.40) changes handedness with changing angle: at sin $(m - n)\varphi > 0$, the field has right-handed elliptic polarization ($S_z > 0$) and at sin $(m - n)\varphi < 0$, the field has left-handed elliptic polarization ($S_z < 0$). At angles that satisfy the condition $\sin(m - n)\varphi = 0$, the field is linearly polarized ($S_z = 0$). Considering that at $n > m$, there holds an inequality $r^{2n} + r^{2m} > 2r^{n+m}$, nowhere across field (6.37) circular polarization ($S_z = \pm 1$) is found.

Next, in a similar way to (6.33), projections of the vector skyrmion field for source vector field (6.40) can be derived in the form:

$$\begin{aligned} \sum\nolimits_x &= \frac{4(m-n)^2\left(r^{2n} - r^{2m}\right)^2 r^{2(n+m-1)}}{\left(r^{2n} + r^{2m}\right)^4}, \\ \sum\nolimits_y &= \frac{16(m-n)^2 r^{2(2n+2m-1)}}{\left(r^{2n} + r^{2m}\right)^4} \cos^2(m-n), \\ \sum\nolimits_z &= \frac{16(m-n)^2 r^{2(2n+2m-1)}}{\left(r^{2n} + r^{2m}\right)^4} \sin^2(m-n). \end{aligned} \tag{6.43}$$

From a comparison of (6.33) and (6.43), both skyrmion fields are seen to be structurally similar, with one skyrmion vector projection being independent of the azimuthal angle and the other two depending on the azimuthal angle in such a way that the sum thereof is azimutal angle-independent. Similar to field (6.33), skyrmion field (6.43) is positively defined, dying out to zero as $r^{-2(n-m+1)}$ and $r^{-4(n-m)-2}$, with the radial variable tending to infinity. The only difference is that for skyrmion

(6.34), the rate of tending to zero is independent of topological charge n, depending on the difference $n - m$ ($n > m$) for skyrmion (6.43). On the optical axis ($r = 0$), the amplitude of skyrmion (6.43) equals zero, unlike skyrmion (6.34), which has the on-axis amplitude $\Sigma_z(r = 0) = 8nf^{-2}$. Along the different axes, the skyrmion numbers can be shown to be the same and equal to $(n - m)/3$:

$$\frac{1}{4\pi}\int_0^{\infty}\int_0^{2\pi}\sum\nolimits_x(r,\varphi)rdrd\varphi = \frac{1}{4\pi}\int_0^{\infty}\int_0^{2\pi}\sum\nolimits_y(r,\varphi)rdrd\varphi$$
$$= \frac{1}{4\pi}\int_0^{\infty}\int_0^{2\pi}\sum\nolimits_z(r,\varphi)rdrd\varphi = \frac{1}{3}(n-m). \quad (6.44)$$

Hence, full skyrmion number (6.23) of superposition (6.40) of two Laguerre-Gaussian beams with mutually orthogonal linear polarizations and topological charges n and m is described by the expression

$$NS = \frac{1}{4\pi}\int_0^{\infty}\int_0^{2\pi}\left(\sum\nolimits_x + \sum\nolimits_y + \sum\nolimits_z\right)rdrd\varphi = n - m. \quad (6.45)$$

Because all skyrmion projections have the same number, Eq. (6.44), it can be asserted that permutations of the skyrmion coordinates produce the same skyrmion. We note that at $m = -n$, the number of skyrmion (6.43) is identical to the number of skyrmion (6.33) and equals $NS = 2n$. Hence, the two skyrmions are topologically identical.

To calculate the number of skyrmion (6.41) based on Eq. (6.38) the following replacements need to be implemented: S_x by S_z, S_y by S_x and S_z by S_y:

$$S_x = \frac{2r^{n+m}}{r^{2n} + r^{2m}}\cos((m-n)\varphi),$$
$$S_y = \frac{2r^{n+m}}{r^{2n} + r^{2m}}\sin((m-n)\varphi), \quad (6.45)$$
$$S_z = \frac{r^{2n} - r^{2m}}{r^{2n} + r^{2m}}.$$

Substituting (6.46) into (6.38) yields (at $n > m$):

$$
\begin{aligned}
NS &= -\lim_{r\to\infty}\frac{\mathrm{Im}}{2\pi}\int\limits_0^{2\pi}\left(\frac{r^{2n}-r^{2m}}{r^{2n}+r^{2m}}\right) \\
&\times\frac{\partial}{\partial\varphi}\left(\frac{2r^{n+m}e^{i(m-n)\varphi}}{r^{2n}+r^{2m}}\right)\left(\frac{2r^{n+m}e^{i(m-n)\varphi}}{r^{2n}+r^{2m}}\right)^{-1}d\varphi \\
&= n-m.
\end{aligned}
\tag{6.46}
$$

The numbers of skyrmions (6.43) calculated using relationships (6.45) and (6.38) are seen to coincide.

6.2.4 Spin Angular Momentum at the Tight Focus of a Poincare Beam

Projections of the E-vector at the tight focus of beam (6.30) can be derived based on a Richard-Wolf formalism [56] in the form ($\varphi_0 = 0$):

$$
\begin{aligned}
E_x &= i^{n-1}e^{-in\varphi}\left(I_{0,n}^{(1)}+e^{i2\varphi}I_{2,n-2}^{(1)}\right) \\
&\quad +i^{n-1}e^{in\varphi}\left(I_{0,n}^{(2)}+e^{-i2\varphi}I_{2,n-2}^{(2)}\right), \\
E_y &= i^{n-1}e^{-in\varphi}\left(I_{0,n}^{(1)}-e^{i2\varphi}I_{2,n-2}^{(1)}\right) \\
&\quad +i^{n-1}e^{in\varphi}\left(-I_{0,n}^{(2)}+e^{-i2\varphi}I_{2,n-2}^{(2)}\right), \\
E_z &= -2i^{n}\left(e^{-i(n-1)\varphi}I_{1,n-1}^{(1)}+e^{i(n-1)\varphi}I_{1,n-1}^{(2)}\right).
\end{aligned}
\tag{6.48}
$$

Entering (6.48) are functions that depend on the radial variable r:

$$
\begin{aligned}
I_{\nu,\mu}^{(1,2)} &= 2kf\int\limits_0^{\alpha}\sin^{\nu+1}\left(\frac{\theta}{2}\right)\cos^{3-\nu}\left(\frac{\theta}{2}\right)\cos^{1/2}(\theta) \\
&\quad\times A_{1,2}(\theta)e^{ikz\cos\theta}J_{\mu}(kr\sin\theta)d\theta, \\
A_1(\theta) &= \frac{f}{\sqrt{r^2+f^2}} = \cos\theta, \\
A_2(\theta) &= \frac{r}{\sqrt{r^2+f^2}} = \sin\theta,
\end{aligned}
\tag{6.49}
$$

where $k = 2\pi/\lambda$ is the wave number of monochromatic light of wavelength λ, f is the focal length of a focusing lens, α is the maximum angle of inclination of rays to the optical axis, which defines the numerical aperture of the aplanatic lens, NA $= \sin\alpha$, $J_\mu(\xi)$ is the Bessel function of the first kind and μ-th order. The functions $A_{1,2}(\theta)$ are real and define the radial symmetry of the source field amplitude which depends on the tilt angle θ of a ray outgoing from a point on the original spherical wavefront and converging to the focal plane center.

Now, using Eq. (6.48), let us find the longitudinal projection of the spin angular momentum (SAM) [57]:

$$\mathbf{SP} = \frac{1}{8\pi\omega}\mathrm{Im}(\mathbf{E}^{*} \times \mathbf{E}). \tag{6.50}$$

We obtain:

$$\begin{aligned} SP_z &= -2\sin 2\varphi\left(I_{0,n}^{(1)}I_{2,n-2}^{(1)} + I_{0,n}^{(2)}I_{2,n-2}^{(2)}\right) \\ &-2\sin 2n\varphi\left(I_{0,n}^{(1)}I_{0,n}^{(2)}\right) \\ &+2\sin(2n-4)\varphi\left(I_{2,n-2}^{(1)}I_{2,n-2}^{(2)}\right). \end{aligned} \tag{6.51}$$

It has been shown [58] that near the optical axis, the integrals with zero first index provide a much higher contribution to the light field than the rest integrals. With this in mind and because a theoretical analysis of Eq. (6.51) is challenging, a simpler approximate relationship for the longitudinal SAM component at the focus of field (6.30) can be analyzed:

$$SP_z \approx -2\sin 2n\varphi\left(I_{0,n}^{(1)}I_{0,n}^{(2)}\right). \tag{6.52}$$

From (6.52) it is seen that meanwhile in source field (6.30) the third Stokes vector projection changes sign with changing radius, longitudinal SAM component (6.52) changes sign with changing angle. When making an arbitrary-radius full circle centered on the optical axis, the longitudinal SAM projection (or third Stokes vector projection) changes sign $4n$ times. Such type of skyrmion has been given the name 'multi-skyrmion' [50].

6.2.5 Numerical Modeling

For the purpose of numerical modeling, field (6.30) is multiplied by the Gaussian function $\exp[-(x^2 + y^2)/w^2]$, where w is the waist radius of the Gaussian beam, assumed to equal $w = 1$ mm upon modeling. Figure 6.9 shows an intensity pattern of Poincare beam (6.30), multiplied by a Gaussian function at $n = 1, f = 1$ mm, $\varphi_0 = 0$. The intensity pattern is depicted in collor level, with white color marking the maximum intensity and black—the minimum. The arrows in Fig. 6.9 show the direction of the transverse Stokes vector in Eq. (6.31), $S_x\mathbf{e_x} + S_y\mathbf{e_y}$, where $\mathbf{e_x}$ and $\mathbf{e_y}$ are unit vectors along the Cartesian coordinates x and y. Sign of the longitudinal

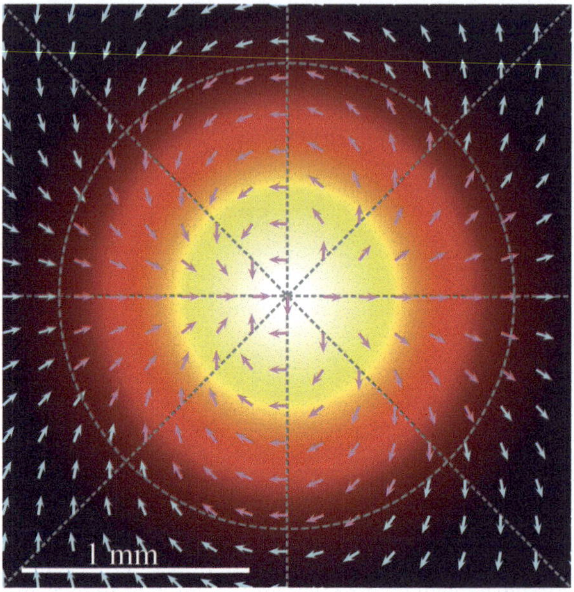

Fig. 6.9 Intensity pattern of field (6.30) (zero intensity marked black, maximum—white), Stokes vector distribution in the source plane of field (6.30): the arrows show the transverse Stokes vector projection in Eq. (6.31), with the arrow color showing the longitudinal projection sign: pink—$S_z > 0$, green—$S_z < 0$. For the illustrative purposes, the region of positive values of S_z is encompassed with a dotted circle. For better visualization, all arrows are same-size. The frame size is 2.5 × 2.5 mm

Stokes vector component in (6.31) is marked in color: pink color for $S_z > 0$, green color for $S_z < 0$. In compliance with the theoretical prediction in Eq. (6.31), the field is right-handed elliptically polarized at the center of the beam cross-section ($S_z > 0$) and left-handed elliptically polarized on the periphery ($S_z < 0$). The intensity pattern of beam (6.30) is shaped as a Gaussian function $I(r) = \exp(-2r^2/w^2)$.

From Fig. 6.9 and Eq. (6.31), on the horizontal axis, the Stokes vectors are seen to lie along the x-axis. On the vertical axis, the Stokes vectors are opposite to the x-axis. On the rays tilted at angles $\varphi = \pi/4$ and $5\pi/4$, the Stokes vectors are directed vertically upwards along the y-axis, being directed vertically downwards oppositely to the y-axis at $\varphi = 3\pi/4$ and $7\pi/4$. At the center of field (6.30), there occurs an on-axis C-point where the field is circularly polarized, for at $r = 0$, the longitudinal Stokes vector in Eq. (6.31) equals unity: $S_z = 1$. This is a polarization singularity point where the direction of axes of the polarization ellipse of field (6.30) is indeterminate. The Poincare-Hopf index [18] of C-point polarization singularity equals the ratio of an integer number of turns of the polarization ellipse axes on a circle centered on the optical axis to the full angle 2π. The Stokes index of polarization singularity is twice as much as the Poincare-Hopf index and equals the ratio of the number of full turns of the transverse Stokes vector on an axis-centered circle to the full angle [59, 60]. From Fig. 6.9, the transverse Stokes vector is seen to rotate by the angle 4π, meaning that the Stokes index is $\sigma = 4\pi/2\pi = 2$ (at $n = 1$). The Stokes index can be derived as the topological charge of a complex Stokes field [61]:

$$S_c = S_1 + iS_2 = \frac{2rf}{r^2 + f^2} \exp(i2n\varphi + i2\varphi_0) \cdot \tag{6.53}$$

The Stokes index, σ, and Poincare-Hopf index, η, are expressed through the topological charge using Berry formula [54]:

$$\sigma = 2\eta = \frac{1}{2\pi} \lim_{r \to \infty} \mathrm{Im} \int_0^{2\pi} d\varphi \frac{\partial S_c(r, \varphi)/\partial \varphi}{S_c(r, \varphi)} = 2n. \tag{6.54}$$

For the situation depicted in Fig. 6.9 at $n = 1$, the Stokes index in (6.54) appears to be equal to the skyrmion number in (6.36) and (6.39).

Figure 6.10 depicts skyrmion field projections (6.33), Σ_x, Σ_y, Σ_z. From a comparison of Stokes field (6.31) and skyrmion field (6.33), it is seen that they are structurally similar, the skyrmion field is positive and approximately defined by the Stokes field squared. With the transverse skyrmion field projections depending on the squared functions cos 2φ and sin 2φ, intensity patterns of the transverse projections have four petals each, rotated relative to each other by $\pi/4$, as seen from Fig. 6.10a and Fig. 6.10b. The longitudinal skyrmion field projection in (6.33) is radially symmetric and has its maximum on the optical axis (Fig. 6.10*c*). Numerically summing up all three projections over the entire image in Fig. 6.9 we find that, in agreement with theoretical prediction (6.35), all three skyrmions have the same numbers, $NS_x = NS_y = NS_z = 2/3$, with the full skyrmion number being $NS = 2$ (Fig. 6.10) at $n = 1$.

Figure 6.11 depicts an intensity pattern of skyrmion (6.30) with arrows marking directions of the transverse skyrmion field projections, i.e. the direction of the field $\Sigma_x \mathbf{e_x} + \Sigma_y \mathbf{e_y}$. Considering that skyrmion field projections (6.33) are positive, angles that the arrow directions in Fig. 6.11 make with the horizontal axis are ranging from zero to 90 degrees. The third projection of skyrmion vector, Σ_z is exclusively directed along the optical axis. Near the optical axis, the projection Σ_z has maximum values (red arrows), decreasing with increasing distance from the axis (blue arrows). The pattern of 3D skyrmion vectors (6.33) resembles hedgehog spikes.

Summing up, we have derived explicit analytical relationships to describe 3D vector fields for two different types of 2D Poincare beams [62]. The skyrmion fields analyzed, Eqs. (5.33) and (6.43), are described by positive definite functions and different numbers of Eqs. (6.36) and (6.45) that have been shown to depend on the topological charges of optical vortices contained in the Poincare beam. A new constructive formula enabling the skyrmion number to be calculated using just Stokes vector projections in Eq. (6.38) has been deduced. The skyrmion numbers obtained using the familiar formula in Eq. (6.28) and the newly proposed formula in Eq. (6.19) have been shown to coincide. Numbers of the individual skyrmion field projections have been shown to the same and equal to one third of the full skyrmion number in Eqs. (6.35), (6.44).

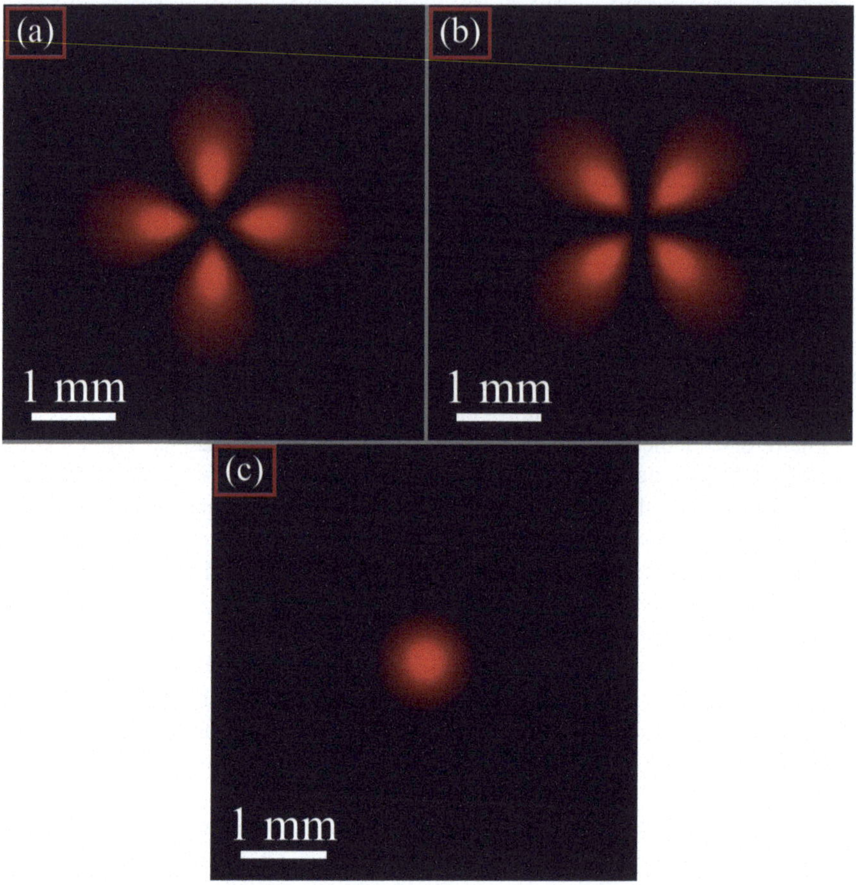

Fig. 6.10 Projections (6.33) of the 3D skyrmion field at $n = 1$: Σx (**a**), Σy (**b**), Σz (**c**). Frame size: 5×5 mm

6.3 Behavior of an Optical Vortex Immediately behind a Spherical Lens Illuminated by Circularly Polarized Light

It is a well-known fact that tightly focused linearly polarized light forms an elliptical focal spot [63–66] and circularly polarized light forms rotating energy flow in the focal plane [67–70]. The latter effect is called spin-orbit conversion [70]. It is also known that using non-uniform linear polarization (radial or azimuthal) it is possible to form a circular focal spot with a minimum size [71, 72]. We will show that, although a circular focal spot is formed during focusing of light with circular polarization and with radial polarization, the shape of the focal spots of the intensity of individual transverse components is elliptical. We will also show that the elliptical

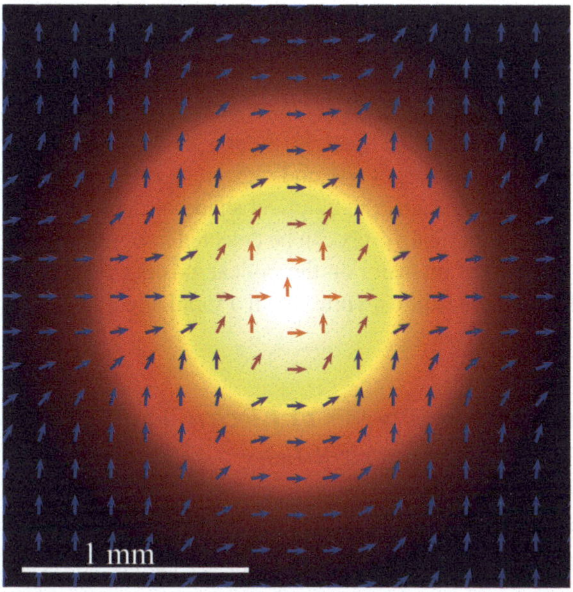

Fig. 6.11 Patterns of intensity (zero marked black and maximum marked white) and arrows of transverse projections of skyrmion vector in Eq. (6.33). The 3D skyrmion vectors resemble hedgehog spikes directed exclusively along the optical axis. The angle of tilt of the skyrmion vector (6.33) is maximum on the optical axis, decreasing with increasing distance from the optical axis, with the on-axis projections of the red arrows being larger than those of the blue ones. Frame size: 2.5 × 2.5 mm

shape of the focal spot during focusing of light with linear polarization is explained by the existence of a second-order optical vortex in each transverse component of the electric field.

In [73], based on the Richards -Wolf theory [64] it was shown for the first time that when focusing circularly polarized light, an optical vortex with a topological charge of +2 (right circular polarization) or −2 (left circular polarization) is formed at the transverse projections of the electric field at the focus. In [73], the existence of optical vortices at the focus was explained by the Berry geometric phase [74] and the spin Hall effect [75]. However, a detailed connection between the optical vortex from the one side and the geometric phase and the optical Hall effect from the other side was not shown in [73]. Based on [64, 76], we will show that optical vortices are formed singly or in pairs with different signs when light with any state of polarization (circular, linear, and cylindrical) passes through a spherical lens. If optical vortices appear in pairs (linear and cylindrical polarizations), they do not create orbital angular momentum. It will be shown that, depending on the azimuthal angle φ, the Cartesian components of the initial transverse light field will change sign four times during one rotation of the angle φ, i.e. they depends from $\sin(2\varphi)$ and $\cos(2\varphi)$.

In [77–79] the azimuthal energy flow in the focus of coherent light was investigated. For example, in [79] it was shown that when focusing circularly polarized light in a tight focus an azimuthal (angular) energy flow is formed in the focal plane. In this work, the existence of this flow is explained by the appearance of transverse and longitudinal optical vortices immediately a spherical lens when the circularly polarized light passes through it.

In this section, we show that: first, a second-order optical vortex is formed in each transverse Cartesian component immediately after circularly polarized light passes through a spherical lens. Second, the existence of this optical vortex immediately behind the spherical lens explains why the light acquires orbital angular momentum immediately behind the spherical lens. Third, this second-order optical vortex also explains the elliptical shape of the intensity of the Cartesian components in the focal plane. Fourth, for linearly polarized light, two transverse optical vortices with topological charges +2 and −2 are formed after the spherical lens. Fifth, their existence explains the elliptical shape of the focal spot for linearly polarized light. Sixth, an optical vortex can be detected in right-hand circularly polarized light by the presence of intensity on the optical axis at the focus when illuminating a spherical lens with an optical vortex with a topological charge of −2.

6.3.1 Components of Electric Field Immediately After the Spherical Lens

Within the Richards -Wolf theory, we can find the components of the electric **E** and magnetic **H** field vectors immediately behind an ideal spherical lens [64, 76]:

$$
\begin{aligned}
E_x &= \left(1 + A(\theta)\cos^2\varphi\right)a(\varphi) + \frac{A(\theta)\sin(2\varphi)}{2}b(\varphi),\\
E_y &= \frac{A\sin(2\varphi)}{2}a(\varphi) + \left(1 + A(\theta)\sin^2\varphi\right)b(\varphi),\\
E_z &= -\sin\theta(a(\varphi)\cos\varphi + b(\varphi)\sin\varphi),\\
H_x &= -\left(1 + A(\theta)\cos^2\varphi\right)b(\varphi) + \frac{A(\theta)\sin(2\varphi)}{2}a(\varphi),\\
H_y &= -\frac{A\sin(2\varphi)}{2}b(\varphi) + \left(1 + A(\theta)\sin^2\varphi\right)a(\varphi),\\
H_z &= -\sin\theta(-b(\varphi)\cos\varphi + a(\varphi)\sin\varphi),\\
A(\theta) &= \cos\theta - 1.
\end{aligned}
\tag{6.55}
$$

In Eq. (6.55), θ is the angle between the optical axis and the segment connecting the point of geometric focus with a point on the converging spherical front immediately behind the lens. In this case, a point on the spherical front located at a distance r from the optical axis is related to the focal length f by the equation: $r = f\sin\theta$, φ is azimuth angle. Next we will find some characteristics of this light field if the initial field is described by the following Jones vectors:

$$\begin{aligned}\mathbf{E}_0(\varphi) &= \begin{pmatrix} a(\varphi) \\ b(\varphi) \end{pmatrix} = \frac{1}{\sqrt{2}}\begin{pmatrix} 1 \\ \sigma i \end{pmatrix}, \\ \mathbf{H}_0(\varphi) &= \begin{pmatrix} -b(\varphi) \\ a(\varphi) \end{pmatrix} = \frac{1}{\sqrt{2}}\begin{pmatrix} -\sigma i \\ 1 \end{pmatrix}.\end{aligned} \tag{6.56}$$

That is, the light illuminating the spherical lens has right-hand (σ = 1) or left-hand (σ = −1) circularly polarization. If the initial field (6.56) is substituted into Eq. (6.55), we obtain the following projections of the electric and magnetic fields:

$$\begin{aligned}
E_x &= \frac{1}{2\sqrt{2}}\left[(1+\cos\theta) - (1-\cos\theta)e^{2\sigma i\varphi}\right], \\
E_y &= \frac{i\sigma}{2\sqrt{2}}\left[(1+\cos\theta) + (1-\cos\theta)e^{2\sigma i\varphi}\right], \\
E_z &= -\frac{1}{\sqrt{2}}\sin\theta e^{\sigma i\varphi}, \\
H_x &= -\frac{i\sigma}{2\sqrt{2}}\left[(1+\cos\theta) - (1-\cos\theta)e^{2\sigma i\varphi}\right], \\
H_y &= \frac{1}{2\sqrt{2}}\left[(1+\cos\theta) + (1-\cos\theta)e^{2\sigma i\varphi}\right], \\
H_z &= \frac{i\sigma}{\sqrt{2}}\sin\theta e^{\sigma i\varphi}.
\end{aligned} \tag{6.57}$$

Eqs. (6.57) show that immediately behind the spherical lens a superposition of a beam with right-hand circularly polarization ($\sigma = 1$) and an optical vortex with a topological charge of 2σ and with left-hand circularly polarization will be formed:

$$\begin{pmatrix} E_x \\ E_y \end{pmatrix} = \frac{1+\cos\theta}{2\sqrt{2}}\begin{pmatrix} 1 \\ \sigma i \end{pmatrix} - \frac{1-\cos\theta}{2\sqrt{2}}e^{2\sigma i\varphi}\begin{pmatrix} 1 \\ -\sigma i \end{pmatrix}. \tag{6.58}$$

Also, an optical vortex with a topological charge σ is formed in the longitudinal field component E_z. Moreover, both vortices (transverse and longitudinal) will be large at large angles of inclination of the beams converging to the focus, i.e. when angle θ is close to $\pi/2$. I.e. the optical vortex will be mainly on the ring, and there will be no optical vortex in the central part of the beam converging to the focus. This optical vortex, in turn, will create an orbital angular momentum (OAM) in the beam immediately behind the spherical lens, although the beam did not have OAM before the lens. Indeed, the longitudinal component of the OAM vector, taking into account Eqs. (6.57), will have the form [79]:

$$\begin{aligned} L_z &= \operatorname{Im}\left(E_x^* \frac{\partial E_x}{\partial \varphi} + E_y^* \frac{\partial E_y}{\partial \varphi} + E_z^* \frac{\partial E_z}{\partial \varphi} \right) = \\ &= \sigma(1 - \cos\theta). \end{aligned} \tag{6.59}$$

The longitudinal component of the spin angular momentum vector (SAM) immediately behind the spherical lens will have the form:

$$S_z = 2\operatorname{Im}\left(E_x^* E_y\right) = \sigma \cos\theta. \tag{6.60}$$

From Eq. (6.60) it is shown that, on the contrary, the longitudinal component of SAM is maximum near the optical axis and decreases to zero as the angle θ increases to $\pi/2$. From Eq. (6.59) and Eq. (6.60) one can find the total SAM and OAM immediately behind the spherical lens:

$$\begin{aligned} \widehat{L}_z &= 2\pi f^2 \int_0^{\theta_0} L_z \sin\theta d\theta = 2W \sin^4(\theta_0/2), \\ \widehat{S}_z &= 2\pi f^2 \int_0^{\theta_0} S_z \sin\theta d\theta = \frac{W}{2} \sin^2(\theta_0), \qquad W = 2\pi f^2. \end{aligned} \tag{6.61}$$

In Eq. (6.61) W is the power of the light beam that passed through the lens (the intensity of the initial beam is equal to unity), f is the focal length of the spherical lens. Eq. (6.61) shows that the maximum OAM is obtained with a numerical aperture *of NA* $= \sin\theta_0 = 1$ and it is equal to half of the beam power, $\widehat{L}_{z,\max} = W/2$. In this case, the longitudinal SAM will decrease by the same amount, $\widehat{S}_{z,\min} = W - W/2 = W/2$. However, this decrease of SAM does not increase the OAM, because it affects the transverse SAM. Similarly, we could obtain the longitudinal projection of the Poynting vector immediately after the spherical lens:

$$P_z = \operatorname{Re}\left(E_x^* H_y - E_y^* H_x\right) = \cos\theta. \tag{6.62}$$

Comparing Eqs. (6.60) and (6.62), we can see that the longitudinal components of SAM and the energy flow coincide up to the sign. The longitudinal component of SAM for left-hand circularly polarization is directed opposite to the optical axis z, and the longitudinal projection of Poynting vector is directed along the positive direction z.

6.3.2 Light with Right Circular Polarization in the Focal Plane

Light field (6.56) in the focal plane have the form [64]:

$$\begin{aligned} E_x &= \frac{-i}{\sqrt{2}}\left(J_{0,0} + e^{2\sigma i\varphi}J_{2,2\sigma}\right), \\ E_y &= \frac{1}{\sqrt{2}}\left(J_{0,0} - e^{2\sigma i\varphi}J_{2,2\sigma}\right), \\ E_z &= -\sqrt{2}e^{i\sigma\varphi}J_{1,\sigma}. \end{aligned} \tag{6.63}$$

In Eq. (6.63) the following function, which depends from the radial variable r, is used:

$$\begin{aligned} J_{\alpha,\beta} = \left(\frac{4\pi f}{\lambda}\right)\int\limits_0^{\theta_0} \sin^{\alpha+1}\left(\frac{\theta}{2}\right)\cos^{3-\alpha}\left(\frac{\theta}{2}\right)\cos^{1/2}(\theta)\times \\ \times B(\theta)e^{ikz\cos\theta}J_\beta(\xi)d\theta, \end{aligned} \tag{6.64}$$

where $k = 2\pi/\lambda$ is the wavenumber, λ is the wavelength, (r, φ, z) are cylindrical coordinates, $\xi = kr\sin(\theta)$. In Eq. (6.64), $J_\beta(\xi)$ is the βth-order Bessel function of the first kind. From Eq. (6.63) we obtain equations for the transverse intensity components:

$$\begin{aligned} I_x &= |E_x(r,\varphi)|^2 = \\ &= \frac{1}{2}\left(J_{0,0}^2 + J_{2,2}^2 + 2J_{0,0}J_{2,2}\cos(2\varphi)\right), \\ I_y &= |E_y(r,\varphi)|^2 = \\ &= \frac{1}{2}\left(J_{0,0}^2 + J_{2,2}^2 - 2J_{0,0}J_{2,2}\cos(2\varphi)\right). \end{aligned} \tag{6.65}$$

Eq. (6.65) will be the same for both left-hand and right-hand circularly polarization. From Eq. (6.65) it follows that the intensity components have the form of an ellipse, which is elongated for I_x along the x axis and for I_y along the y axis. These elliptical intensity distributions can be measured experimentally, and thus we can be sure that an optical vortex with a topological charge of 2 is present at the focus. However, as shown in the contribution to the intensity of the second terms in Eq. (6.63) is much smaller than the contribution of the first terms, so this ellipse will be a slightly distorted circle.

It is more efficient to use an additional vortex to compensate for the optical vortex in Eq. (6.9) and obtain a weak intensity at the center of the focal plane. Indeed, for an initial field of the form

$$\mathbf{E}_0(\varphi) = \frac{e^{in\varphi}}{\sqrt{2}}\begin{pmatrix} 1 \\ \sigma i \end{pmatrix},$$
$$\mathbf{H}_0(\varphi) = \frac{e^{in\varphi}}{\sqrt{2}}\begin{pmatrix} -\sigma i \\ 1 \end{pmatrix}. \tag{6.66}$$

In [79] an expression for the projections of the electric vector in the plane of focus was obtained:

$$\begin{aligned} E_{x,n} &= \frac{i^{n-1}}{\sqrt{2}} e^{in\varphi}\left(J_{0,n} + e^{2\sigma i\varphi} J_{2,n+2\sigma}\right), \\ E_{y,n} &= \frac{i^{n}}{\sqrt{2}} e^{in\varphi}\left(J_{0,n} - e^{2\sigma i\varphi} J_{2,n+2\sigma}\right), \\ E_{z,n} &= -\sqrt{2} i^{n} e^{i(n+\sigma)\varphi} J_{1,n+\sigma}. \end{aligned} \tag{6.67}$$

Therefore, at $n = -2\,\sigma$ in the plane on the optical axis there will be a local intensity maximum proportional to $J_{2,0}^2$. This intensity maximum proves the presence of a transverse optical vortex with a topological charge of +2 at the focus.

6.3.3 Linearly Polarized Light in Tight Focus

It is known that when focusing a linearly polarized light, the focal spot acquires an elliptical shape. This is explained by the fact that after the spherical lens for linear polarization, two optical vortices with topological charges 2 and −2 with left-hand and right-hand circularly polarizations are formed. These vortices introduce asymmetry into the focal spot. Indeed, the Jones vector of the initial field for an optical vortex with linear polarization has the form:

$$\mathbf{E}_n(\varphi) = \exp(in\varphi)\begin{pmatrix} 1 \\ 0 \end{pmatrix}. \tag{6.68}$$

The projections of the electric vector in the plane of the tight focus for the initial field (6.68) have the form:

$$\begin{aligned} E_x &= \frac{i^{n-1}}{2} e^{in\varphi}\left(2J_{0,n} + e^{2i\varphi} J_{2,n+2} + e^{-2i\varphi} J_{2,n-2}\right), \\ E_y &= \frac{i^{n}}{2} e^{in\varphi}\left(e^{-2i\varphi} J_{2,n-2} - e^{2i\varphi} J_{2,n+2}\right), \\ E_z &= i^{n} e^{in\varphi}\left(e^{-i\varphi} J_{1,n-1} - e^{i\varphi} J_{1,n+1}\right). \end{aligned} \tag{6.69}$$

In Eq. (6.69) the maximum contribution near the optical axis is given by the term $J_{0,n}$. The remaining terms are several times smaller [19]. From Eq. (6.69) it follows that at the focus at $n = 0$ two optical vortices with topological charges 2 and −2 at the transverse components, and two optical vortices with charges 1 and −1 at the longitudinal component of the field are formed. Using Eqs. (6.69) for $n = 0$, we obtain expressions for the intensity of individual components at the focus for light with linear polarization along the horizontal axis:

$$\begin{aligned} I_x &= |E_x(r,\varphi)|^2 = \\ &= J_{0,0}^2 + J_{2,2}^2\cos^2(2\varphi) + 2J_{0,0}J_{2,2}\cos(2\varphi), \\ I_y &= |E_y(r,\varphi)|^2 = \\ &= \frac{1}{2}J_{2,2}^2(1 - \cos(4\varphi)), \\ I_z &= |E_z(r,\varphi)|^2 = \\ &= 2J_{1,1}^2(1 + \cos(\varphi)). \end{aligned} \tag{6.70}$$

From Eqs. (6.70) it follows that all intensity components depend on the azimuthal angle φ and are not radially symmetric. But in terms of magnitude, the main contribution to the intensity near the optical axis, where the focal spot is located, is given by the term $J_{0,0}$, which forms a round focal spot. The second largest term contributing to the intensity is the term $2J_{0,\,0}J_{2,\,2}\cos(2\varphi)$. This term has elliptical shape, it is elongated along the horizontal axis, since it has two local maxima at $\varphi = 0$ and $\varphi = \pi$. According to the sum of these two terms, the shape of the focal spot for light with initial linear polarization will have the form of an ellipse elongated along the direction of the polarization vector. Thus, it is the two optical vortices with charges 2 and −2, which are formed during focusing, that are responsible for the elliptical shape of the focal spot.

From Eq. (6.69), we can obtain the SAM density at the focus for a first-order optical vortex ($n = 1$) with linear polarization:

$$\begin{aligned} S_z = 2\mathrm{Im}\left(E_x^* E_y\right) = -\frac{1}{2}(J_{2,1} + J_{2,3})\times \\ (-J_{2,1} + J_{2,3} + J_{0,1}\cos 2\varphi). \end{aligned} \tag{6.71}$$

From Eq. (6.71), it follows that for $\varphi = 0$ and $\varphi = \pi$, the SAM will be negative ($S_z < 0$), while for $\varphi = \pi/2$ and $\varphi = 3\pi/2$, the SAM will be positive ($S_z > 0$). The longitudinal projection of the SAM reveals the presence of local regions with right- and left-handed elliptical polarizations at the focus. In the initial plane, the light field (6.69) had no spin angular momentum ($S_z = 0$), while regions with SAMs of opposite signs formed at the focus. This demonstrates the presence of the spin Hall effect (6.71).

6.3.4 Radially Polarized Light in Tight Focus

The Jones vector for a radially polarized light field is

$$\mathbf{E}(\varphi) = (\cos\varphi \sin\varphi). \tag{6.72}$$

In the plane of the focus, the components of electric vector for the initial field (6.72) have the form:

$$\begin{aligned} E_x(r,\varphi) &= \cos(\varphi)(I_{0,1} - I_{2,1}), \\ E_y(r,\varphi) &= \sin(\varphi)(I_{0,1} - I_{2,1}), \\ E_z(r,\varphi) &= 2iI_{1,0}. \end{aligned} \tag{6.73}$$

From Eqs. (6.73) it is seen that the radial polarization (6.72) is preserved at the focus, however, it is modulated by a function depending on the radial variable $(I_{0,\,1} - I_{2,\,1})$. It is also seen that the intensity of each transverse coordinate will have the shape of an ellipse elongated along the corresponding coordinate axis, although the total transverse intensity will have a circular shape.

6.3.5 Numerical Simulation

Simulation parameters were as follows: wavelength $\lambda = 532$ nm, focal length $f = 20\lambda$, annular aperture with $\theta_{min} = \arcsin 0.95 \le \theta \le \text{asin}(0.99) = \theta_{max,}$ the amplitude inside the ring is uniform, i.e. $A(\theta) \equiv 1$, the polarization is right-hand and left-hand circularly, the Jones vector is $[1, \pm i]^{\text{T}}$, computational domain $|x|$, $|y| \le R = 5$ μm. Figure 6.12 shows distribution of intensity (Fig. 6.12a,b) and their components (Fig. 6.12c) at the focus.

From Fig. 6.12 it follows that the focal spot I_x (excluding side lobes) has the shape of an ellipse elongated along the horizontal axis, in accordance with the first formula in Eqs. (6.65). Similarly, the focal spot I_y (excluding side lobes) has the shape of an ellipse elongated along the vertical axis, in accordance with the second formula in Eqs. (6.65). In this case, the transverse distribution of intensity at the focus (Fig. 6.12c) has a radially symmetric shape.

Figure 6.13 shows the transverse components of intensity at the focal plane for an initial optical vortex with a topological charge $n = 2$ with left-hand (Fig. 6.13a) and right-hand (Fig. 6.13b) circularly polarization.

From Fig. 6.13 it follows that in the picture $I_x + I_y$ the main ring (shown in Fig. 6.13 by vertical lines) is of the same radius, but with left-hand circularly polarization there is another ring inside (the dark area in the center is small)—this is the vortex ring with topological charge $n - 2 = 1$. With right-hand circularly polarization there is no ring inside. There is a ring outside, this is a vortex ring with

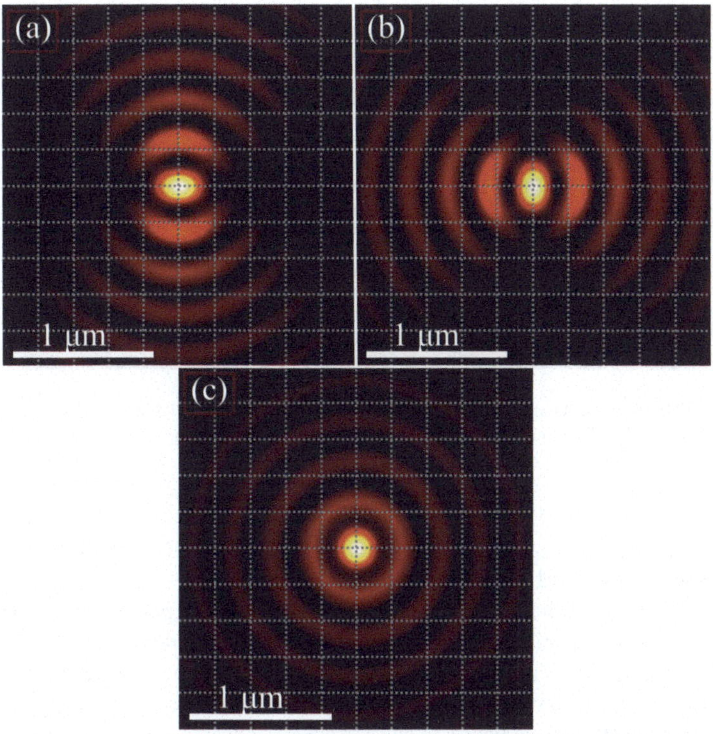

Fig. 6.12 Intensity at the focus for I_x (**a**), I_y (**b**) and $I_x + I_y$ (**c**) for the initial field with circular polarization Eq. (6.56)

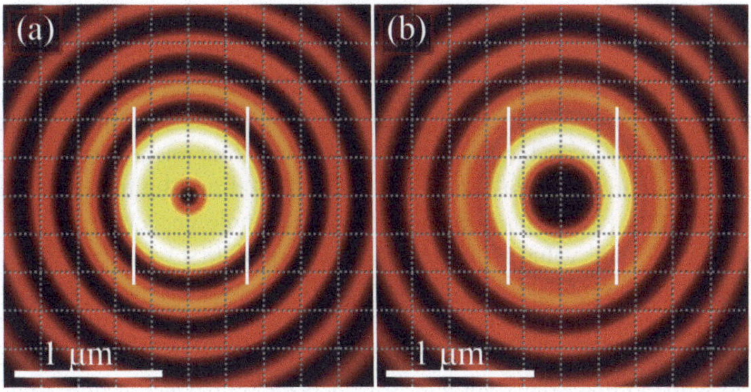

Fig. 6.13 Transverse intensity $I_x + I_y$ at the focal plane of the initial field (6.66) at $n = 3$, and $\sigma = -1$, left-hand (**a**) and $\sigma = 1$, right-hand (**b**) circularly polarization

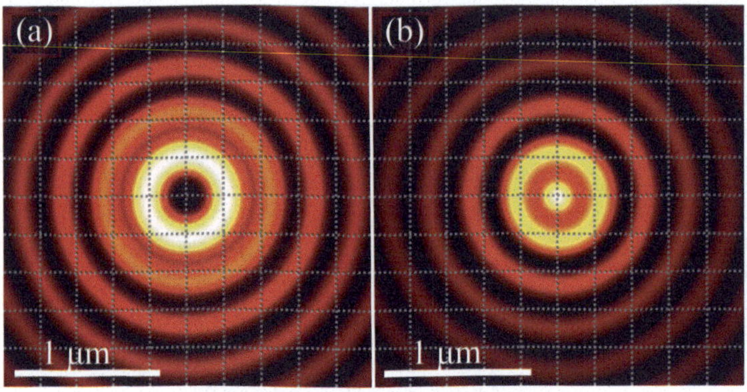

Fig. 6.14 Transverse intensity $I_x + I_y$ at the focal plane of the initial field (6.66) with the same parameters as in Fig. 6.13 but for $n = -2$ with $\sigma = -1$, left-hand (**a**) and $\sigma = 1$, right-hand (**b**) circularly polarization

topological charge $n + 2 = 5$; however, it is not detectable because of the side lobes of the vortex of the third order.

From Fig. 6.14 it follows that in the transverse intensity pattern $I_x + I_y$ with left-hand circularly polarization only a ring is formed, since the field consists of vortices of orders −2 and −4. With right-hand circularly polarization inside the ring there is an intensity peak on the optical axis, since the field consists of two terms: vortex $I_{0,2}e^{-2i\varphi}$ and radially symmetric $I_{2,0}$.

Figures 6.12, 6.13 and 6.14 prove that when circularly polarized light passes through a spherical lens, a second-order optical vortex ($n = 2\sigma$) is formed. Although this vortex is not visible in the phase distribution at the focus (Fig. 6.15). Only the first-order optical vortex ($n = \sigma$) is visible in the phase of the longitudinal component of the field (6.63).

Figure 6.15 shows the phases at the focus for all components of the electric field obtained using Eqs. (6.63). From Fig. 6.15c it is seen that an optical vortex with topological charge $n = 1$ is formed at the longitudinal component of the electric field at the focus for light with right-hand circularly polarization. But in Figure 6.15a,b it is difficult to see the phase of the optical vortex with topological charge $n = 2$. This is explained by the small value of the second terms in the first two Eqs. (6.63).

The particles captured on the light ring is shown experimentally in [22, 23] to rotate in a circle counterclockwise ($\sigma = 1$) and clockwise ($\sigma = -1$) in the absence of an optical vortex in the sharp focus of the axicon $\exp(iar)$. Here a is the axicon parameter, with circular polarization $(1, i\sigma)^T$, T is the transposition sign. Figure 6.16 shows two identical ring intensity distributions (a, c) in the sharp focus (NA = 0.95) of an ideal spherical lens with the axicon initial amplitude $B(\theta) = \exp(ia\tan\theta)$. And the OAM density distribution on the intensity ring: for the right-hand circularly polarization (b) and the left-hand circularly polarization (d). Numerical simulation parameters are as follows: wavelength $\lambda = 532$ nm is used, the focal length is $f = 20\lambda$, the axicon parameter $a = 10$, the calculation region size is 8×8 μm.

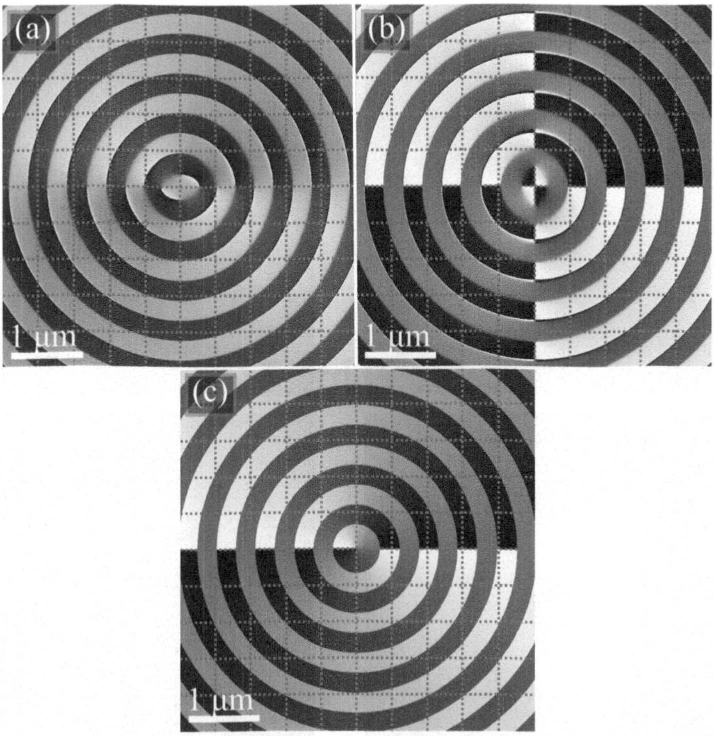

Fig. 6.15 Phase distribution in the focal plane of the initial field (6.56): arg E_x (**a**), arg E_y (**b**) and arg E_z (**c**). Black color—phase 0, white color—2π

Due to spin-orbit conversion the OAM density is seen from Fig. 6.16 to change sign when the direction of rotation of the polarization vector changes. The following results were obtained in the section. It was shown that when focusing circularly polarized light immediately behind a spherical lens, the transverse light field is divided into two light fields converging to the focus. One of these fields is light with the same circular polarization. The second one (smaller in magnitude) is a field of a second-order optical vortex with circular polarization of the opposite direction. This additional optical vortex is responsible for the existence of the orbital angular momentum of the light field immediately behind the spherical lens. However, both the optical vortex and the orbital angular momentum appear not in the entire field, but only in a certain part of the field, both light fields (both vortex and non-vortex) propagate together, and they cannot be separated by the distribution of the total intensity at the focus. The optical vortex manifests itself in the elliptical shape of the intensities of individual transverse field components in the focal plane, and it also manifests itself in the existence of a local intensity peak on the optical axis if a second-order optical vortex with the opposite sign is present in the initial field. The work also shows that when focusing light with linear polarization, two second-order optical vortices with plus and minus signs are formed at the transverse components

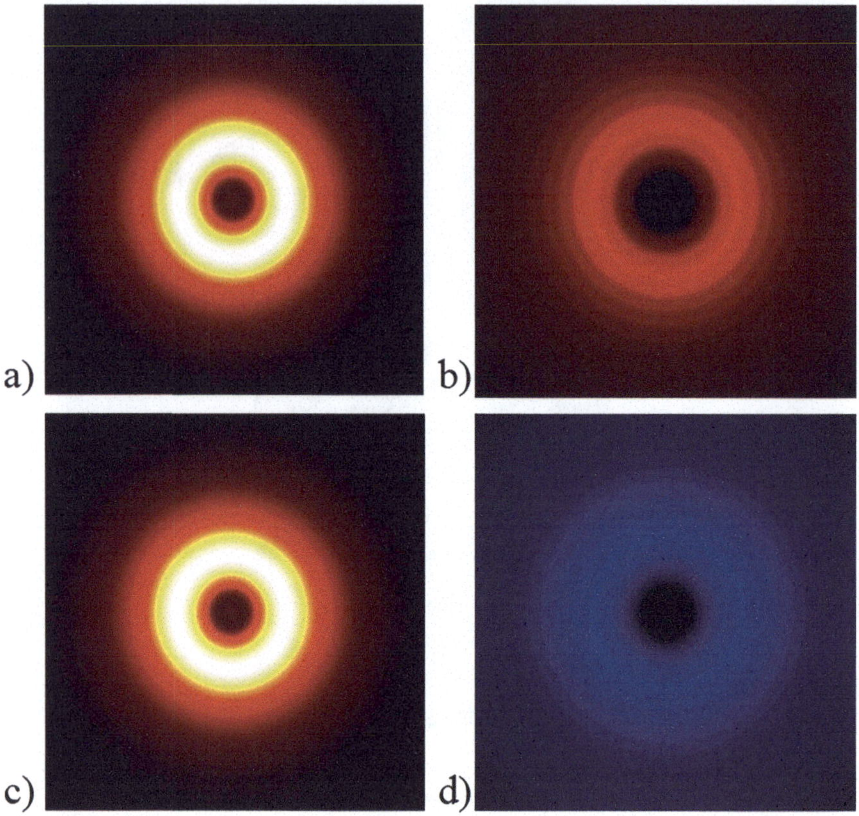

Fig. 6.16 Intensity distributions (**a**, **c**) and OAM densities (**b**, **d**) in the sharp focus of the field with right-hand (**a**, **b**) and left-hand (**c**, **d**) circularly polarization. Red and blue colors (**b**, **d**) represent positive and negative values, respectively

of the light field. It is shown that these optical vortices determine the elliptical shape of the focal spot. Moreover, the elliptical focal spot is determined by the initial transverse component and it is elongated along the Cartesian axis of the initial linear polarization direction. The results obtained in the work can be used to rotate microparticles in a circle without preliminary formation of an optical vortex, but using only a spherical lens and an axicon.

References

1. Q. Zhan, Adv. Opt. Photon. **1**, 1 (2009)
2. V.V. Kotlyar, S.S. Stafeev, E.S. Kozlova, A.G. Nalimov, Sensors **21**, 6424 (2021)
3. H.-T. Wang, X.-L. Wang, Y. Li, J. Chen, C.-S. Guo, J. Ding, Opt. Express **18**, 10786 (2010)
4. I. Freund, Opt. Commun. **201**, 251 (2002)

5. A.A. Kovalev, V.V. Kotlyar, S.S. Stafeev, Micromachines **14**, 777 (2023)
6. J. Wang, Photonics Res. **4**, B14 (2016)
7. J. Wang, J. Liu, S. Li, Y. Zhao, J. Du, L. Zhu, Nanophotonics **11**, 645 (2022)
8. Y. Zhao, L. Zhou, X. Jiang, L. Zhu, Q. Shi, Nanomaterials **14**, 691 (2024)
9. Y. Zhang, M. Li, S. Yan, Y. Zhou, W. Gao, R. Niu, X. Xu, B. Yao, Small **20**, 2309395 (2024)
10. A. Porfirev, S. Khonina, D. Porfirev, N. Ivliev, Appl. Opt. **63**, 3779 (2024)
11. A.M. Beckley, T.G. Brown, M.A. Alonso, Opt. Express **18**, 10777 (2010)
12. S. Chen, X. Zhou, Y. Liu, X. Ling, H. Luo, S. Wen, Opt. Lett. **39**, 5274 (2014)
13. V.V. Kotlyar, S.S. Stafeev, V.D. Zaitsev, A.M. Telegin, Photonics **9**, 969 (2022)
14. E.J. Galvez, S. Khadka, W.H. Schubert, S. Nomoto, Appl. Opt. **51**, 2925 (2012)
15. V.V. Kotlyar, S.S. Stafeev, A.G. Nalimov, Phys. Rev. A **99**, 033840 (2019)
16. H. Li, C. Wang, M. Tang, X. Li, Opt. Express **28**, 18607 (2020)
17. C. Leyder, M. Romanelli, J.P. Karr, E. Giacobino, T.C.H. Liew, M.M. Glazov, A.V. Kavokin, G. Malpuech, A. Bramati, Nat. Phys. **3**, 628 (2007)
18. V.V. Kotlyar, A.A. Kovalev, A.M. Telegin, E.S. Kozlova, Photonics **11**, 1 (2024)
19. D. Li, S. Feng, S. Nie, C. Chang, J. Ma, C. Yuan, J. Appl. Phys. **125**, 073105 (2019)
20. Z. Gu, D. Yin, F. Gu, Y. Zhang, S. Nie, S. Feng, J. Ma, C. Yuan, Sci. Rep. **9**, 15301 (2019)
21. M. Liu, P. Huo, W. Zhu, C. Zhang, S. Zhang, M. Song, S. Zhang, Q. Zhou, L. Chen, H.J. Lezec, A. Agrawal, Y. Lu, T. Xu, Nat. Commun. **12**, 2230 (2021)
22. J. Mun, M. Kim, Y. Yang, T. Badloe, J. Ni, Y. Chen, C.-W. Qiu, J. Rho, Light Sci. Appl. **9**, 139 (2020)
23. M. Babiker, J. Yuan, K. Koksal, V.E. Lembessis, Opt. Commun. **554**, 130185 (2024)
24. K.A. Forbes, Opt. Commun. **561**, 130499 (2024)
25. J. Yuan, K. Koksal, V.E. Lembessis, O. Communications **130185**, 1 (2024)
26. M. Babiker, K. Koksal, V.E. Lembessis, J. Yuan, J. Opt. **26**, 115601 (2024)
27. B. Richards, E. Wolf, Proc. R. Soc. A mMath. Phys. Eng. Sci. **253**, 358 (1959)
28. S.F. Pereira, A.S. van de Nes, Opt. Commun. **234**, 119 (2004)
29. D.R. Abujetas, J.A. Sánchez-Gil, ACS Photonics **7**, 534 (2020)
30. N. Mackinnon, J. Opt. **21**, 125402 (2019)
31. K.A. Forbes, G.A. Jones, J. Opt. **23**, 115401 (2021)
32. K.A. Forbes, G.A. Jones, Phys. Rev. A **103**, 53515 (2021)
33. M. Rashid, O.M. Maragò, P.H. Jones, J. Opt. A Pure Appl. Opt. **11**, 065204 (2009)
34. A.A. Kovalev, V.V. Kotlyar, A.M. Telegin, Photonics **10**, 719 (2023)
35. V.V. Kotlyar, S.S. Stafeev, Y. Liu, L. O'Faolain, A.A. Kovalev, Appl. Opt. **52**, 330 (2013)
36. V. Kotlyar, S.S. Stafeev, A.G. Nalimov, L. O'Faolain, Appl. Phys. Lett. **114**, 141107 (2019)
37. M.A. Alonso, J. Opt. **13**, 64016 (2011)
38. F.S. Roux, Opt. Commun. **223**, 31 (2003)
39. K.A. Forbes, J. Opt. **26**, 125401 (2024)
40. X. Yi, Y. Liu, X. Ling, X. Zhou, Y. Ke, H. Luo, S. Wen, D. Fan, Phys. Rev. A **91**, 023801 (2015)
41. S.S. Stafeev, V.D. Zaitsev, V.V. Kotlyar, Opt. Commun. **583**, 131803 (2025)
42. T.H.R. Skyrme, Proc. R. Soc. Lond. A Math. Phys. Sci. **260**, 127 (1961)
43. S. Gao, F.C. Speirits, F. Castellucci, S. Franke-Arnold, S.M. Barnett, J.B. Götte, Phys. Rev. A **104**, 49901 (2021)
44. Y. Shen, E.C. Martínez, C. Rosales-Guzmán, ACS Photonics **9**, 296 (2022)
45. C. Cisowski, C. Ross, S. Franke-Arnold, Adv. Photonics Res. **4**, 2200350 (2023)
46. R. Gutiérrez-Cuevas, E. Pisanty, J. Opt. **23**, 24004 (2021)
47. Y. Shen, Q. Zhang, P. Shi, L. Du, X. Yuan, A.V. Zayats, Nat. Photonics **18**, 15 (2024)
48. S. Cao, L. Du, P. Shi, X. Yuan, Opt. Express **32**, 4167 (2024)
49. S. Gao, F.C. Speirits, F. Castellucci, S. Franke-Arnold, S.M. Barnett, J.B. Götte, Phys. Rev. A **102**, 53513 (2020)
50. A. McWilliam, C.M. Cisowski, Z. Ye, F.C. Speirits, J.B. Götte, S.M. Barnett, S. Franke-Arnold, Laser Photon. Rev. **17**, 2300155 (2023)

51. M.V. Berry, J. Opt. A Pure Appl. Opt. **6**, 259 (2004)
52. V.V. Kotlyar, A.A. Kovalev, S.S. Stafeev, V.D. Zaitsev, Bull. Russ. Acad. Sci. Phys. **86**, 1158 (2022)
53. H. Teng, J. Zhong, J. Chen, X. Lei, Q. Zhan, Photonics Res. **11**, 2042 (2023)
54. S.M. Barnett, L. Allen, Opt. Commun. **110**, 670 (1994)
55. A.A. Kovalev, V.V. Kotlyar, Micromachines **14**, 494 (2023)
56. I. Freund, Opt. Lett. **26**, 1996 (2001)
57. V.V. Kotlyar, A.A. Kovalev, V.D. Zaitsev, Photonics **9**, 298 (2022)
58. V.V. Kotlyar, A.A. Kovalev, A.M. Telegin, S.S. Stafeev, Opt. Mem. Neural Networks **34**, 347 (2025)
59. T. Ando, Y. Ohtake, T. Inoue, H. Itoh, N. Matsumoto, N. Fukuchi, Appl. Phys. Lett. **92**, 21116 (2008)
60. A. Boivin, E. Wolf, Phys. Rev. **138**, B1561 (1965)
61. J.W.M. Chon, X. Gan, M. Gu, Appl. Phys. Lett. **81**, 1576 (2002)
62. T.A. Nieminen, A.B. Stilgoe, N.R. Heckenberg, H. Rubinsztein-Dunlop, J. Opt. A Pure Appl. Opt. **10**, 115005 (2008)
63. Z. Bomzon, M. Gu, J. Shamir, Appl. Phys. Lett. **89**, 241104 (2006)
64. Y. Wu, P. Yu, Y. Liu, Z. Wang, Y. Li, L. Gong, J. Light. Technol. **41**, 2252 (2023)
65. K.Y. Bliokh, F.J. Rodríguez-Fortuño, F. Nori, A.V. Zayats, Nat. Photonics **9**, 796 (2015)
66. S. Quabis, R. Dorn, M. Eberler, O. Glöckl, G. Leuchs, Opt. Commun. **179**, 1 (2000)
67. R. Dorn, S. Quabis, G. Leuchs, Phys. Rev. Lett. **91**, 233901 (2003)
68. Z. Bomzon, M. Gu, Opt. Lett. **32**, 3017 (2007)
69. M.V. Berry, Proc. R. Soc. London. A. Math. Phys. Sci. **392**, 45 (1984)
70. M. Onoda, S. Murakami, N. Nagaosa, Phys. Rev. Lett. **93**, 083901 (2004)
71. M. Zhu, S. Fu, Z. Man, Opt. Express **30**, 41048 (2022)
72. B. Ghosh, A. Daniel, B. Gorzkowski, R. Lapkiewicz, Optica **10**, 1217 (2023)
73. V.V. Kotlyar, A.G. Nalimov, S.S. Stafeev, J. Opt. Soc. Am. B **36**, 2850 (2019)
74. K.Y. Bliokh, E.A. Ostrovskaya, M.A. Alonso, O.G. Rodríguez-Herrera, D. Lara, C. Dainty, Opt. Express **19**, 26132 (2011)
75. V.V. Kotlyar, A.A. Kovalev, E.S. Kozlova, A.M. Telegin, Micromachines **14**, 788 (2023)
76. V.V. Kotlyar, S.S. Stafeev, A.A. Kovalev, V.D. Zaitsev, Appl. Sci. **12**, 12218 (2022)
77. X.-L. Wang, J. Chen, Y. Li, J. Ding, C.-S. Guo, H.-T. Wang, Phys. Rev. Lett. **105**, 253602 (2010)
78. S.-Y. Huang, G.-L. Zhang, Q. Wang, M. Wang, C. Tu, Y. Li, H.-T. Wang, Optica **8**, 1231 (2021)
79. V.V. Kotlyar, A.A. Kovalev, S.S. Stafeev, A.M. Telegin, Optik **338**, 172505 (2025)

Conclusion

The book contains results on the canonical energy flux in laser beams obtained by the authors in 2025. We will briefly list these results.

We explore the diffraction of a plane wave by an opaque rectangular screen and show that at any on-axis distance from the obstacle plane, a multitude of diffraction fringes occur in a screen-parallel plane, characterized by a (negative) canonical backflow. Remarkably, the localized fringe recurrence interval is found to decrease with larger distance from the screen edge, with the first fringe of the canonical energy backflow departing from the edge with increasing distance from the optical axis. The energy backflow is shown to occur in the diffraction pattern areas characterized by subwavelength values of phase and amplitude modulation, i.e. where the local wave-vector expressed through the phase gradient is larger than the incident wave wave-vector.

We have demonstrated both analytically and numerically that in the cross-section of a 2D paraxial accelerating Airy beam propagating on a parabolic trajectory there are areas where a canonical energy backflow occurs. These areas arise in the far sidelobes of the Airy beam characterized by subwavelength local period (superoscillations areas). At an arbitrary propagation distance from the initial plane, there is a threshold transverse coordinate, such that at all smaller negative values of the transverse coordinate, the longitudinal component of the canonical energy flow is negative. For a nonparaxial Airy beam, we derive an explicit analytical expression for the canonical energy flow near the initial plane. The energy backflow is revealed to occur near the intensity null where a phase jump of π takes place. Near the intensity null, the phase derivative with respect to the longitudinal coordinate takes large negative values, leading to the longitudinal wave-vector projection becoming larger than the wave-number of light and directed oppositely. The maximum energy backflow is found to be about one fifth of the direct maximum energy flow.

In this book, several 2D non-paraxial light fields near the initial plane (sinc-beam, Gaussian beam and Hankel beam) are considered. These TE-polarized beams are

V. Kotlyar et al., *Inverse Canonical Energy Flow in Laser Beams*,
https://doi.org/10.1007/978-3-032-27289-8

shown theoretically and numerically to have a reverse canonical energy flux in the initial plane. The negative value of the longitudinal projection of the canonical flux vector means that the longitudinal derivative of the phase is negative, and hence the local wave vector is directed against the positive direction of the optical axis. The reverse flux occurs in the presence of evanescent waves. Although running along the initial plane evanescent waves themselves do not transfer energy along the optical axis perpendicular to the initial plane, they participate in the formation of the reverse flux. Near the initial plane, the amplitude of the light field consists of a linear combination of amplitudes of the propagating and evanescent waves. Therefore, evanescent waves participate in the formation of the phase function of the complex amplitude of the entire field and affect the sign of the longitudinal phase derivative. For a sinc-beam, the backflow occurs in the first intensity side lobes, where the standing evanescent wave has odd antinodes.

We have demonstrated that non-paraxial 2D TE-polarized light fields form 2D optical vortices near intensity zeros, with the amplitude having the form $x + iz$, where x and z are the transverse and longitudinal Cartesian coordinates. Near the intensity zeros, the longitudinal projections of the wave vector also have gigantic values of both signs. Negative values of the longitudinal projections of the wave vectors indicate that a reverse energy flow occurs near the intensity zeros. Around the intensity zeros, the energy flow circulates both clockwise (if the topological charge of the optical vortex is -1) and counterclockwise (if the topological charge of the optical vortex is $+1$). The gigantic values of the wave vectors near the intensity zeros indicate that the wavelength of the light is small, and therefore the phase velocity of rotation of light around the intensity zero is also small, compared to the speed of light in a vacuum. A giant wave vector, energy flow circulation and energy return flow are formed in a substantially subwavelength region around intensity zeros of fractions of a wavelength in size, demonstrating the presence of superoscillations.

We derive an explicit analytical expression for the distribution of longitudinal component of the canonical energy flux vector near the initial plane (in the near field). We consider a nonparaxial light beam with linear polarization and with an amplitude of the plane waves spectrum in the form of a circle. This distribution has the shape of an ellipse elongated along the direction of the linear polarization vector. If the plane waves spectrum includes evanescent waves, then in the near field, a reverse canonical energy flux arises. This reverse flux almost coincides with the first side lobe of the intensity distribution in the near field. The reverse flux arises near intensity zeros. Computer simulations is performed using two independent methods: plane-wave spectral decomposition and Rayleigh-Sommerfeld integral. The simulation results agree with theoretical predictions.

Employing the Richards-Wolf formalism that adequately describes an electromagnetic field near the sharp focus of an ideal spherical lens, we demonstrate that certain light fields (linearly polarized optical vortex, cylindrical vector fields of an arbitrary order) have in the focus plane a reverse canonical energy flow. Maximal magnitude of the reverse energy flow amounts for the numerical aperture 0.95 nearly 1% of the maximal magnitude of the direct energy flow. Distribution of the reverse canonical flow in the focus plane can have the shape of concentric rings or only arcs of the concentric rings. For certain light fields, for instance, for an azimuthally polarized light field, longitudinal component of the canonical energy flow vector coincides with the longitudinal component of the Poynting vector. It is shown that a circularly polarized optical vortex does not have the reverse flow at the focus.

In this book, we investigated the behavior of the phase and local wave vector of a tightly focused optical vortex with linear polarization. It was shown that in contrast to a scalar optical vortex, there are some centers of phase singularity points of optical vortices under tight focusing condition. If the topological charge of the focused vortex is equal to one, then the center of the focused vortex is located on the optical axis. If the topological charge is equal to two, then two optical vortices with unit topological charges are observed in the focal plane. This singularity points are shifted from the optical axis and bounded the area of energy backflow. If the topological charge exceeds two, then three optical vortices are formed in the focal plane, two of them have a unit topological charge and are shifted from the optical axis. It was analytically shown that with increasing of topological charge the centers of shifted optical vortices move away from the optical axis.

For the well-known linear electric Hertzian dipole, oriented along the Cartesian axis z, we derive explicit analytical expressions for the radial and meridional components of the canonical energy flow vector in spherical coordinates in the near field. We demonstrate theoretically that, in the near field, the dipole has energy backflow. The backflow area is bounded by a curve of zero radial energy flow. A line of forward energy flow outgoes from the dipole along a radius, then it bends and intersects the curve of zero radial energy flow, and then the energy flow line returns back to the dipole. This returning energy flow, outgoes from the dipole at meridional angles from 55° till 125°. Outside these angles, the energy flow is forward and the energy flow lines, despite bending, go away from the dipole to infinity. At meridional angles from 0° to 90°, energy flow lines bend clockwise, while at angles from 90° to 180°, energy flow lines bend counterclockwise. It is interesting that the linear dipole has the azimuthal component of the spin angular momentum vector. At $z < 0$, the spin vector is directed along the azimuthal coordinate, while at $z > 0$—oppositely. Due to nonzero meridional component of the orbital energy flow, the dipole has the azimuthal component of the orbital angular momentum. It is directed oppositely to the direction of the azimuthal component of the spin angular momentum.

We discuss two source vector fields of Poincare-beam type that can be looked upon as optical skyrmions, i.e. topological quasiparticles. We derive explicit analytical relationships that describe projections of a three-dimensional (3D) skyrmion vector field in the source plane and skyrmion numbers, which are shown to be pro-portional to the topological charges of constituent optical vortices of the Poincare beams. We also propose a new constructive formula as an effective tool for calculating the skyrmion number via normalized Stokes vector projections rather than skyrmion vector field projections. The skyrmion numbers calculated using the familiar and newly proposed formulae coincide. Numbers of each pro-jection of a 3D skyrmion vector field are shown to com-prise a third of the full skyrmion number. The theoretical conclusions are validated by a numerical simulation. The non-uniform linear polarization in the skyrmion cross-section depends on the azimuthal angle and can be used to form a spiral microrelief due to the mass transfer of molecules on the surface of the material.

We introduce here new physical quantities. By analogy with the conventional orbital angular momentum (OAM), they are called polarization OAM and hybrid OAM. It is known that the conventional OAM indicates the presence of an optical vortex in the light field and is equal to the topological charge. The same way, the polarization OAM indicates the presence of a "vortex of the linear polarization direction" in the vector field and is equal to the singularity index of the vector field. Thus, it shows how many full turnovers makes the linear polarization vector along a closed contour around the optical axis. The hybrid OAM demonstrates that along a closed contour around the optical axis, the vector field changes not only by the direction of major axis of the polarization ellipse, but also by the ratio between the major and minor ellipse axes. The hybrid OAM is equal to the number of full turnovers of the polarization ellipse. These new OAMs are derived for several typical initial vector fields to demonstrate their functionality. The initial electric field has only two transverse components, which determine the longitudinal component of the OAM vector. However, at the tight focus, all three components of the E-vector arise, and therefore, longitudinal component of the OAM vector is defined by all three electric field components. Thus, if one of these three OAMs is zero in the initial field, then at the focus, this OAM can become nonzero due to appearing longitudinal field component, contributing to this OAM. We believe that these newly introduced OAMs can be applicable to other vector fields, allowing more complete description of their properties.

The manufacturer's authorised representative in the EU is Springer Nature Customer Service Centre GmbH, Europaplatz 3, 69115 Heidelberg, Germany. If you have any concerns regarding our products, please contact ProductSafety@springernature.com

Printed and bound by CPI Group (UK) Ltd, Croydon, CR0 4YY

07/10/2026

02229357-0003